Animal Welfare

Volume 24

The Animal Welfare series has been designed to help contribute towards a culture of respect for animals and their welfare by producing academic texts addressing how best to provide for the welfare of the animal species that are managed and cared for by humans. Books in the series do not provide a detailed blue-print for the management of each species, but they do describe and discuss the major welfare concerns, often in relation to the wild progenitors of the managed animals. Welfare has been considered in relation to animals' needs, concentrating on nutrition, behaviour, reproduction and the physical and social environment. Economic effects of animal welfare provision were also considered where relevant, as were key areas where further research is required.

Barbara Padalino • Bernard Faye
Editors

Dromedary Camel Behavior and Welfare

Camel Friendly Management Practices

Co-ordinating Editor: Carly I. O'Malley

Springer

Editors
Barbara Padalino
University of Bologna
Bologna, Italy

Bernard Faye
Campus International de Baillarguet
CIRAD-ES
Montpellier, France

ISSN 1572-7408
Animal Welfare
ISBN 978-3-031-48599-2 ISBN 978-3-031-48600-5 (eBook)
https://doi.org/10.1007/978-3-031-48600-5

This Springer imprint is published by the registered company Springer Nature Switzerland AG
The registered company address is: Gewerbestrasse 11, 6330 Cham, Switzerland

Printed on acid-free paper

By Carlo Pastrana

Preface

Concern for animal welfare is not a new phenomenon in the history of the coexistence of humans and animals. Even the principle of domestication is based on the natural selection of animals that learn to not fear humans and develop less aggressive behavior, helping to reduce their stress no longer. During this multi-millennial history of coexistence between humans and domestic animals, protection rules have been established by various authorities to avoid animal suffering. Regarding farm animals, the concept of welfare raised new questions in the context of the emergence worldwide of industrial farms marked by profitability and intensification with all the excesses that we know. However, concerning the large camelids, dromedaries or Bactrians, most people believe that until recently their breeding and management was too extensive, and therefore too close to the animals' "natural" way of life, therefore the question of their well-being was not as closely considered as for other species that are more commonly intensively farmed for livestock production, such as dairy and beef cattle, pigs, and chickens.

Nothing could be further from the truth. Living in a particularly hostile environment, camels in the desert are far from being respected by their owners (for example, when they are destined to carry excessively heavy loads in caravan activities) or by their environment (drought, heat, low pastoral resources). Certainly, their remarkable adaptive capacities to these environmental contexts are well known. But for all that, the services that the people of the desert expect from the camels in the harsh desert conditions may have given priority to their survival, the question of welfare being linked firstly to the ability for people and animals to survive in an environment characterized by its low resources and climatic constraints.

Elsewhere, the relatively backward-looking image of the camel, the nomad's animal in harmony with nature, has been significantly tarnished in recent decades, because, obviously, camel breeding is experiencing a real revolution that has been emphasized by various authors. Indeed, the emergence of intensive and specialized breeding systems (dairy farms, feed-lots, racing stables, intensive tourism, etc.) being characterized by a limitation or even the cessation of herd mobility, by richer and less diversified diets, by more systematic use of veterinary inputs, the breeding context is moving considerably away from the "natural" way of life mentioned above. Therefore, the question of well-being arises for camels in the same way as other farm species subject to the same constraints.

Moreover, another phenomenon is emerging, i.e., the geographical expansion of the species, not only on the margins of arid zones as we can observe in Africa, but also in countries far from desert areas, in Western Europe and in the United States for example, as part of the diversification of agriculture. Faced with these new environments, camel biology is therefore subject to other climatic constraints (for example, wet and cold) and human constraints (lack of knowledge of the animal by these new farmers). The society rightly wonders about the consequences of these developments on the well-being of large camelids.

Welfare is about "quality of life" and scientists are working to develop protocols able to measure the mental and physical states of the animals, to identify the most reliable animal-based and environmental-based measures and to recommend preventive and corrective measures of poor welfare. However, knowledge about the behavioral and physiological needs of the camels, kept both extensively and intensively, are scant, and many husbandry and management systems of camels are based on traditions and beliefs. Consequently, there are many gaps of knowledge in what can make the camels feel well and "happy."

The aim of this edited book is to provide the reader with a review of the current literature on the welfare of camels, and hopefully this book can be used not only by academic and students, but also by all camel lovers, keepers, and owners who care about their camels. To achieve this, the book begins with the domestication of camels, with a focus on how the role of this animals has changed to date, and how modern genomics could help in the selection of welfare-related traits. It then considers how to assess welfare of dromedary camels, and there is a chapter dedicated to each welfare principle, namely "good feeding," "good housing," "good health," and "appropriate behavior." Each of these chapters explains how the welfare principle can be achieved while rearing camels and presents a list of recommendations to protect and enhance camel welfare. The book closes with four chapters in which specific problems related to camel training, reproductive management, transportation, slaughtering, and killing are discussed.

It is therefore the ambition of this book to take into account all the dimensions of animal welfare for a species so necessary to the life of people in arid regions and beyond. To achieve this, we have called on geneticists as well as nutritionists, behaviorists, and veterinarians to address all facets of well-being and knowledge to ensure the best quality of life for the camels and provide the most up-to-date research evidence to our readers. This was the ambition of this work, in many ways innovative because it is the first of its kind for the camel species.

We are extremely grateful to all authors of the chapters in this book, to the series editors, Dr Clive Phillips who proposed this idea to us, and Dr Carley O'Malley who carried out the final edit, and to our families who supported us in this adventure.

Montpellier, France Bernard Faye
Bologna, Italy Barbara Padalino

Animal Welfare Series Preface

Animal welfare is attracting increasing interest worldwide, and the knowledge and resources are available to, at least potentially, provide better management systems for farm animals, as well as companion, zoo, laboratory, and performance animals. The key requirements for adequate food, water, a suitable environment, companionship, and health are important for animals kept for all of these purposes.

The attention given to animal welfare in recent years derives largely from the fact that the relentless pursuit of financial reward and efficiency, to satisfy market demands, has led to the development of intensive animal management systems that challenge the conscience of many consumers, particularly in the farm and laboratory animal sectors. Livestock are the world's biggest land users, and the farmed animal population is increasing rapidly to meet the needs of an expanding human population. This results in a tendency to allocate fewer resources to each animal and to value individual animals less than the group. In these circumstances, the importance of each individual's welfare is diminished.

Increased attention to welfare issues is just as evident for zoo, companion, sport, and wild animals. Of growing importance is the ethical management of breeding programs since genetic manipulation is now technically advanced. There is less public tolerance of the breeding of extreme animals if it comes at the expense of animal welfare (e.g., brachycephalic dogs). The quest for producing novel genotypes has fascinated breeders for centuries. Dog and cat breeders have produced a variety of deformities that have adverse effects on their welfare, but nowadays the breeders are just as active in the laboratory, where the mouse is genetically manipulated with equally profound effects.

In developing countries, human survival is still a daily uncertainty for many, so that provision for animal welfare has to be balanced against human welfare. Animal welfare is usually a priority only if it supports the output of the animal, be it food, work, clothing, sport, or companionship. However, in many situations the welfare of animals is synonymous with the welfare of the humans that look after them, because happy, healthy animals will be able to assist humans best in their struggle for survival. In principle, the welfare needs of both humans and animals can be provided for, in both developing and developed countries, if resources are properly husbanded. In reality, the inequitable division of the world's riches creates physical and psychological poverty for humans and animals alike in many parts of the world.

The intimate connection between animals and humans that was once so essential for good animal welfare is rare nowadays, having been superseded by technologically efficient production systems where animals on farms and in laboratories are tended by increasingly few humans in the drive to enhance labor efficiency. With today's busy lifestyles, companion animals too may suffer from reduced contact with humans, although their value in providing companionship, particularly for certain groups such as the elderly, is beginning to be recognized. Animal consumers also rarely have any contact with the animals that are kept for their benefit.

In this estranged, efficient world, people struggle to find the moral imperatives to determine the level of welfare that they should afford to animals within their charge. A few people, and in particular many companion animal owners, strive for what they believe to be the highest levels of welfare provision, while others, deliberately or through ignorance, keep animals in impoverished conditions in which their health and well-being can be extremely poor. Today's multiple moral codes for animal care and use are derived from a broad range of cultural influences, including media reports of animal abuse, guidelines on ethical consumption and campaigning and lobbying groups.

This series has been designed to contribute toward a culture of respect for animals and their welfare by producing learned treatises about the provision for the welfare of the animal species that are managed and cared for by humans. The early species-focused books were not detailed management blue-prints; rather they described and considered the major welfare concerns, often with reference to the behavior of the wild progenitors of the managed animals. Welfare was specifically focused on animals' needs, concentrating on nutrition, behavior, reproduction, and the physical and social environment. Economic effects of animal welfare provision were also considered where relevant, as were key areas where further research is required.

This volume addresses the behavior and welfare of dromedary camels, for the first time in this Springer series. It discusses the changing role of camels from riding and pack animals that were managed extensively in their natural habitat to more intensive management practices globally for dairy, meat, racing, zoos, circuses, and camel shows. The intensive management of dromedary camels has raised welfare concerns related to feeding, housing, health, behavior, reproduction, transport, and slaughter. The welfare and behavior of camels in intensive systems is relatively understudied compared to more common livestock species. The volume addresses current gaps in knowledge and outlines recommendations for consideration for camel management and presents an animal welfare assessment protocol specific to dromedary camels and is sure to be a key reference for those working with or interested in camels.

Wilmington, MA Carly O'Malley
Perth, WA, Australia Clive Phillips

Contents

About the Authors

Dawood Sulaiman Al-Ajmi is a highly qualified and experienced researcher in the field of animal husbandry and food safety. With a PhD in Animal Husbandry and a Master's degree in Animal Science, he has a strong foundation in the field and has focused his work on the gut microbiology of animals and the impact of diet on the proliferation of pathogenic microorganisms. Dr. Al-Ajmi has made significant contributions to the field, including the successful isolation of *E. coli* O157 from camel fecal samples in the UAE for the first time. In addition to his expertise in animal and microbial molecular biology, he also has knowledge on animal feed formulations and animal husbandry. Currently serving as an Assistant professor at the College of Food and Agriculture at UAEU, Dr. Al-Ajmi is dedicated to promoting our understanding of the complex interplay between animal health and food safety.

Issa S. Al-Amri is an Associate Professor at the Department of Biological Sciences and Chemistry, and Dean of College of Arts and Sciences at the University of Nizwa, Oman. He received his BSc in Biomedical Sciences from Glasgow Caledonian University, MSc in Biological Electron Microscopy from the University of Wales, Aberystwyth, and PhD in Biological Sciences from the University of Portsmouth in the UK. Dr. Al-Amri worked at the Department of Pathology, College of Medicine & Health Sciences, Sultan Qaboos University at the capacity of Director of Diagnostic Electron Microscopy Unit for more than 24 years before joining the University of Nizwa in 2012. He published more than 113 articles in international peer-reviewed journals, book chapters, and conference proceedings. Dr. Al-Amri experience and research interests include histopathology, stress physiology, reproductive biology and endocrinology, cancer biology, life sciences, nanotechnology, electron microscopy, and microanalyses

Abdulaziz AlKindi is a Professor of Physiology at the Department of Biological Sciences and Chemistry, Vice Chancellor of Academic Affairs and the Secretory General of the Boards of Trustees at the University of Nizwa, Oman. He received his BSc and MSc from the University of Arizona in USA and his PhD from the

University of Exeter in the UK. Prof. AlKindi worked as an Associate Professor at the Department of Biology and Dean of College of Science at Sultan Qaboos University between 1988 and 2009. He published more than 90 articles in international peer-reviewed journals and conferences. Prof. AlKindi main research interests focused on stress and reproductive physiology and endocrinology.

Mohammed Hocine Benaissa has completed his doctorate in veterinary medicine at the High National Veterinary School of Algiers (Algeria), then his MSci in animal reproduction from the same school, has received his PhD in Veterinary Theriogenology (ENSV Algiers), then the postdoctoral degree "habilitation to manage research teams" (HDR) at the same school. He joined the Scientific and Technical Research Centre for Arid Areas (CRSTRA- Algeria) as head of the camel research programs, and Head of Experimental station. Currently, he is working as research director and Head of the division of biological resources in arid regions and scientific project manager in animal production sector. He is a member of several national and international scientific societies including ISOCARD, AVN, AASEA, etc. He is serving as an editorial member of several international reputed impact factors and non-impact factor journals. He has authored more than 40 international peer-reviewed scientific articles and has presented his research at many international scientific conferences worldwide. His research interests span from theriogenology and Veterinary epidemiology to camel farming system and animal reproduction.

Pamela Burger graduated in 2004 as a veterinarian at the Vetmeduni Vienna with a special focus on mitochondrial disorders in cheetahs and snow leopards. Immediately afterward, she started her first project about the genetic origin and domestication of Old World camel. Main questions were to identify the genetic origin and species status of the last wild two-humped camels in Mongolia and China, and the process of domestication in dromedaries and Bactrian camels. In 2015, she founded the International Camel Consortium for Genomic Improvement and Conservation together with colleagues from Italy, Saudi Arabia, France, and Kuwait. In 2021, Pamela Burger acquired her Habilitations (equivalent to Associate Professorship) in "Animal Genetics and Conservation" and is head of the Population Genetics and Conservation group at the Research Institute of Wildlife Ecology, Vetmeduni Vienna; her current projects investigate the immune genome of camels and felids in response to economical relevant and zoonotic diseases.

Elena Ciani graduated in 2001 in Animal Production Science at University of Pisa (Italy) and got a PhD in Animal Breeding and Genetics in 2006, as a joint title between University of Pisa and University of Limoges (France). Her main early research interest has been in sheep genetic diversity and evolution, also thanks to her active participation within the International Sheep Genomic Consortium. Since 2009, she has been involved in EU-funded projects (PROCAMED, CARAVAN, CAMEL-SHIELD) addressing the development of the camelid sector in Northern

Africa. In 2019, she was appointed, on behalf of the International Camel Consortium for Genetic Improvement and Conservation, winner of the Illumina® Agricultural Greater Good Initiative, a whole-genome sequencing project of worldwide dromedaries to better understand genetic diversity and evolution of this species as well as to develop a commercial genotyping tool (SNP array) to foster genomic-based breeding management.

Asim Faraz secured the degrees of DVM, MSc. (Hons.) Livestock Management and PhD Livestock Management from University of Agriculture Faisalabad (UAF) Pakistan. Currently, he is serving as Assistant Professor in Department of Livestock & Poultry Production, Faculty of Veterinary Sciences, Bahauddin Zakariya University Multan Pakistan. Moreover, he is performing the duties of Technical Advisor Dairy and Cattle Farmers Association (DCFA) Pakistan. He is Secretary General Camel Association of Pakistan (CAP) & Pakistan Camel Club (PCC). He is also Member Executive & Public Relations Officer (PRO) International Society of Camelid Research and Development (ISOCARD). He is acting as Focal Person for International Camel Organization (ICO). He is the Chief Editor of the International Journal of Camel Science, and editor of a special issue in Animals (Trends in Camels Health & Production). Furthermore, he is Promoting Camels for Life & World.

Meriem Fatnassi is an Animal Production Researcher in the Division of Animal Production Research of the Department of Agricultural at Ministry of Municipality Qatar. She graduated in Animal Production Engineer, and she got a Master in Genetics and Animal Reproduction, before completing her PhD at the University of Agronomic Science of Sousse (Tunisia) and Arid Lands Institute Medenine (Tunisia). She has 10 years of experience in scientific research and teaching. She has published more than 10 articles in journals with impact factors, participated in several national and international meetings, and supervised the research work of many post-graduate students. Her research areas are animal reproductive physiology and camel behavior.

Bernard Faye Is a veterinarian, specialized in tropical veterinary medicine, PhD Paris University and HDR Montpellier University. He lived in Africa for 8 years (Ethiopia, Niger) for research and development activities (1975–1983) before joining the National Institute of Agricultural Research (INRA) as director of ecopathology laboratory. Joining CIRAD (Centre for International Cooperation in Agricultural Research for Development) as Head of Animal Productions Program (1996), then as Scientific project manager (2007). His interest in camels started in Ethiopia in 1975, studying mineral metabolism. Gradually, through his multiple research programs in cooperation and his international network of camel scientists (North, Western and Horn of Africa, Middle-East, India, Central Asia, Latin America), he founded the International Society of Research and Development on Camelids (ISOCARD). From 2010 to2015, he worked in Saudi Arabia as FAO consultant in a camel research center. He is currently emeritus researcher at CIRAD and independent international camel expert. He is also an author of more than 450 scientific publications, 340 communications in national and international conferences, 45 books and chapters in book, 210 technical reports, and 12 scientific editions.

Syeda Maryam Hussain got her PhD in Animal Production and Quality from University of Sao Paulo, Brazil. She also worked on Animal toxicity and Biochemistry in Poland and USA. Currently, she is serving as an Assistant Professor in Department of Livestock Production and Management, Faculty of Veterinary and Animal Sciences, Pir Mehr Ali Shah-Arid Agriculture University Rawalpindi. Moreover, she is member of Camel Association of Pakistan (CAP). Her research specially focuses on Animal welfare, behavior, and stress models, mainly due to management. She is currently running projects with focus on reducing the environmental stress in rabbits and a housing model for reducing the diseases incidence in milking goats. Moreover, she is enthusiastic toward developing special welfare protocols for cattle and camels in Pakistan.

Rafat Al Jassim Rafat Al Jassim is an animal scientist with specialisation in Nutrition Biochemistry and Gut Microbiology. During the past 28 years, the focus of his research programs was on the impact of commensal gut microorganisms in large animal health and disease and the role of gut microbial ecology in the nutrition of domesticated and wild herbivores. Interaction between diet and the microbial community of the intensively managed animals has been of particular interest.

Dr Rafat Al Jassim holds a BSc in Animal Science from Baghdad University, an MSc in Animal Nutrition from the University of Wales UK and a PhD in Animal Nutrition from the University of New South Wales, Australia. Before joining UQ in 2000, Dr Al Jassim was a Senior Research fellow at the University of New England (1997–2000), visiting Scientist at Ian Clunies Ross Laboratories CSIRO Sydney (1995–1997), Associate Professor at Jordan University of Science and Technology (1991–1995), and Associate Professor at Baghdad University (1985–1991). Dr Al Jassim has retired from The University of Queensland in 2016 after 16 years of service at Gatton Campus and 38 years in academia. As of Jan 2013, Dr Al Jassim started a Professorial position at the African Sustainable Agriculture Research Institute (ASARI), Mohammed VI Polytechnic University (UM6P), Morocco.

Judit Juhasz is a veterinarian, received her degree in 1988 from the University of Veterinary Medicine in Budapest, Hungary. She worked as a reproductive specialist at the Experimental Station of the University where she developed the equine artificial insemination program. She received her PhD degree in 2002 on Equine andrology. Judit has been working with dromedary camels over 20 years. In 2000–2002, she worked in the Sultanate of Oman to develop an embryo transfer and artificial insemination program for racing camels. Since 2003, she has been in

Dubai, UAE to develop, then to manage the world first large-scale camel milking farm (EICMP, Camelicious®) where she has been the Chief Veterinary Officer since 2006. Judit has special interest in sustainable animal welfare and large-scale farm management. She is also one of the founders and co-organizer of the ICAR Satellite Meetings on Camelid Reproduction.

Isam T. Kadim is a professor in the Department of Biological Sciences and Chemistry, College of Arts and Sciences at University of Nizwa, Sultanate of Oman. He obtained his Ph.D. from Massey University, New Zealand. He has more than 40 years of teaching and research experience at different international institutes. He has supervised many M.Sc. and Ph.D. students; published more than 120 refereed articles, four books, and 17 book chapters; and presented 85 papers at scientific conferences. Muscle biology is a major focus of his research and publications. He has obtained more than US $1.4 million funding for his research projects

Laura Menchetti is currently a temporary researcher at the School of Biosciences and Veterinary Medicine of Camerino (Italy) where she teaches physiology, ethology and animal welfare, and Animal adaptation to environmental conditions. She graduated in Animal Science and obtained a PhD degree in Animal health, livestock production and food safety at the University of Perugia, Italy. She also obtained the Master's degree in Biostatistics for Clinical Research and Scientific Reporting at the University of Padova, Italy. She previously worked as a researcher and adjunct professor at the University of Perugia, Bologna, and Milano (Italy). Laura Menchetti's research focuses on animal physiology and, in particular, on the relation between nutrition, welfare, and productivity in rabbits, and behavior and welfare in dogs and dromedary camels. She is the author of more than 90 papers published in international indexed journals and is involved in several research projects. She is currently the national secretary of Italian Federation of Doctors in Animal Science (FIDSPA) and member of the Italian Society of Veterinary Physiology (SoFiVet).

Coralie Le Meur is a trainer. She is based in France and she is training many dromedaries for fairs and riding purposes. She has many years' experience in animal handling and training. She practices camel races in France to promote camel activities on French soil. More information on www.dromasud.fr

El Khasmi Mohammed is a professor in the Department of Biology, Ben M'Sik Faculty of Sciences, Hassan II University of Casablanca, MOROCCO. Holder of a PhD on phosphocalcic and bone metabolism of the dromedary, the author has more than 30 years of experience in scientific research, teaching, and community services. He has published a large number of articles in indexed journals, participated in several national and international meetings, and supervised the research work of a large number of master's and doctoral students. His main interests are metabolism, endocrinology, and the impact of stress on the homeostasis and quality of camel products.

Peter Nagy Is a veterinarian, graduated from the University of Veterinary Medicine in Budapest, Hungary in 1990. He worked in his alma mater until 1999 as assistant professor on large animal reproduction, theriogenology, and endocrinology. He obtained his PhD degree in 1998 on Equine reproduction. Peter has been working with dromedary camels over 20 years. In 2000–2002, he developed an embryo transfer and artificial insemination program for dromedary camels in the Sultanate of Oman. Since 2003, he has been in Dubai, UAE to develop and manage the world's first large-scale camel milking farm (EICMP, Camelicious®). Peter is a founding diplomat of the European College of Animal Reproduction (ECAR), an honorary professor at the Széchenyi István University in Hungary and chairman of the Camelid Working Group of EAAP. He is one of the founders and co-organizer of the ICAR Satellite Meetings on Camelid Reproduction.

Barbara Padalino is an associate professor in the Division of Animal Sciences of the Department of Agricultural and Food Sciences, University of Bologna, Italy. She graduated in veterinary medicine from the University of Bari (Italy) and completed her PhD at the University of Sydney (Australia). She is a diplomate of The European College in Animal welfare, ethics and law and a member of the Animal Health Animal Welfare Panel of the European Food Safety Authority. She has more than 20 years of experience in scientific research, teaching, and community services. She has published more than 130 articles in indexed journals, participated in several national and international meetings, and supervised the research work of a large number of master's and doctoral students. Her main research interests are behavior, health, and welfare of different species, including dromedary camels, equine internal, and sports medicine.

Carlos Iglesias Pastrana is a veterinary doctor and predoctoral researcher at the University of Cordoba, Spain. He is working on the genetic and cultural diversity of endangered animal resources. His work focuses specifically on dromedary cognition, social behavior, ecology, welfare, and other functional traits with relevance to animal breeding and sustainable conservation. Carlos additionally participates in international projects aiming at the study of wildlife (primates and ungulates) and donkey behavior.

J. A. (Lulu) Skidmore graduated with a BSc (Hons) in Animal Science (Wye College, University of London, UK. 1985), before spending 5 years at the Equine Fertility Unit in Cambridge, as a research associate specializing in equine reproduction. In 1991, she joined the Camel Reproduction Centre as a post-graduate research student and successfully completed her PhD, "Reproduction in the dromedary camel," in 1994 (University of Cambridge, UK). Subsequently, she has remained as Scientific Director of the Camel Reproduction Centre in Dubai for the last 30 years leading a team of scientists and has published over 130 publications covering all aspects of camel reproduction physiology including embryo transfer, artificial insemination, cryopreservation of embryos and semen, and hybridization of New and Old World camelids. She has also been Chairman and co-organizer of the ICAR Satellite Meetings on Camelid Reproduction.

Martina Zappaterra is a senior researcher at the Department of Agricultural and Food Sciences (DISTAL) of Alma Mater Studiorum University of Bologna, Italy. She began her career in the field of animal breeding and genetics, and then went on to work in the field of animal science, applying different statistical and laboratory methodologies in studies evaluating animal production sustainability. At the time of writing this chapter, she has been involved in more than 50 scientific papers published in peer-reviewed journals. She has also co-authored studies on camel welfare and the provision of shade to dromedary camels.

The Camels: From Domestication to the Twenty-First Century

1

Bernard Faye

Abstract

The domestication of large camelids is relatively recent in the history of mankind (5–6000 years) and was motivated primarily by their role in the transport of goods and humans, for war or trade. Indeed, milk and meat were probably only "by-products" of the carrying activities. The expansion of large camelids throughout history has brought them to new territories, on the margins of the Sahara, to other desert regions of the world, and recently in Western countries, i.e., in environments to which they are not necessarily adapted. Due to the changes in use and the intensification of animal husbandry, the welfare of these animals may become a concern, and welfare standards for this species are needed.

Keywords

Large camelids · Domestication · Multipurpose animal · Intensification · Adaptation

1.1 The Camel Domestication

The Camelidae family originated in North America about 40 million years ago (Burger 2016). From this original nucleus, two migrations, one to South America through the Isthmus of Panama and the other to the Asian continent via the Bering Strait, gave rise to the two current large branches of the camelid family, respectively, the small Andean camelids (Lamini group) and the large camelids (Camelini group). The divergence between these two groups would have occurred 11 million years ago

B. Faye (✉)
UMR SELMET, CIRAD-ES, Montpellier, France

Al-Farabi National Kazakh University, Almaty, Kazakhstan

B. Padalino, B. Faye (eds.), *Dromedary Camel Behavior and Welfare*, Animal Welfare 24, https://doi.org/10.1007/978-3-031-48600-5_1

in North America, and the ancestor of the small camelids would have arrived in the Andean mountains approximately 2 million years ago (Wheeler et al. 2006). For its part, the ancestor of the large camelids would have migrated to Asia around 8 million years ago. The large camelids in turn divided into dromedary (one-humped camel) and Bactrian (two-humped camel), 4–5 million years ago; the former migrating to the warmer lands of the Arabian Peninsula and the latter to the colder lands of central Asia. A final divergence occurred less than 1 million years ago between the Bactrian camel and the one that some call the "Tartarian camel" whose remnant is the current wild camel (Hare 1999).

This long history has therefore structured the current family of camelids which today includes three genera and seven species. Two genera of *Camelus* (the **dromedary** (*C. dromedarius*) also called Arabian camel or single-humped camel and the **Bactrian** (*C. bactrianus*) or double-humped camel, sometimes called Asian camel) were domesticated. The third genus, the **Tartarian camel**, longtime regarded as a feral Bactrian camel remaining wild, was recently recognized as a different species (*C. bactrianus ferus*) as a result of molecular genetic studies showing a clear divergence by making a full-fledged genotype (Li Yi et al. 2017). It is therefore a cousin and not a direct ancestor of Bactrian.

Domestication would have occurred for Bactrian 5–6000 years ago probably in an area more Western than previously thought, toward Uzbekistan and present-day West Kazakhstan, rather than toward Mongolia (Sala 2017). The name "Bactrian" comes from a region located between Afghanistan, Iran, and present-day Kazakhstan. The domestication of the dromedary would be more recent (3–4000 years) and in all probability in the south-east of the Arabian Peninsula (current Sultanate of Oman, United Arab Emirates, and southern Saudi Arabia) (Uerpmann and Uerpmann 2002). According to current data (Digard 2009), large camelids are therefore among the last large species domesticated by humans (Fig. 1.1).

However, the camelid family has expanded since domestication with the creation of crossbreed and hybrids. Formally, the crossing of the dromedary and the Bactrian is a crossbreeding rather than hybridization, both belonging to the same genus. Nevertheless, this practice, common since antiquity along the trade routes of the Asian continent ("Silk Roads"), consisted in obtaining hybrids combining the strength of Bactrian and the endurance of the dromedary, qualities very useful in caravan activity (Dioli 2020).

Nowadays, hybridization is actively implemented in two contexts: (i) in Central Asia, and especially in Kazakhstan, to obtain females producing more milk than the parents and higher in fat, and (ii) in Türkiye, to obtain males reputed in the festivals of "camel wrestling" (Figs. 1.2 and 1.3). In fact, Kazakh and Turkish breeders have set up several hybridization schemes depending on whether the male is one or two humps (Faye and Konuspayeva 2012; Dioli 2020).

The domestication process mainly consists in the reduction of human fear, i.e., tameness (Jensen 2014), and leads to anatomical, physiological, and behavioral

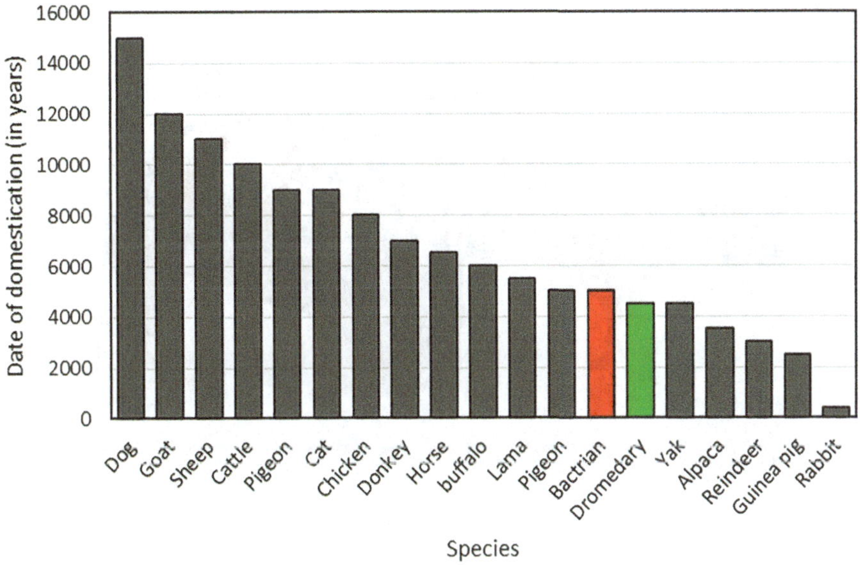

Fig. 1.1 Approximate dates of domestication of animals (source: B. Faye according to the data retrieved from https://fr.mahnazmezon.com/articles/science/animal-domestication-table-of-datesand-places.html on 10/07/23)

Fig. 1.2 Hybrid F1 for the milk production – Kazakhstan (Credit: B. Faye)

Fig. 1.3 Hybrids F1 for camel wrestling – Türkiye (Credit: B. Faye)

changes that Wilkins et al. (2014) have described as the “domestication syndrome”.[1] In a recent study (Fitak et al. 2020), the underlying genes responsible for such “domestication syndrome” were investigated both in dromedary and Bactrian camels showing that several genes associated with neural crest cell deficiencies (linked to behavioral changes in domesticated animals) and altered thyroid hormone-based signalling (linked to tameness) common to other domesticated mammals were present in camels. While domestication of camels is visible through their genomic signatures, the main reason for this domestication is still hypothetical.

Indeed, the oldest evidence of camel use is related to some petroglyphs in Central Asia or North Africa and in Sumerian frescoes. In all the cases, those representations show camels used for riding during hunting (Fig. 1.4) or for military purposes.

One of the oldest Sumerian frescoes (4000 BP) is showing a military raid achieved by Arab tribes in the Euphrates valley, the Arabs riding camels with their spears and bows (Fig. 1.5).

Such “military vocation” is perpetuated until nowadays through the “camel corps,” maintained in the army of desertic countries. Camel riding is also maintained among fans of falcon or eagle hunting in Middle East and Central Asia. However, only camels with slender conformation were adapted to such activity, while stocky camels were selected for packing. In other words, at the beginning of the domestication, camels were submitted to hard work, either they were forced to run long distances carrying fighters with their weapons or occasionally to carry a high quantity of goods on their back for long distances.

[1]Domestication syndrome refers to a set of hereditary phenotypic, physiological, and behavioral traits that are common to domesticated animals, but not in their respective wild species variants.

Fig. 1.4 Petroglyphs showing antelope hunting by camel riders (Alashan Desert, Inner Mongolia, China) (Credit B. Faye)

Fig. 1.5 Sumerian frescoes showing camel raid by Arab tribes in the second millennium BC (museum of Doumat-Al-Jandal, Saudi Arabia) (Credit: B. Faye)

1.2 The Domestic Camel Migrations: Wars of Conquest and Trade Routes

From the places of origin of their domestication, the large camelids spread into the ancient arid and semiarid world by two main routes: war and trade. If their use in the war of conquest was formerly limited to the margin of Arabian Peninsula in ancient times, the role of camels in the Muslim conquest in the north of Africa after Hegira is not deeply documented. Regarding Bactrian camel, the conquests of Genghis Khan in the thirteenth century were achieved thanks to the Mongol horse rather than to the camel. Most probably, the venue of the dromedary camels in Africa is linked to his role as packing and carrying animals (Agut-Labordère and Redon 2020) rather than as an auxiliary of warriors. The merchants all along the trade routes took advantage of the camel's ability to survive without water or abundant food for several days and even weeks.

The weight of packages on the back of the animals could be highly variable. According to the literature, the load capacity could vary from 100 to 680 kg depending on the distance to travel and animal weight (Schwartz 1986). However, these values mainly attest to the ability of caravaners to overload animals. Various regulations in the colonial era limited the loads to about 200 kg or even less (Faye 1997). In caravans, animals are tied one behind the other in one or two columns, the rope usually passed through a nasal ring being attached to the tail of the animal in front. The package, which varies according to the goods, is based on a wooden frame adapted to the morphology of the animal. The load is usually distributed symmetrically on each side of the hump. However, the friction of the pack on the back skin leads to wounds, occasionally extensive. Harness wounds are the more common disorders in packing animals (Muhammad et al. 2006), difficult to cure if the animal is not exempt from porterage (Fig. 1.6).

The practice of packing requires training animals. This training begins at the age of 4–5 years, but the full load is assigned to the dromedary only from 6–8 years. The career of a "carrier" camel can last 12 years. Nowadays, even if long caravans crossing the Sahara or Central Asia (silk road) disappeared, short-distance caravans are still available for carrying non-perishable goods in remote places such as salt or fodder (Brachet 2004), and they still participate in trade or barter, for example, between salt and cereals (Bernus and Bernus 1972; Museur 1977). Notably, they are used also in ways of cross-border smuggling, i.e., on potentially more dangerous routes away from water points, for example (Alary and Faye 2016).

1.3 From Extensive to Intensive Farming: From the Desert to the Barn

Dedicated, by its physiology oriented toward survival in the harsh conditions of the desert, to pastoral mobility, the camel has been confronted in recent decades with a profound change in its farming mode, gradually oriented toward more intensive systems (Faye 2016a).

Fig. 1.6 Harness wound in packing camel at the vet clinic of Dire-Dawa, Ethiopia (Credit, B. Faye)

For a long time, either the camel walked in the desert spaces from one oasis to another along the trade routes or it valued these same spaces by ensuring the production of milk and more rarely meat for nomadic populations (Faye et al. 2017). In all cases, its farming model was characterized by its mobility and sobriety. In a context of scarce and dispersed resources (fodder and water), only mobility allowed him to meet his nutritional requirements. Its grazing behavior is characterized by high selectivity and the large diversity of the plants ingested (Slimani et al. 2013).

A study carried out in Sudan showed that a camel could graze between 15 and 22 different plants in a determined environment, from herbaceous tufts at ground level to Sahelian trees several meters high. This is much more compared to sheep able to graze only from 12 to 15 different plants, at no more than 30 cm height (Rutagwenda et al. 1989). Fairly gregarious, the camel rarely concentrates with its congeners in a small area. On the contrary, it disperses spatially moving up to more than 20 km daily to ensure its diet. Browser rather than grazer, he constantly practices an "ambulatory pasture" consisting of grabbing here and there the different plant strata from the ground up to 2.50 m height without overgrazing any of them (Moussa et al. 2020).

The current trend toward more intensive systems marks the end of mobility. Settlement, therefore, puts an end to pastoral diversity. While in pastoral areas, the camel spent more than 8 to 12 h a day enjoying a diversified but often low nutritive food, in intensive systems where it is distributed at best twice a day a rich ration (typically alfalfa hay + concentrates), it spends little more than 2 h consuming its daily diet (Breulmann et al. 2007). From animal welfare point of view, this trend toward settlement and intensification of production systems results in a shift from work in open space to less activities in limited space and from a diversified diet but

Fig. 1.7 The camel from desert (**a**) to stable (**b**)

with low protein-energy concentration to an enriched but monotonous diet (Fig. 1.7a and b). Such a change is not without effect on dietary status, sedentary lifestyle leading to obesity, and altered digestive physiology with the appearance of metabolic disorders (Faye 2016a). The camel gains easy access to feed, but its physiology widely oriented toward sobriety, endurance, and anticipation of difficult periods (drought, undernourishment) can be disturbed by these new living conditions (protein overeating, sedentary lifestyle, limited space, boredom) (see Chap. 4 on good feeding).

Another consequence of intensification and more generally of the changes in camel production systems is the transformation of the camel from "carrier" animal to "carried" animal. Nowadays, his role is less to carry goods and humans on his back or to pull carts than to be transported himself from one point to another for going to slaughterhouse, market, or for export. However, transport conditions are often sources of stress with physical consequences (injuries, various traumas) on the living animal (Bauer and Havenstein 2017) or on meat quality (El-Khasmi et al. 2011) (see Chaps. 10 and 11).

Such changes in camel farming have implications also for the human-animal interactions. Element of Bedouin identity (Faye 2016b) and more widely of nomadic identity, the camel may have been an "idealized" animal (Faye and Brey 2005) because it was ideal in his relationship with the environment and in his ability to satisfy the requirements of man in the desert as a multipurpose animal. Sung by poets, adulated by the cameleers, sometimes adorned and made up in festivals and beauty contests, and venerated by the Qur'an (Al-Jassim and Al-Ani 2015), it is often at the top of the animal hierarchy (Cabalion 2013).

Both at the milking time (manual), during care, watering, or guiding to pasture, the cameleer (owner or shepherd) constantly accompanies the animal, weaving with their camels an almost fusional relationship (Breulmann et al. 2007).

Contrariwise, in specialized intensive systems (dairy farms, feed-lots, racing stables), the camel is subjected to a more utilitarian design, intended to produce for the market, and for which productivity becomes a parameter to be particularly taken into consideration. The development of mechanical milking and reproductive biotechnologies, the automated distribution of the diet, the ad libitum watering, and

Fig. 1.8 Wire mesh roller containing straw or hay used for spreading out feeding time

the presence of more abundant technical staff in large production units tend to "anonymize" the animal and "technologize" human-camel relations (Faye 2016b).

Assisted reproduction or mechanical milking requires a restraint system (forced animal in corridor of milking parlor or in restraint cage) and a set of manipulations, requiring invasive practices such as the fitting of milking sleeves or transrectal examination (Nagy and Juhasz 2015) or even electroejaculation used for semen collection in male (Tibary and Anouassi 1997). Those practices can be stressful for the animal, in particular when they are untrained (see Chap. 9).

Sedentarization and intensification, therefore, tend to distend the human-animal relationship to be included in a more impersonal process imposed by the technical contributions of modernity in camel breeding. Some alternatives were suggested and applied in modern intensive farms as regular walking out of the paddocks for at least 2 h/day, distribution of hay in wire mesh roller (Fig. 1.8) and concentrates in automatic dispenser for imposing regular feeding all along the day, and sensitization of the staff for manipulating camel (entry into corridor or restraint cage, use of milking machine, weighing animals, etc.) without using sticks or other too coercive methods (see Chap. 8 on behavior and training).

1.4 New Locations and New Uses

Until recently, the geographical expansion of large camelids from their domestication area (Arabian Peninsula, Central Asia) was achieved mainly through military and commercial activity in the arid and semiarid regions of the old world as mentioned above. Toward the end of the twentieth century, climate change contributed to a continuation of African expansion on the Sahara margins (Faye

2020). But this conquest of new territories and spaces by the large camelids remained limited to regions marked by their aridity, i.e., to environments to which the camel is perfectly adapted. Similar considerations can be made about the introduction of camel in the desert area of Namibia or Botswana in southern Africa (Wilson 2012; Seifu et al. 2019). However, those new camel farmers are not accustomed to the animal and could achieve occasionally mistakes in the management.

Anyway, for securing their production system following recurrent droughts affecting more sub-humid regions, some pure cattle breeders such as Fulani in Western Africa or Maasai in Eastern Africa have switched their bovines to camels. Therefore, the camel herd can be confronted with climate factors favorable to the emergence of new diseases as has been observed many times in the Horn of Africa (Roger et al. 2000; Dirie and Abdurahman 2003; Khalafalla 2017) (see Chap. 6 on good health).

The introduction of camel in the USA (Baum 2011) and overall, in Australia (McKnight 1969) started in the nineteenth century, but contrary to Southern Africa, the importation of camel herds in those new locations was accompanied by cameleers, native from Arab countries or Pakistan (Jones and Kenny 2010) able to manage the animals in their new arid environment. During the twentieth century, the main part of the Australian camel population returned to the wild. The favorable environmental conditions in the desert area of the inland facilitated their growing demography, leading to increased pressure on the local resources, especially after the recent drought. A population of around half a million wild Australian camels was regarded by the authorities as excessive density requiring strict management control (Hart and Edwards 2016), including shooting from helicopters. Facing international critics regarding the brutality of the control methods applied in such a context, a new project aiming to achieve such control under high standards of animal welfare was implemented (Hampton et al. 2016). The two main methods used for controlling the camel population, i.e., aerial shooting and live capture (mustering and transport to the slaughterhouse), were based on transparent procedures through an audit system assessed by independent observers.

The implementation of camel farming in Europe is not necessarily new, as it was revealed by archeological findings along the Roman ways up to the northern regions of the Roman Empire (Henrotay and Pigière 2012) where camels were used as packing animals. But, contrary to the formerly described introduction in the USA and Australia, the European location did not correspond to the desert or arid environment to which the camel is adapted. Moreover, for a long time mainly confined to zoological gardens and circuses (Faye and Brey 2005), his use for tourist attraction and more recently for milk production (Smits and Montety 2009) is increasing (Fig. 1.9), contributing to the diversification of the livestock production in the European Union (Faye et al. 1995).

In this new environmental and managerial context, the question of well-being is often raised by the European public more sensitive to this issue than in most countries of the South. The question of adaptation to a wetter and colder environment with abundant resources is legitimate. However, it should be recalled that their

Fig. 1.9 New camel dairy farm implemented in France

original environment is characterized by the following: (i) a strong thermal amplitude in the day (hot day, cold night sometimes freezing) or seasonal, especially for Bactrian (+45 °C in summer, up to −30 or even −40 °C in winter); (ii) the scarcity of food resources, which forces them to travel long distances to feed (up to 40 km per day) and to spend a lot of energy for their walking; (iii) the scarcity of water, which forces them to stay without drinking for several days or even weeks depending on the quality of the fodder; (iv) a more or less aggressive soil for the feet, either sand (ergs) that does not temper the thermal amplitudes (burning during the day, freezing at night) or pebbles (regs) traumatic for an animal without hoof; and (v) a harsh health environment marked by two major parasitic diseases that affect more than 30% of animals on average (and up to 100% in more intensive systems), trypanosomosis and scabies. Added to this is the diarrhea of baby camels which leads to a high mortality rate, up to 20% in a year (Higgins and Kock 1984).

Obviously, camels are adapted to survive in such conditions, where most other species would not survive, but it does not mean that they cannot cope with a better environment, or that they can have a good welfare status without being in a survival mode. Indeed, when the conditions are generally much more favorable, in terms of thermal comfort (see Chap. 5 on good housing), resources (abundant grass, absence of significant water shortage, see Chap. 4 on good feeding), and sanitary coverage (dense network of veterinarians, see Chap. 5), camels are able to adapt to their new environment having an enhanced welfare status. Their longevity is also higher, and the sanitary pressure much less. Moreover, due to the lack of specialized slaughterhouses, European camels are not culled and are living fully their life (Faye et al. 1995; Karpyta 2021).

The management of these animals in Europe and beyond nevertheless requires specific wariness: (i) livestock buildings adapted in terms of space (surface, ceiling height) and sufficiently protective against bad weather and wet season; (ii) a diet based on roughage to avoid excessive fat deposit (obesity) conferred by diets too rich in crude proteins (spring grass), and on concentrates in reasonable quantities to fill the energy-deficient periods (winter period which also corresponds to the breeding season, calving, and mating); (iii) special care for metabolic disorders of food origin (laminitis, hepatorenal insufficiency) while the main infectious or parasitic diseases (trypanosomosis in particular) specific to camels in Africa and Asia are normally absent from European continent; (iv) more specific training for veterinarians with camel farms in their clientele; (v) good quality of riding saddle for tourist activities to avoid harness wounds already mentioned above; (vi) properly maintain callosities (sternum, knees, and elbows) that allow animals not to suffer on erosive surfaces such as asphalt or concrete; (vii) facilitate interactions with people to avoid stress; and (viii) train new camel owners to handle their animals.

Concerning the new uses of camels, they only extend the formidable range of traditional uses that embed the camel as the archetype of a multipurpose animal (Hjort af Örnas and Ali Hussein 1993). Beyond its use for zootechnical production (milk, meat, wool, leather, manure), transport (packing, carting), and leisure activities (racing, camel trek), its carrying activities continue to diversify as evidenced by the development of "bibliocamels" (see, e.g., https://www.globalcitizen.org/en/content/camel-library-remote-villages-covid-19-ethiopia) or "camel-buses" for urban transport; he also increasingly plays the role of auxiliary of agriculture (plowing, harrowing, noria) and participates in the Gulf countries in beauty competitions. These uses can strongly solicit the animal beyond its working capacities or subject it to a stressful or even dangerous practice (e.g., doping). Recently, the injection of botox subcutaneously on camels participating in beauty contests in Saudi Arabia caused a scandal (see, e.g., https://www.theguardian.com/world/2021/dec/08/camels-enhanced-with-botox-barred-from-saudi-beauty-contest) indicating the urgency in these circumstances to establish rules of good practice for the well-being of these animals.

Overall, due to the large and different use of camels worldwide, it is crucial to develop some objective tool to measure their welfare status, so that it is possible to collect pieces of evidence on which to base new regulations for camel farming, transport, and killing (see Chap. 3) as emphasized, for example, for dairy production (Smits et al. 2023).

1.5 Conclusion

Since the domestication of the great camelids, emblematic animals of the old-world deserts, humans gradually contributed to bringing them out of their original environment to which their anatomy and physiology were adapted notwithstanding their welfare conditions. This "exit" is accelerated in recent decades, both geographically with territorial expansion beyond the deserts and zootechnically, the camel production system moving toward breeding intensification. These deep changes are characterized by their relative speed and raise new questions about animal welfare, the ability of camels to adapt to other environmental contexts, and the effects of these changes on metabolism and health. Data on the welfare status of camels both in their original habitat and in others are needed to objectively create welfare standards for camels worldwide.

References

Agut-Labordère D, Redon B (2020) Dromadaires et chameaux de l'Asie centrale au Nil dans les mondes anciens (IVe millénaire av J C – premiers siècles de notre ère) In: « Les vaisseaux du désert et des steppes: les camélidés dans l'Antiquité (*Camelus dromedarius* et *Camelus bactrianus*). Archéologies 2 MOM Editions, Lyon France, p 9–20

Alary V, Faye B (2016) The camel chains in East Africa-importance of gaps between the data and the apparent reality. J Camelid Sci 9:1–22

Al-Jassim R, Al-Ani A H (2015) Did Prophet Muhammad advise his followers to drink camel urine? In Proc of 4th Conference of ISOCARD 'silk road camel: the camelids main stakes for sustainable development' 8–12 June (2015 Almaty Kazakhstan) Konuspayeva G (ed). Special issue of Scientific and Practical Journal Veterinariya #2 (42) (2015 poster presentation 460–461)

Bauer H, Havenstein J (2017) The welfare of dromedary camels during road transport in the Middle East. Animals' Angels, Francfort Germany. https://www.animals-angelsde

Baum D (2011) The status of the camel in the United States of America. In: "Camel conference SOAS camel cultures: historical traditions present threats and future prospects" London 24–25 May 2011: 75–80

Bernus E, Bernus S (1972) Du sel et des dattes: Introduction à l'étude de la communauté d'In Gall et de Tegidda-n-Tesemt. Etudes nigériennes Publ Centre Nigérien d'Etudes en Sciences Humaines Niamey (Niger) 31: 130

Brachet J (2004) Le négoce caravanier au Sahara central: histoire évolution des pratiques et enjeux chez les Touaregs Kel Aïr (Niger). Les Cahiers d'outre-mer 226–227:117–136

Breulmann M, Boer B, Wernery U, Wernery R, El-Shaer H, Alhadrami G, Gallacher D, Peacock J, Chaudhary SA, Brown G, Norton J (2007) The camel from tradition to modern times. UNESCO Doha Publ, Doha (Qatar)

Burger PA (2016) The history of old world camelids in the light of molecular genetics. Trop Anim Health Prod 48:905–913. https://doi.org/10.1007/s11250-016-1032-7

Cabalion S (2013) Le « système domesticatoire » touareg (Tagaraygarayt Niger). vols 1 et 2 (Thesis in social anthropology and ethnology EHESS Paris)

Digard JP (2009) L'homme et le animaux domestiques. Publ Fayard, Paris France, p 328

Dioli M (2020) Dromedary (*Camelus dromedarius*) and Bactrian camel (*Camelus bactrianus*) crossbreeding husbandry practices in Turkey and Kazakhstan: an in-depth review. Pastoralism: Res Policy Pract 10:6. https://doi.org/10.1186/s13570-020-0159-3

Dirie MF, Abdurahman O (2003) Observations on little known diseases of camels (*Camelus dromedarius*) in the horn of Africa. Rev Scient Techn (International Office of Epizootics) 22(3):1043–1049. https://doi.org/10.20506/rst.22.3.1456

El Khasmi M, Riad F, Safwate A, Tahri EH, Farh M, El Abbadi N, Coxam V, Faye B (2011) Effects of preslaughter stress on meat quality and phosphocalcic metabolism in camels (*Camelus dromedarius*). J Camelid Sci 3:33–38

Faye B (1997) Guide de l'élevage du dromadaire. Ed Sanofi, Libourne, France, p 126

Faye B (2016a) The camel new challenges for a sustainable development. Trop Anim Health Prod 48:689–692. https://doi.org/10.1007/s11250-016-0995-8

Faye B (2016b) Des dromadaires et des hommes au Moyen-Orient: Identité et modernité. Anthropol Middle East 11(1):51–65

Faye B (2020) How many large camelids in the world? A synthetic analysis of the world camel demographic changes. Pastoralism: Res Pol Pract 10(1):25. https://doi.org/10.1186/s13570-020-00176-z

Faye B, Brey F (2005) Les relations entre chameaux et société: entre marginalisation et idéalisation. Revue Ethnozootechnie 77:43–50

Faye B, Konuspayeva G (2012) The encounter between Bactrian and dromedary camels in Central Asia. In: Knoll E-M, Burger P (eds) Camels in Asia and North-Africa-interdisciplinary perspectives on their past and present significance. Austrian Academy of Sciences press, Wien, Austria, pp 27–33. (photos p 248–250)

Faye B, Jouany JP, Chacornac JP, Ratovonanahary M (1995) L'élevage des grands camélidés Analyse des initiatives réalisées en France. INRA Prod Anim 8:3–17

Faye B, Senoussi H, Jaouad M (2017) Le dromadaire et l'oasis: du caravansérail à l'élevage périurbain. Cah Agric 26:14001

Fitak RR, Mohandesan E, Corander J et al (2020) Genomic signatures of domestication in Old World camels. Commun Biol 3:316. https://doi.org/10.1038/s42003-020-1039-5

Hampton JO, Jones B, Perry AL, Miller CJ, Hart Q (2016) Integrating animal welfare into wild herbivore management: lessons from the Australian feral camel management project. Rangel J 38(2):163–171. https://doi.org/10.1071/RJ15079

Hare J (1999) The lost camels of Tartary: a quest into forbidden China. Abacus Publ, London UK, p 272

Hart Q, Edwards G (2016) Outcomes of the Australian feral camel management project and the future of feral camel management in Australia. Rangel J 38(2):201–206. https://doi.org/10.1071/RJ15087

Henrotay D, Pigière F (2012) Camels in the northern provinces of Roman Empire. J Archaeol Sci 39:1531–1539. https://doi.org/10.1016/j.jas2011.11.014

Higgins AJ, Kock R (1984) The camel in health and disease. Br Vet J 140(5):482–504

Hjort af Ornäs A, Ali Hussein M (1993) Camel herd dynamics in southern Somalia: long term development and milk production implications. In: Hjort af Ornäs A (ed) The multi-purpose camel: interdisciplinary studies on pastoral production in Somalia. Uppsala University, pp 31–42

Jensen P (2014) Behavior genetics and the domestication of animals. Annu Rev Anim Biosci 2:85–104. https://doi.org/10.1146/annurev-animal-022513-114135

Jones P, Kenny A (2010) Australia's Muslim cameleers: pioneers of the inland 1860s–1930s. South Australian Museeum Wakefield Press, Cambridge USA

Karpyta F (2021) Le dromadaire future vache à lait ? Ca m'intéresse 488(Oct 21):28–31

Khalafalla AI (2017) Emerging infectious diseases in Camelids. In: Bayry J (ed) Emerging and re-emerging infectious diseases of livestock. Springer, Cham. https://doi.org/10.1007/978-3-319-47426-7_20

McKnight TL (1969) The camel in Australia. Melbourne University Press, Australia

Moussa M, Ouologuem B, Coulibaly L, Coulibaly N, Teme N, Nialibouly O, N'Diaye M, Traoré MD, Aune JB (2020) Comportement des dromadaires sur pâtures Sahélienne et Subhumide au Mali: 2 Espèces végétales et quantités de fourrage ingérées. Agron Afr 32(2):199–206

Muhammad G, Jabbar A, Iqbal Z, Athar M, Saqib M (2006) A preliminary passive surveillance of clinical diseases of cart pulling camels in Faisalabad metropolis (Pakistan). Prev Vet Med 76(3–4):273–279. https://doi.org/10.1016/j.prevetmed.2006.05.006

Museur M (1977) Un exemple spécifique d'économie caravanière: l'échange sel-mil. J Afr 47(2): 49–80

Nagy P, Juhasz J (2015) Present knowledge and future questions regarding intensive camel Milk production. In Proc of 4th Conference of ISOCARD 'silk road camel: the camelids Main stakes for sustainable development' 8–12 June 2015 Almaty Kazakhstan, Konuspayeva G (ed) special issue of Scientific and Practical Journal Veterinariya #2 (42) 2015 Key-note presentation: 51–57

Roger F, Yigezu LM, Hurard C, Libeau G, Mebratu GY, Diallo A, Faye B (2000) Investigations on a new pathological condition of camels in Ethiopia. J Camel Pract Res 7:163–165

Rutagwenda T, Lechner-Doll M, Kaske M, Engelhardt WV, Schultka W, Schwartz HJ (1989) Adaptation stratégies of camels on a thombush savannah pasture comparison with other domestic animals. Options Méditerranéennes, Série Séminaires 2. CIHEAM:69–73

Sala R (2017) The domestication of camel in the literary archaeological and petroglyph record. J Arid Land Studies 26(4):205–211. https://doi.org/10.14976/jals.26.4_205

Schwartz HJ (1986) The potential of the camel (*Camelus dromedarius*) as a transport and draught animal. In: "FAO-MINEADEP Workshop on "Camels" Kuwait Oct.1986 111: 2–25

Seifu E, Madibela OR, Teketay D (2019) Camels in Botswana: herd dynamics and future development implications Botswana. J Agric Appl Sci 13(1):12–25

Slimani N, Chehma A, Faye B, Huguenin J (2013) Régime et comportement alimentaire du dromadaire dans son milieu naturel désertique en Algérie. Livest Res Rural Dev 25(12) http://lrrd.org/lrrd25/12/slim25213html

Smits MG, Montety GJ (2009) Ammonia emission from camel dairy in The Netherlands. J Camel Pract Res 16(2):139–142

Smits M, Joosten H, Faye B, Burger PA (2023) The flourishing camel milk market and concerns about animal welfare and legislation. Animals 13:47. https://doi.org/10.3390/ani13010047

Tibary A, Anouassi A (1997) Theriogenology in camelidae. Actes Editions IAV Hassan II Rabat Maroc)

Uerpmann HP, Uerpmann M (2002) The appearance of the domestic camel in SE-Arabia. J Oman Studies 12:235–260

Wheeler JC, Chikhi L, Bruford MW (2006). Genetic analysis of the origins of domestic south American camelids. Documenting domestication: new genetic and archaeological paradigms, 329–341

Wilkins AS, Wrangham RW, Fitch WT (2014) The "domestication syndrome" in mammals: a unified explanation based on neural crest cell behavior and genetics. Genetics 197:795–808. https://doi.org/10.1534/genetics.114.165423

Wilson RT (2012) The one-humped camel in southern Africa: imports to and use in South-West Africa/Namibia. J Camel Pract Res 19(1):1–6

Yi L, Ai Y, Ming L, Hai L, He J, Guo FC, Qiao XY, Ji R (2017) Molecular diversity and phylogenetic analysis of domestic and wild Bactrian camel populations based on the mitochondrial ATP8 and ATP6 genes. Livest Sci 199:95–100. https://doi.org/10.1016/j.livsci.2017.03.015

How Early Domestication and Modern Genomics Contribute to Camel Welfare

2

Elena Ciani, Pamela Burger, Martina Zappaterra, and Carlos Iglesias Pastrana

Abstract

This book chapter explores the impacts of both early domestication and historical breeding practices on camel welfare-related traits, emphasizing the need for a holistic approach to enhance the well-being of these animals in view of the contemporaneous rise of socio-economic interest in camel breeding. Under the so-called "domestication syndrome", several genomic regions mostly controlling neural crest development and thyroid function have been identified in Old World camels. However, these animal species remain less explored than other conventional livestock species for the genetic mechanisms underlying domestication-related traits such as behaviour, coat colour, and social communication. In this scenario, the standardization of phenotype characterization activities and the ulterior use of genomic tools for further study of genotype-phenotype associations becomes prominent to increase the knowledge on the genetic architecture of welfare-related traits, thus their potential inclusion on marker-assisted selection schemes in the camel species. Additionally, precision livestock farming technologies can significantly support herd and breeding management under the

E. Ciani (✉)
Department of Biosciences, Biotechnologies and Environment, University of Bari 'Aldo Moro', Bari, Italy
e-mail: elena.ciani@uniba.it

P. Burger
Research Institute of Wildlife Ecology, Vetmeduni Vienna, Vienna, Austria
e-mail: pamela.burger@vetmeduni.ac.at

M. Zappaterra
Department of Agricultural and Food Sciences, Bologna University, Bologna, Italy
e-mail: martina.zappaterra2@unibo.it

C. I. Pastrana
Faculty of Veterinary Sciences, Department of Genetics, University of Cordoba, Cordoba, Spain

B. Padalino, B. Faye (eds.), *Dromedary Camel Behavior and Welfare*, Animal Welfare 24, https://doi.org/10.1007/978-3-031-48600-5_2

concerns of animal welfare for maintaining highly productive and healthy animals.

Keywords

Genomics · Marker-assisted selection · Welfare-related traits · Old World camels

2.1 Early Livestock and Camel Domestication from an Animal Welfare Perspective

The process of animal domestication that started 10,000 years ago (e.g. rev. in Frantz et al. 2020) continues shaping the relationship between animals and humans. Yet, a definition of domestication is still widely discussed with lack of consensus among scientists. A rather holistic concept was offered by Melinda Zeder (2015) considering both participants in the process, the domesticator and the domesticate:

> Domestication is a sustained multigenerational, mutualistic relationship in which one organism assumes a significant degree of influence over the reproduction and care of another organism in order to secure a more predictable supply of a resource of interest, and through which the partner organism gains advantage over individuals that remain outside this relationship, thereby benefiting and often increasing the fitness of both the domesticator and the target domesticate.

But what does this mean for the domesticated animal? Adopting Zeder's concept, which clearly goes beyond perceiving domestication as one-way subjection of animals by humans, in this chapter we will discuss (camelid) domestication and selection from an animal welfare point of view.

Across scientists and across species, there is a consensus that reduced fear and increased tolerance towards humans were central and at the basis of the domestication process. A similar suite of modifications in response to domestication including behaviour and behaviour-related phenotypes have been described in different mammal and bird domesticates (Jensen 2014). With the notion that selection for tameness would have been at the beginning of domestication, we need to take a closer look at the possible pathways to become a domestic animal. From initial management to intensive selection and breed formation, three types can broadly be classified (Zeder 2012, Larson and Burger 2013): (i) the commensal pathway, where the domesticate first moves into human habitat and benefits from a mutual relationship; (ii) the prey or harvest pathway, in which hunter and gatherers initially target a resource and later enhance its production traits according to their needs; and (iii) the directed pathway, where humans deliberately bring animals under their control in captivity and subject them to intensive selection and breed management. The pre-required behavioural traits in animals following the first two pathways would make them more suitable candidates for domestication, like high curiosity or docility and low flight distance as seen in cats and dogs, or pigs, sheep, goats, and New World camelids (Larson and Burger, 2013). The directed pathway, however, would require more knowledge and skills from humans to overcome domestication barriers like high flight distances or aggressive behaviour as we would find in horse, donkey, dromedary, Bactrian camel, rabbit, or mink, fox, and chinchilla. This last pathway comprises more recent domestication processes in which previously acquired techniques were transferred or more advanced technologies (e.g. artificial insemination or genetic manipulation) were applied to the wild target species (Zeder 2015).

2.2 Genetic Mechanisms of Domestication Related to Behaviour and Other Traits

The domestication process, no matter which pathway was involved, has always come with a change in phenotype and genotype in the domesticated species. In particular, behavioural changes with increased docility and tameness; anatomical changes with reduction in muzzle, teeth, ears, and brain size; and alterations to the endocrine and reproductive system have been described in the context of domestication and were summarized under the so-called "domestication syndrome" (Wilkins et al. 2014, 2021). Pioneering experimental work on silver foxes and rats has shown that strong selection for tameness (Trut et al. 2009; Albert et al. 2009 and 2011) can modify physiological, morphological, and behavioural phenotypes within a few generations. Underlying genetic networks influencing tameness included the synthesis of the neurotransmitter serotonin (tryptophan hydroxylase 1; *TPH1*) as well as the gamma-aminobutyric acid type A receptor subunit alpha5 (*GABRA5*) gene, which encodes an inhibitory neurotransmitter (Albert et al. 2011). Low fear of humans was associated with higher body weight and growth rates as well as reduced brain size in red junglefowls (*Gallus gallus*; ancestor of domestic chicken), that had been selected for eight generations either for high or low fear levels towards humans (Katajamaa and Jensen 2021). Gut-brain module analysis by metagenomics further revealed enrichment in microbial synthesis and degradation potential of metabolites associated with fear extinction in low-fear red junglefowls (Puetz et al. 2021).

Not only reduced fear but also the ability to cooperate and communicate with humans has been favoured in animals during domestication. For example, domestic pigs and ferrets showed some communicative skills that are usually attributed only to dogs, suggesting that this might be a general domestication trait (Jensen 2014).

Using ancient DNA analysis, selected genomic regions were identified in early domestic horses that were enriched for genes involved in cognition and social behaviour, like response to fear, guidance and learning ability, neural growth, but also locomotion, development of the skeleton and limbs, and the regulation of blood pressure (Schubert et al. 2014).

Another important domestication-related trait is coat colour, which has been a target for selection starting from the early phases of domestication (MacHugh et al. 2017). Genomic mechanisms underlying coat colour have been detected in different livestock species using also ancient genomic analysis. For example, chestnut (Agouti signalling peptide gene; *ASIP*) and black (Melanocortin 1 receptor; *MC1R*) colouration appeared at a very early time point during the process of domestication, ca. 500 years after the start of horse domestication (MacHugh et al. 2017). Genome-wide data from ancient aurochs and modern taurine and indicine cattle revealed selected genes (e.g. diacylglycerol O-acyltransferase 1; *DGAT1*) associated with lactation traits. Furthermore, selection candidate genomic loci included genes related to neurobiology, muscle development and growth, metabolism, and immunity, showing that behaviour and meat traits were key domestication targets in cattle (Park et al. 2015).

2.3 Selection during Domestication in Camelids from an Animal Welfare Perspective

New World camelids had a key role in the transition from hunter-gatherer to a mixed pastoral-agronomy in the Andean communities. The large wild herbivores guanaco (*Lama guanicoe*) and vicugna (*Vicugna vicugna*) were dominant in the upper Andean ecosystem and had great importance for the subsistence of early human populations until the domestication of the llama (*L. glama*) and alpaca (*V. pacos*) around 7000 years ago (Diaz-Maroto et al. 2021 and references therein). These species were an inherent part of the production system, culture, and cosmic views of the past Andean pastoralists and highly valued for their meat and fibre, which was also used for religious artefacts. Interestingly, the different behaviour of the wild species is also reflected in their domestication pathway – (ii) prey or harvest (Larson and Burger 2013). While vicugna are gregarious animals and form large herds of 100 individuals or more, guanacos tend to live in family groups of five to ten animals including one male and one to two females plus their offspring. Such differences in their behaviour would make vicunas a preferential hunting target for the upper Andean population, as they were found in larger quantities, while guanacos might have been easier to handle in a domestic setting of smaller breeding units (Diaz-Maroto et al. 2021).

The domestication process in Old World camels followed most likely the directed pathway (iii; Larson and Burger 2013) as humans had already gathered experience on the domestication of large mammals like horses or cattle. This knowledge was consecutively transferred to the wild ancestors of Bactrian camels that were domesticated 4000–6000 years ago (Bulliet 1975; Benecke 1994). This happened either in the historical region of "Bactria" (eponymous for Bactrian camels), located in north-eastern Iran and stretching until the Kopet Dagh mountains in south-western Turkmenistan (Benecke 1994), or further to the East, in the steppes of Mongolia and Kazakhstan within the historical distribution range of wild two-humped camels (Peters and von den Driesch 1997). Later, the practice of camel domestication was also applied to dromedaries (*Camelus dromedarius*), which were domesticated 3000–4000 years ago in the Southeast Coast of the Arabian Peninsula (Almathen et al. 2016). Notably, a third Old World camel species is recognized today, the wild two-humped camel (*Camelus ferus*), which has never been domesticated and diverged from the ancestors of modern domestic Bactrian camels around 1.2 million years ago (Mohandesan et al. 2017, Jirimutu et al. 2012; Wu et al. 2014).

Re-sequencing of multiple genomes (Fitak et al. 2020) showed positive selection signals during domestication of Old World camels for candidate genes consistent with the domestication syndrome hypothesis (Wilkins et al. 2014). Specifically, genes underlying neural crest deficiencies and altered thyroid hormone signalling associated with domestication-related behaviour were identified. It was hypothesized that a pan-domestication set of overlapping genes might be shared between different livestock species in connection with the selection for tameness during the early phase of the domestication process (Fitak et al. 2020).

2.4 The "Modern Genomics" Toolbox and its Use in Understanding the Genetic Architecture of Welfare-Related Traits

Welfare-related traits such as morphology, body condition and feed efficiency, stress and disease tolerance, and behaviour are complex in nature. Their phenotypic variation is continuously distributed, their underlying genetic architecture is polygenic, and their phenotypic manifestation can be even highly impacted by environmental variability factors. The above elements have long slowed progress in identifying their molecular basis, also because only single genes, or single small genomic regions, could be affordably investigated in the past.

The availability of methodologies allowing the investigation of the genome as a whole has been one of the greatest advances in the understanding of the genetic architecture of complex traits. Since the availability of the early DNA markers (e.g. microsatellites) amenable to genotyping in a genome-wide manner, various approaches have been developed allowing identification of genome regions under selective pressure. These approaches either relies on the availability of phenotype measurements for the considered complex traits, such as mapping of quantitative trait loci (QTL) and genome-wide association studies (GWAS), or focus on specific patterns of variation observable at the genomic level (e.g. patterns of heterozygosity, allele diversity and linkage disequilibrium), the latter methods being known as detection of "selection signatures" (Brøndum et al. 2015; Olschewsky and Hinrichs, 2021; Zhang et al. 2012). Since relying on the analysis of co-segregation of alleles at genome-wide markers and of quantitative phenotypes within a given family, QTL mapping is subordinate to the availability of reliable individual identification and genealogical recording systems, as well as on the availability of families sizes able to assure statistical power in the segregation analysis, which is typically the case in species and breeds where artificial insemination is routinely adopted.

In the last 15 years, short tandem repeat (STR) markers have been superseded by the more abundant, automatable, and more reproducible single nucleotide polymorphism (SNP) markers. SNP genotyping has been revolutionized with the advent of SNP array technologies (Nicolazzi et al. 2015) that allow fast, reproducible, and cost-affordable simultaneous genotyping of tens to hundreds of thousands of *loci* in a genome. Nowadays, various array formats are commercially available for domestic animals, from low-density (generally 1 to 20 k loci), to medium-density (commonly

50–60 k loci), to high-density (>500 k loci). These different formats are in part the heritage of the historical process behind SNP array design and development, with low- and medium-density arrays being the earlier version, while high-density formats generally coming later on, as new whole-genome re-sequencing data become available and allow the identification of millions of novel SNP *loci* in domestic animal species. But different formats are also justified by different applications, with low-density arrays being mainly used for DNA-based parentage tests, individual identification and product traceability, medium-density arrays mainly used for population genetic diversity studies and detection of selection signatures, and high-density arrays mainly adopted for genome-wide association studies (GWAS) and genomic selection (GS, see paragraphs below). Although biotech companies offer the option for the development of custom SNP arrays, the design and manufacturing costs are still high, and most of the commercially available arrays have been developed within the frame of large international collaborative projects and consortia (ISGC et al., 2010; Elsik et al. 2009). For minor species, where SNP arrays are not yet fully available, an interesting option for SNP genotyping is represented by a set of methods known as genotyping-by-sequencing (GBS) technologies (Gorjanc et al. 2015; Gurgul et al. 2019). These methods are based on the sequencing, using next generation sequencing (NGS) technologies, of a limited part of the whole genome, such as in the case of the sequencing of the exome, or the sequencing of the extremities of DNA fragments randomly generated via restriction enzyme-mediated digestion of the whole genome (restriction site-associated DNA sequencing (RAD) and double digest restriction associated DNA sequencing (ddRAD)) (Davey and Blaxter 2010; Peterson et al. 2012), or the sequencing of a region flanking a primer (single primer enrichment technology (SPET)) (Barchi et al. 2019). Exome sequencing and the SPET technology still suffer from a relatively high cost due to the sequencing library preparation. On the other hand, ddRAD has emerged as a reliable and affordable methodology for medium-density SNP genotyping in domestic and wild animals (Lado et al. 2020; Sooriyabandara et al. 2023). Compared to SNP arrays, a ddRAD experiment can be competitive in terms of genotyping costs. However, ddRAD experiments can be successfully applied if a good quality reference genome at the chromosomal level is available for the considered species and the chromosome position of the genotyped *loci* can be inferred through bioinformatic analysis. Furthermore, in any case, ddRAD cannot assure the *a priori* defined evenness of spacing of the genotyped SNP loci that is usually guaranteed by SNP arrays (some regions of the genome may have less dense SNPs, and this may represent a negative aspects when ddRAD data are to be used in GWAS studies).

The ultimate level of SNP genotyping approach is represented by the re-sequencing of whole genomes (WGS). While the drop in NGS costs has significantly paved the way for this approach to represent a more common option in the future, the sequencing costs are still about ten times higher than a medium-density array genotyping cost. Hence, WGS is today mainly employed for SNP discovery projects and for whole-genome SNP genotyping of relevant animals (Bai et al. 2012) rather than being used for routine SNP genotyping. More recently, NGS technologies that allow single molecule sequencing and thus definition of gametic phase from sequence data instead of its statistical inference from pedigree or populations genotype data have become available (Shendure and Ji 2008).

The availability of dense genome-wide SNP data has marked the possibility to move from phenotype/genotype association studies relying on exploiting genetic recombination while they occur along the few generations of the considered pedigrees (linkage mapping), towards approaches exploiting historical recombination events, occurred along the many generations in the past that separate not related individuals in a population (linkage disequilibrium mapping). Moving from pedigree-based approaches towards population-based approaches (such as GWAS) brings a number of advantages, including the possibility to perform phenotype/genotype association studies also in species where the unavailability, or the poor use, of artificial insemination does not allow to obtain large-sized families and where pedigree records are not available, such as in natural, feral, or poorly managed populations. On the other hand, it imposes to collect very large population samples, in order to account for a higher risk of false positive signals. In addition, approaches based on the detection of, and correction for, genetic stratification of the collected population samples have to be implemented. Furthermore, optimal GWAS design and choice of the bioinformatic analysis methods, mainly in the case of phenotypes affected by rare alleles, have to be taken into consideration (Zuk et al. 2014).

Based on the above, while it appear that a number of constraints, notably the lack of individual identification and genealogical recording systems and the very limited use of artificial insemination, may significantly hamper the adoption of family-based QTL mapping approaches in camelids, population-based GWAS approaches look more promising, provided that efforts are put in place to implement national/international individual identification systems, collection, and storage of metadata (e.g. sex, age, farm of origin, parity, lactation stage, reproductive history, and routine and sound phenotype recording). Indeed, from the genotyping technology point of view, the recent or expected availability of medium-density SNP arrays from different commercial providers (Bitaraf Sani et al. 2021; Gray et al. 2023; Guo et al. 2020; Hedayat-Evrigh et al. 2020) paves the way for GWAS to become the main approach in camelids for disentangling the genetic architecture of complex traits.

2.5 Perspectives for Application of Genomic Selection to Welfare-Related Traits

The genomic selection (GS) approach has been theoretically conceived at the very beginning of the present century (Hayes and Goddard 2010; Meuwissen et al. 2013; Meuwissen et al. 2021) though it was practically implemented in livestock species only a few years later, once medium- and high-density (HD) SNP arrays became commercially available. The method is based on exploitation of the linkage disequilibrium between causative variants/QTLs and the SNPs in a genotyping array and on estimating individual SNP effects using a representative subset of animals (ideally several thousands) for which both phenotypes for the traits of interest have been collected and genotypes at genome-wide SNP *loci* have been generated. Predicted SNP effects are then used to infer the genomic estimated breeding value (GEBV) of single animals within the considered breed. GS is now routinely used in breeding programmes for a range of livestock species, including cattle (dairy and beef), pigs, poultry, sheep, and goats (Meuwissen et al. 2016; Rupp et al. 2016; Wolc et al. 2016) and for several aquaculture species (Boudry et al. 2021; Houston et al. 2020). Indeed, GS has been shown to remarkably increase the rate of genetic improvement for most traits, particularly for those with low heritability (such as fertility, longevity, and resistance to specific diseases) (García-Ruiz et al. 2016; Meuwissen et al. 2016; Rexroad et al. 2019). Moreover, GS allows genomic predictions with good accuracy to be generated for genotyped animals with little, if any, phenotypic data themselves or on their relatives. This is a particularly attractive benefit for traits that are traditionally difficult to measure, such as many welfare-related traits, and under extensive and highly mobile camelids rearing systems. While the use of an HD SNP array, or even of the whole genome sequence, was shown to assure higher selection accuracies in GS, the still higher cost hampers their widespread use. Statistical approaches have been developed that allow accurate imputation of genotypes for medium-density arrays, provided good whole-genome sequenced reference populations are available (e.g. in cattle, the 1000 Bull Genomes Project (Hayes and Daetwyler 2019)).

The requirement for large training populations makes successful implementation of GS a more difficult challenge for breeds or populations in which the number of available phenotypes for a trait of interest is small due to either small population sizes or recording difficulties. Consequently, the potential value of multibreed training populations, for across-breed prediction is attracting considerable interest. Expanding phenotype availability via greater utilization of on-farm data, increased integration of data collected in different parts of the supply chain (e.g. slaughterhouses) or as part of government-led animal health testing, coupled with adoption of new precision livestock farming (PLF) data collection systems may bring further benefits.

In camelids, especially in the Old World ones, the lack of large genotyped and phenotyped reference populations is currently hindering intensively GWAS analysis and GS implementation not only for welfare-related traits but also for more strictly productive traits (milk, meat and wool production, racing and beauty performances). Despite these factors, which severely limit the application of GS in camelids, the scientific literature concerning association studies between genetic variability and welfare-related traits in camels is expanding. One of the first studies to associate camel temperament with variations at the DNA level was performed by Ramadan et al. (2018) on 138 dromedary camels belonging to 4 Egyptian breeds. Behavioural tests (the novel object and the exposure to an unfamiliar person tests) were coupled with the search for genetic variability in monoamine oxidase A (*MAOA*) and androgen receptor (*AR*) camel genes. A polymorphism with different glutamine repeats in the first exon of *AR* gene was found associated with the camels' response to behavioural tests, as camels carrying a shorter *AR* genotype in homozygosis were more fearful when compared with the other individuals (Ramadan et al. 2018). Despite the small sample size limiting the significance of the study, the results identified by Ramadan et al. could still be of interest as genes involved in the synthesis of sex hormones and in the control of their effects at the transcription level (such as the *AR* gene) were associated with changes in aggressiveness and temperament in dogs and humans (Konno et al. 2011; Aluja et al. 2011). Temperament and leadership in social hierarchies are complex multifactorial traits (Ramos et al. 2018), determined by sex-related hormones, physical fitness, body size, age, and phaneroptic characteristics (Iglesias Pastrana et al. 2021). Some of these traits are highly or completely dependent on genetics, such as traits concerning coat and eye colour. Interestingly, a study carried out on a population of Canarian dromedary camels showed that within a herd the dominant camels tended to have a darker coat colour, and the lighter-coated camels with delimited white-haired zones and blue-eyed tended to reach lower positions in the herd hierarchy (Iglesias Pastrana et al. 2021). Camels with a variable proportion of white coat are generally reported by camel herders to be the least aggressive but also the most fearful and submissive (Launois et al. 2002), in agreement with observations in foxes (Trut et al. 2004) and dogs (Pérez-Guisado et al. 2006). In animals, the extension of white spots in the coat is often associated with deafness and visual deficit, and blue-eyed subjects may suffer from visual impairment or being less reactive to visual stimuli. These features may have contributed to determining a lower ranking of light-haired and/or blued-eyed camels in the herd social network (Iglesias Pastrana et al. 2021). These results are of particular interest and represent a first step towards a greater understanding of the heritability of welfare-related traits in camels. Further genetic studies are however highly necessary to further elucidate the aetiological complexity of welfare-related traits and temperament in camelids.

The study of genes involved in temperament and coping abilities is still at an early stage in livestock animal species, but there is growing interest in the application of GS for improving animal welfare. The application of GS for temperament has already been proposed in cattle (reviewed in Haskell et al. 2014) and pigs (Brajon et al. 2020), and it is to be hoped that something similar will also happen in camels, although more investment is needed to speed up this process. While the availability of large capital in the hands of the major camel owners in Gulf countries may allow predicting implementation of GS as private business initiatives, in our opinion the adoption of GS in other areas (notably, in African countries for Old World camelids and in South American countries for New World camelids) may take benefit from government-supported nation-wide initiatives. In addition, given the low genetic differentiation highlighted so far in dromedaries (Lado et al. 2020), and the lack, in most cases, of officially recognized dromedary breeds (due to the lack of modern selective breeding), adoption of across-countries reference populations, while requiring relevant organizational efforts and willfulness, possibly coordinated through international intergovernative bodies, may allow sharing costs and risks related with GS implementation. In dromedaries, the availability of about 700 whole-genome sequences from about 20 countries (yet unpublished), partly generated within the frame of the Illumina's Agricultural Greater Good Initiative, represents a solid foundation for haplotype reconstruction and genotype imputation in animals typed at medium density SNP arrays. The main challenge for the coming times, besides the many and relevant organizational ones, will be the ability to incrementally and/or disruptively adapt PLF devices to the extensive breeding systems most of the camelids will be hopefully still confronted in the future. As what concerns the contribution of modern genomics to the improvement of welfare-related traits, this will also require a real paradigm shift, with concepts and methodological approaches developed so far by the animal welfare science to be reconsidered and updated in order to be integrated into extant and/or adapted conceptual and methodological approaches of modern animal breeding science.

2.6 Conclusions and Outlook

Integrating tameness, tranquillity, and high tolerance towards humans as criteria into future breeding and genomic selection programs will be key to maintaining highly productive and healthy, well-cared camels. There has always been a strong and often highly personal relationship between camels and their owners. In some countries over the global camel distribution, these animals are even considered part of the family, highly appreciated, priced, and valued. This close connection between camels and humans should be acknowledged and strengthened by future scientific evidence in the fields of behavioural genomics, animal welfare, and genotype-phenotype interaction studies. The upcoming field of precision livestock farming that connects physiological, individual movement and environmental monitoring with modern digital tools and surveillance can support effective herd and breeding management including the important aspect of animal welfare.

References

Albert FW et al (2009) Genetic architecture of tameness in a rat model of animal domestication. Genetics 182(2):541–554

Albert FW et al (2011) Targeted resequencing of a genomic region influencing tameness and aggression reveals multiple signals of positive selection. Heredity 107(3):205–214

Almathen F et al (2016) Ancient and modern DNA reveal dynamics of domestication and cross-continental dispersal of the dromedary. Proc Natl Acad Sci 113(24):6707–6712

Aluja A, García LF, Blanch A, Fibla J (2011) Association of androgen receptor gene, CAG and GGN repeat length polymorphism and impulsive-disinhibited personality traits in inmates: the role of short–long haplotype. Psychiatr Genet 21:229–239. https://doi.org/10.1097/YPG.0b013e328345465e

Bai Y et al (2012) Current status and future perspectives for sequencing livestock genomes. J Anim Sci Biotechnol 3(1):1–6

Barchi L et al (2019) Single primer enrichment technology (SPET) for high-throughput genotyping in tomato and eggplant germplasm. Front Plant Sci 10:1005

Benecke N (1994) Der Mensch und seine Haustiere. Theiss Verlag, Stuttgart, p 470

Bitaraf Sani M et al (2021) Genomic prediction for growth using a low-density SNP panel in dromedary camels. Sci Rep 11(1):7675

Boudry P et al (2021) Current status and potential of genomic selection to improve selective breeding in the main aquaculture species of International Council for the Exploration of the sea (ICES) member countries. Aquac Rep 20:100700
Brajon S, Ahloy-Dallaire J, Devillers N, Guay F (2020) The role of genetic selection on agonistic behavior and welfare of gestating sows housed in large semi-static groups. Animals 10(12): 2299. https://doi.org/10.3390/ani10122299
Brøndum R et al (2015) Quantitative trait loci markers derived from whole genome sequence data increases the reliability of genomic prediction. J Dairy Sci 98(6):4107–4116
Bulliet RW (1975) The camel and the wheel. Columbia University Press, New York, p 327
Consortium ISG et al (2010) The sheep genome reference sequence: a work in progress. Anim Genet 41(5):449–453
Davey JW, Blaxter ML (2010) RADSeq: next-generation population genetics. Brief Funct Genomics 9(5–6):416–423
Diaz-Maroto P et al (2021) Ancient DNA reveals the lost domestication history of South American camelids in Northern Chile and across the Andes. Elife 10:e63390
Elsik CG et al (2009) Bovine genome sequencing and analysis consortium. The genome sequence of taurine cattle: a window to ruminant biology and evolution. Science 324:522–528
Fitak RR et al (2020) Genomic signatures of domestication in Old World camels. Commun Biol 3(1):316
Frantz LAF et al (2020) Animal domestication in the era of ancient genomics. Nat Rev Genet 21(8): 449–460
García-Ruiz A et al (2016) Changes in genetic selection differentials and generation intervals in US Holstein dairy cattle as a result of genomic selection. Proc Natl Acad Sci 113(28):E3995–E4004
Gorjanc G et al (2015) Potential of genotyping-by-sequencing for genomic selection in livestock populations. Genet Sel Evol 47:1–14
Gray N et al (2023) Identification of six genomic regions associated with pigment intensity in alpacas. Animal Production Science
Guo F et al (2020) A genome-wide association study identifies quantitative trait loci affecting hematological traits in camelus bactrianus. Animals 10(1):96
Gurgul A et al (2019) Genotyping-by-sequencing performance in selected livestock species. Genomics 111(2):186–195
Haskell MJ, Simm G, Turner SP (2014) Genetic selection for temperament traits in dairy and beef cattle. Front Genet 5:368. https://doi.org/10.3389/fgene.2014.00368. PMID: 25374582; PMCID: PMC4204639
Hayes BJ, Daetwyler HD (2019) 1000 bull genomes project to map simple and complex genetic traits in cattle: applications and outcomes. Annu Rev Anim Biosci 7:89–102
Hayes B, Goddard M (2010) Genome-wide association and genomic selection in animal breeding. Genome 53(11):876–883
Hedayat-Evrigh N et al (2020) Genome-wide identification and analysis of variants in domestic and wild bactrian camels using whole-genome sequencing data. Int J Genomics 2020(4):1–9
Houston RD et al (2020) Harnessing genomics to fast-track genetic improvement in aquaculture. Nat Rev Genet 21(7):389–409
Iglesias Pastrana C, Navas González FJ, Ciani E, Arando Arbulu A, Delgado Bermejo JV (2021) The youngest, the heaviest and/or the darkest? Selection potentialities and determinants of leadership in Canarian dromedary camels. Animals (Basel) 11(10):2886. https://doi.org/10.3390/ani11102886
Jensen P (2014) Behavior genetics and the domestication of animals. Annu Rev Anim Biosci 2: 85–104
Jirimutu, Wang Z., Ding G., Chen G., Sun Y., Sun Z., et al. (2012). Genome sequences of wild and domestic bactrian camels. Nat. Commun. 3:1202. https://doi.org/10.1038/ncomms2192
Katajamaa R, Jensen P (2021) Tameness correlates with domestication related traits in a red Junglefowl intercross. Genes Brain Behav 20(3):e12704

Konno A, Inoue-Murayama M, Hasegawa T (2011) Androgen receptor gene polymorphisms are associated with aggression in Japanese Akita Inu. Biol Lett 7:658–660. https://doi.org/10.1098/rsbl.2011.0087

Lado S et al (2020) Genome-wide diversity and global migration patterns in dromedaries follow ancient caravan routes. Commun Biol 3(1):387

Larson G, Burger JA (2013) A population genetics view of animal domestication. Trends Genet 29(4):197–205

Launois M, Faye B, Kriska MA (2002) Le Dromadaire Pédagogique. FAO, Rome, Italy

MacHugh DE et al (2017) Taming the past: ancient DNA and the study of animal domestication. Annual Review of Animal Biosciences 5:329–351

Meuwissen T et al (2013) Accelerating improvement of livestock with genomic selection. Annu Rev Anim Biosci 1(1):221–237

Meuwissen T et al (2016) Genomic selection: a paradigm shift in animal breeding. Anim Front 6(1): 6–14

Meuwissen T et al (2021) On the use of whole-genome sequence data for across-breed genomic prediction and fine-scale mapping of QTL. Genet Sel Evol 53(1):1–15

Mohandesan E et al (2017) Mitogenome sequencing in the genus Camelus reveals evidence for purifying selection and long-term divergence between wild and domestic Bactrian camels. Sci Rep 7(1):9970

Nicolazzi E et al (2015) Software solutions for the livestock genomics SNP array revolution. Anim Genet 46(4):343–353

Olschewsky A, Hinrichs D (2021) An overview of the use of genotyping techniques for assessing genetic diversity in local farm animal breeds. Animals 11(7):2016

Park SD et al (2015) Genome sequencing of the extinct Eurasian wild aurochs, Bos primigenius, illuminates the phylogeography and evolution of cattle. Genome Biol 16:234

Pérez-Guisado J, Lopez-Rodríguez R, Muñoz-Serrano A (2006) Heritability of dominant–aggressive behaviour in English cocker spaniels. Appl Anim Behav Sci 100:219–227. https://doi.org/10.1016/j.applanim.2005.11.005

Peters J, von den Driesch A (1997) The two-humped camel (Camelus bactrianus): new light on its distribution, management and medical treatment in the past. J Zool 242:651–679

Peterson BK et al (2012) Double digest RADseq: an inexpensive method for de novo SNP discovery and genotyping in model and non-model species. PLoS One 7(5):e37135

Puetz LC et al (2021) Gut microbiota linked with reduced fear of humans in red Junglefowl has implications for early domestication. Adv Genet 2(4):2100018

Ramadan S, Nowier AM, Hori Y, Inoue-Murayama M (2018) The association between glutamine repeats in the androgen receptor gene and personality traits in dromedary camel (Camelus dromedarius). PloS one 13(2):e0191119. https://doi.org/10.1371/journal.pone.0191119

Ramos A, Manizan L, Rodriguez E, Kemp YJ, Sueur C (2018) How can leadership processes in European bison be used to improve the management of free-roaming herds. Eur J Wildl Res 64: 16. https://doi.org/10.1007/s10344-018-1175-0

Rexroad C et al (2019) Genome to phenome: improving animal health, production, and Well-being–a new USDA blueprint for animal genome research 2018–2027. Front Genet 10:327

Rupp R et al (2016) Genomic application in sheep and goat breeding. Anim Front 6(1):39–44

Schubert M et al (2014) Prehistoric genomes reveal the genetic foundation and cost of horse domestication. Proc Natl Acad Sci 111(52):E5661–E5669

Shendure J, Ji H (2008) Next-generation DNA sequencing. Nat Biotechnol 26(10):1135–1145

Sooriyabandara M et al (2023) Genetic features of Sri Lankan elephant, Elephas maximus maximus Linnaeus revealed by high throughput sequencing of mitogenome and ddRAD-seq. PLoS One 18(6):e0285572

Trut LN, Plyusnina I, Oskina I (2004) An experiment on fox domestication and debatable issues of evolution of the dog. Russ J Genet 40:644–655. https://doi.org/10.1023/B:RUGE.0000033312.92773.c1

Trut L et al (2009) Animal evolution during domestication: the domesticated fox as a model. BioEssays 31(3):349–360

Wilkins AS et al (2014) The "domestication syndrome" in mammals: a unified explanation based on neural crest cell behavior and genetics. Genetics 197(3):795–808

Wilkins AS et al (2021) The neural crest/domestication syndrome hypothesis, explained: reply to Johnsson, Henriksen, and Wright. Genetics 219(1):iyab098

Wolc A et al (2016) Implementation of genomic selection in the poultry industry. Anim Front 6(1): 23–31

Wu H et al (2014) Camelid genomes reveal evolution and adaptation to desert environments. Nat Commun 5:5188

Zeder MA (2012) The domestication of animals. J Anthropol Res 68:161–190

Zeder MA (2015) Core questions in domestication research. Proc Natl Acad Sci 112(11): 3191–3198

Zhang H et al (2012) Progress of genome wide association study in domestic animals. J Anim Sci Biotechnol 3(1):1–10

Zuk O et al (2014) Searching for missing heritability: designing rare variant association studies. Proc Natl Acad Sci 111(4):E455–E464

3 Welfare Assessment in Dromedary Camels

Laura Menchetti and Barbara Padalino

Abstract

The assessment of welfare requires a multidimensional approach determining the actual welfare of animals, including both physical and mental states, using environmental- and animal-based measures. This chapter aims at critically reviewing the literature to present the last updates on how to assess welfare in dromedary camels. A protocol for assessing their welfare status when kept in semi-intensive and intensive systems was recently developed, adapting the AWIN protocol to the uniqueness of dromedary camels and adding some positive welfare indicators. In the protocol, data related to housing, feeding, health, and behaviour of dromedary camels are collected at three levels of assessment: caretaker, herd, and animal. Data are then aggregated to obtain overall assessment indices and classify the camel units/pens/farms according to their welfare level. Using this protocol, welfare issues can be identified and recommendations to enhance camel welfare provided. The protocol, however, requires further refinement and validation. To date, space allowance, shaded space, cleanliness of bedding, water management, and infectious disease prevention were the major welfare hazards for camels kept in intensive systems, while poor health and bad handling were reported for camels kept in extensive systems. Overall, more evidence-based studies are needed before being able to suggest welfare standards for dromedary camels.

Keyword

Welfare · Dromedary camels · Protocol · ABMs · EBMs

L. Menchetti · B. Padalino (✉)
Department of Agricultural and Food Sciences, University of Bologna, Bologna, Italy
e-mail: Barbara.padalino@unibo.it

B. Padalino, B. Faye (eds.), *Dromedary Camel Behavior and Welfare*, Animal Welfare 24, https://doi.org/10.1007/978-3-031-48600-5_3

3.1 Animal Welfare: Background and Camel Caretakers' Perspectives

Animal welfare is a complex matter, and despite the growing global interest, there is still not a single or universally accepted definition of what animal welfare is. Scientists have emphasized different aspects of animal physiology and pathology as well as the animal-environment and animal-human interactions (Mellor et al. 2010). Changes over time in scientific knowledge and husbandry systems may partially explain the different approaches to animal welfare. Human cultures, traditions, religious faiths, economic, and practical contexts also contribute to emphasizing some facets while dampening others (Szucs et al. 2012; Dioli 2022). Finally, individuals, with their wealth of experience and beliefs, tend to develop their concepts of what animal welfare is. To date, there is only one study on the definition of animal welfare in people working with camels, which was carried out in a market in Qatar (Menchetti et al. 2020). This survey documented the perception of animal welfare by camel caretakers asking them what animal welfare is and what criteria they used to identify poor welfare in their animals. A noteworthy finding was that camel caretakers experienced difficulty understanding questions related to what is animal welfare so the interviewer had to intervene to introduce the concept and obtain a valid response. This suggests that the concept of animal welfare could be very far from the daily life of camel caretakers and that they do not receive any training, formal or informal, on animal welfare and on how to identify stress in animals. This type of training and certificates are compulsory for all people who work with livestock (i.e. cattle, pigs) in Europe.

The different approaches and concerns about animal welfare can be generally framed within three different conceptions: (i) the basic health and functioning of animals, (ii) their affective state, and (iii) natural behaviour and living conditions (Fraser et al. 1997). Camel caretakers working at a market in Doha were mainly oriented toward the first concept as they defined animal welfare in terms of appropriate nutrition, water supply, and health (Menchetti et al. 2020). However, age differences emerged because for the younger caretakers, welfare also meant treating the animals gently. Considering that camels are found in many different countries, the results of this study cannot be generalized as other orientations may prevail in other cultural and regional contexts. The husbandry system could also strongly influence the views on animal welfare. It was assumed that among nomadic pastoralists, natural living conditions might not emerge as a welfare concern because those animals are free to express many behavioural needs: they are free-range, have a natural mother-calf relationship, and can mate naturally (Dioli 2022). Conversely, in these systems, the welfare concerns should focus on the functioning of animals and their affective status since it has been reported that dromedary kept in pastoralism conditions often have poor health and husbandry conditions, particularly due to the limited professional healthcare access, irregular forage and water availability, and lack of facilities (Dioli 2022). Regardless of environmental and husbandry context, however, Fraser and Broom pointed out that animal welfare requires a multidimensional approach and the evaluation of all three aspects, namely, physical health, affective states, and natural life (Broom 1996; Fraser 2009). The approach of Fraser and Broom seems to reconcile different ethical frameworks but the question of how to fit the different views about what constitutes a good life for animals, animal practices, and welfare standards remained unsolved (Fraser 2009). This is the reason why animal welfare scientists are pushing toward very objective methods to measure the quality of life of a living animal as perceived by the animals and not based on the perception and belief of humans. Consequently, a good definition of welfare is the following by Pond and Bazer (2011): "Welfare is a term that describes a potentially measurable quality of a living animal at a particular time and in a scientific concept". This is the only possible approach which may be useful to define welfare standards to protect animal welfare.

A first concrete attempt to capture and summarize the key insights arising from the scientific research on animal welfare was made by the World Organisation for Animal Health with the Terrestrial Animal Health Code. The Code implements standards of animal health and welfare and public health from a veterinary point of view and is now embraced worldwide (World Organization for Animal Health 2022). Section 7 of the mentioned code provides specific recommendations on how to protect animal welfare on farm, during transport, and at slaughter in different animal species. The recommendations listed in the Code could also be applied to camelids but no specific chapters address their welfare aspects while their peculiarities are mentioned only sporadically, such as in Chap. 7.3 on transport by land. Moreover, although it is a world reference point for setting the generic standards of animal welfare, the Terrestrial Animal Health Code does not provide practical and scientific tools for animal welfare assessment, but mainly guidelines and general recommendations to protect welfare from farm to fork.

Based on a scientific approach, practical tools for welfare assessment were instead developed by two major Europe projects, the Welfare Quality project® and the Animal Welfare Indicators Project (AWIN), which started in 2004 and comprised a partnership of many institutions. These projects defined four principles for the evaluation of animal welfare that extended and concretized the concept of the Five Freedoms. The novelty of these practical tools was that for each welfare principle, they had two or more welfare criteria (Fig. 3.1), and for each criterion one or more welfare indicators (e.g. good feeding, absence of prolonged hunger, Body Condition Score).

The welfare assessment proposed by the Welfare Quality should rely on complementary measures covering all these dimensions. This multidimensional approach did not seem to be perceived by the camel caretakers of the market as their animal welfare perspective usually included only one or two principles (Menchetti et al. 2020). Over the years, Welfare Quality® and AWIN projects have developed practical and feasible tools for assessing the welfare of different farm species but, once again, the camel was not considered.

Although the Five Freedoms and the four principles of the Welfare Quality project® are today considered cornerstones in the assessment of animal welfare, some scientists have highlighted several limitations (Mellor 2016). Mellor claimed that these approaches are overly focused on negative welfare states and that the

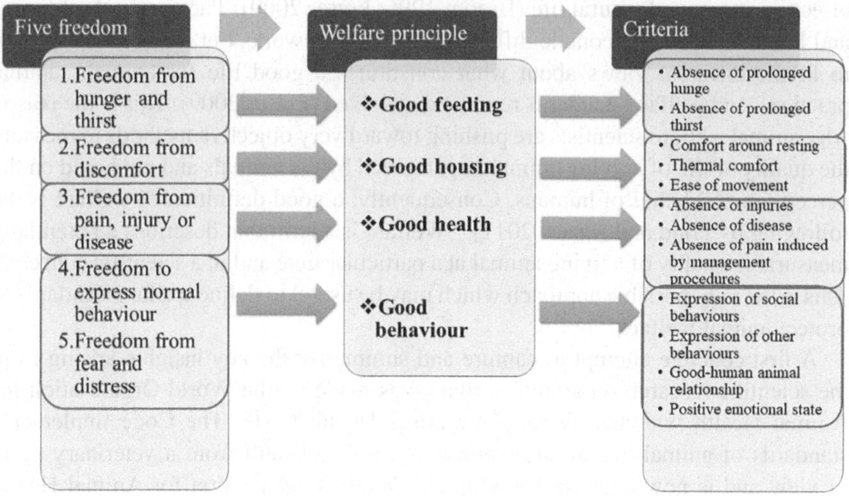

Fig. 3.1 From the Five Freedoms to the Welfare principles and criteria proposed by AWIN and Welfare Quality (modified from Dalla Costa et al. 2014)

freedom from circumstances that cause these negative states only ensures survival for the animal and not "a life worth living" (Mellor and Beausoleil 2015). The model for welfare assessment proposed by Mellor consisted of five domains. Besides the domains that follow the four principles (i.e. Nutrition, Physical Environment, Health, and Behaviour), a fifth domain related to the animals' Mental state was added. According to Mellor, only the latter could ultimately determine the overall welfare state of the animals as inputs associated with the first four domains are subjectively processed by the brain resulting in negative or positive experiences (Mellor et al. 2020). Mellor's concepts, from a practical point of view, suggest that a greater emphasis on positive states and human-animal relationships should be given in the protocols for animal welfare assessment. An appropriate welfare assessment is crucial to collect research evidence, on which regulations on the protection of animal welfare should be based.

In conclusion, welfare is a term that describes the quality of living of an animal at a particular time and in a scientific concept. The assessment of welfare requires a multidimensional approach and aims at determining the actual welfare of animals. The tools for welfare assessment should be holistic and consider all the aspects included in the four welfare principles while also emphasizing positive experiences, both in extensive and intensive husbandry systems. In addition to being specific according to the species, the animal category and the breeding system, the welfare protocol should also be adapted to the sociocultural and economic context. Literature on camel caretakers' perspectives is scarce and suggests that their understanding of animal welfare is limited and one-dimensional. On the other hand, despite the great economic and social role that dromedary camels play in several countries, their territorial expansion, and the shift towards intensive farms (see Chap. 1 for details), to date, this species has been neglected by the most important international projects for the evaluation of animal welfare and by current regulations and recommendations to protect animal welfare from farm to fork.

3.2 Scientific Tool to Assess Welfare on Farm

To date, the only protocol that has been proposed for the assessment of welfare in dromedary camels kept in intensive and semi-intensive systems that applies a multidimensional approach was developed by Padalino and Menchetti (2021). This protocol revisited the protocols used to assess the welfare of livestock in Europe, i.e. AWIN and Welfare Quality® protocols (Welfare Quality Network 2009; AWIN 2015b), and adapted them for use in dromedary camels. It, therefore, foresees the on-farm collection of both animal-based (ABMs) and environmental-based measures (EBMs) relating to the four fundamental welfare principles (good feeding, good housing, good health, and appropriate behaviour) but at three levels of investigation (i.e. caretaker, herd, and animal level) (Fig. 3.2). It was also inspired by Mellor's recommendations (Mellor et al. 2020) and as such, includes indicators of positive states and human-animal relationship.

For each welfare principle, ABMs and EBMs need to be collected at three levels of assessment, for a total of 105 measures (Table 3.1).

The welfare assessment should be carried out at a fixed time, for example, 10: 00 a.m., respecting the farm's routine practices, and some measures would be from

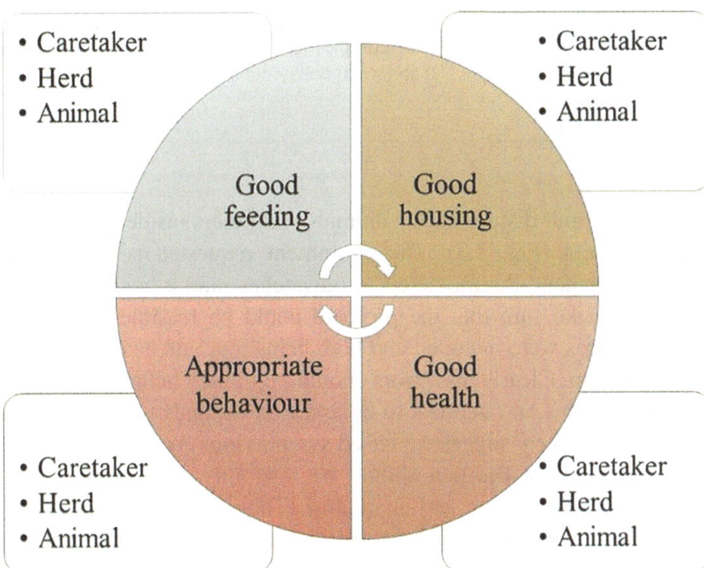

Fig. 3.2 An example of the data collection of the protocol for camel welfare assessment proposed by Padalino and Menchetti (2021). Modified by Padalino and Menchetti (2021) and AWIN (2015b)

Table 3.1 Environmental- and animal-based measures included in the protocol are divided by welfare principle and level of investigation (adapted from Menchetti et al. 2021b)

Welfare principle	Level of investigation		
	Caretaker	Herd	Animals
Good feeding	Feeding and watering management	Feeding and water points Feed and water availability Feed and water quality Feeding and water space per animal Presence of salt Drinking, eating, and ruminating camels	Body condition score Thirst index
Good housing	Caretaker's experience in working with animals Number of animals handled by the caretaker in the busiest week	Space allowance Shaded areas Fence condition Bedding Rubbish Hobbled/tethered camels	Resting behaviour Location (under the sun/in shade) Insects (quality, quantity) Tethering Hobbled
Good health	Past camel disease observed Camel health check Medical treatments	Camels with disease, physical injuries, scars from hobbles, cauterization, nose ring Camels in pain	Camels with disease, physical injuries, scars from hobbles, cauterization, nose ring Camels in pain
Appropriate behaviour	Experience in camel handling Skills in identifying distress Reported behavioural problems	Camels resting, standing quietly, aggressive Camels showing stereotypies and other abnormal behaviour	Social interaction Stereotypies Abnormal behaviour Feeding and rumination Approaching test

outside the pen without disturbing the animals and others inside the box/pen where the animals are kept (Fig. 3.3). The equipment requested is minimal, namely, weather stations, meters, thermometers, stopwatches, and a bucket with fresh and clean water. To make sure that the protocol could be feasible in the field, only noninvasive practices were included and both behaviour and health can be assessed by direct observation. Clearly, assessors should be trained before applying it.

The number of pens and animals to be assessed depends on how big the camel farm is, and it has been suggested based on previous Welfare Quality (AWIN 2015a). The selection of the pen should be randomly conducted excluding the pens used as infirmary, culling, and quarantine (Table 3.2).

The number of animals to be assessed for each pen/paddock/herd should be chosen following the rules proposed by AWIN for goats' selection assuming a 50% prevalence, a confidence interval of 95%, and an accuracy of 10%. However,

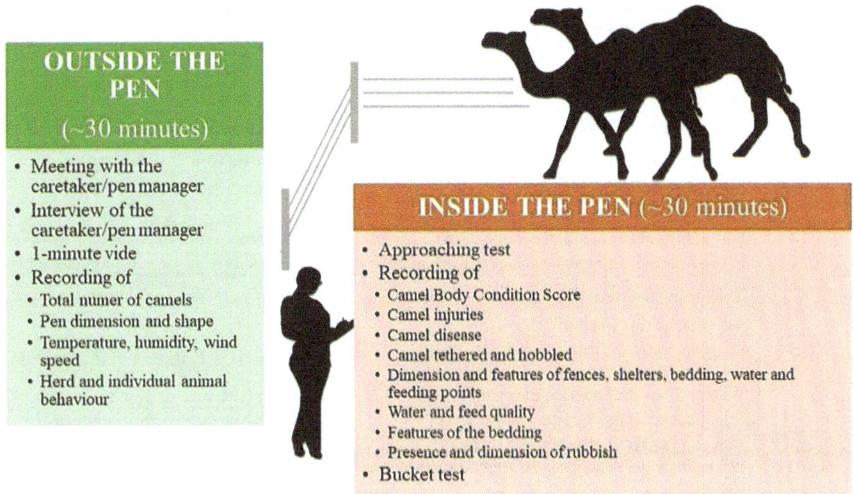

Fig. 3.3 The main steps of the protocol for the assessment of dromedary camels kept in intensive and semi-intensive systems proposed by Padalino and Menchetti (2021). Adapted from Menchetti et al. 2021b

Table 3.2 Number of pens to be assessed

Number of pens[a] in the farm	Number of pens to be assessed
1–2	All pens
3–7	2 pens
8–10	3 pens
>10	25% of the pens

[a]Paddock/house facilities

Table 3.3 Number of camels per pen/paddock/herd to be assessed

Number of camels in the pen/paddock/herd	Number of camels to be assessed
<15	All animals
15–29	13–19
30–49	21–28
50–99	29–39
100–149	41–44

to minimize the impact on camels, non-restrictive criteria, such as a level of confidence of 90% or less, or rules of thumb could be adopted (Table 3.3).

Based on the number of pens and animals to assess on each farm, it may take a different amount of time. However, small and medium farms will not require more than 1 day to collect the data at all three levels.

The caretaker level is a face-to-face interview, which should be carried out in a friendly manner, and lasts about 10–15 min. It includes questions related to the

Date ____________ **Assessor** ____________ **Farm** ____________ **ID caretaker** ____________

DEMOGRAPHICS	How old are you?	_ _ _ _ _ _ years
	How long have you worked with camels?	_ _ _ _ _ _ years
	What other species have you worked with?	_ _ _ _ _ _ _ _
GOOD FEEDING	How often do you feed camels?	_ _ _ _time(s)/day □ Ad libitum
	How often do you water camels?	_ _ _ _time(s)/day □ Ad libitum
GOOD HOUSING	In your busiest week of the year, approximately how many camels are reared at the farm?	_ _ _ _ _ _ _ _ camels
	Do you keep other animal species at the farm?	□ Yes (specify) _ _ _ _ _ _ □ No
	What is the rearing purpose of your camels?	□ Meat □ Milk □ Other: _ _ _ _
	Are the camels exercised?	□ Yes □ No
	Do you change the management/housing according to season?	□ Yes □ No
GOOD HEALTH	Who assesses the health of the camels?	□ Vet □ Non-vet □ Not conducted
	Who treats the camel when it is sick?	□ Vet □ Non-vet □ Not conducted
	Who administers vaccinations to camels?	□ Vet □ Non-vet □ Not conducted
	Who administers endoparasite treatments to camels?	□ Vet □ Non-vet □ Not conducted
	Who administers ectoparasite treatments to camels?	□ Vet □ Non-vet □ Not conducted
	Which health problems have you observed in camels over the last year?	□ None □ Colic □ Injuries (e.g. cuts, bruises) □ Skin problems □ Muscular problems □ Diarrhoea □ Respiratory problems □ Overheating/sunstroke □ Other _ _ _ _
	What criteria do you use to identify a camel in pain or distress?	_ _ _ _ _ _ _ _ _ _ _ _ _ _
APPROPRIATE BEHAVIOR	How many years of experience in camel handling do you have?	_ _ _ _ _ _ years
	Do your camels show behavioral problems?	□ Yes □ No
	If yes, what behavioral problems do camels show?	□ Aggression □ Biting □ Kicking □ Anxiety or escaping from the pen □ Other _ _ _ _ _ _ _ _
	How do you grade your ability in identifying a camel in distress/pain?	□ Low □ Some □ Moderate □ High □ Very high
	What criteria do you use to identify a camel in pain or distress?	_ _ _ _ _ _ _ _ _ _ _ _ _ _
	How do you rank your understanding of animal welfare?	□ Low □ Some □ Moderate □ High □ Very high

Fig. 3.4 Camel welfare recording sheet at caretaker level (from Padalino and Menchetti 2021)

caretaker's background and then some specific questions for each welfare principle (Fig. 3.4).

The herd level is composed mainly of EBMs (including resource- and management-based indicators) for the principle of good feeding and good housing, while mainly of ABMs for the principle "good health" and "appropriate behaviour" (Fig. 3.5). The herd level starts with some observations from outside the pens (i.e. the number of animals and all behavioural observations) and then the assessors must go

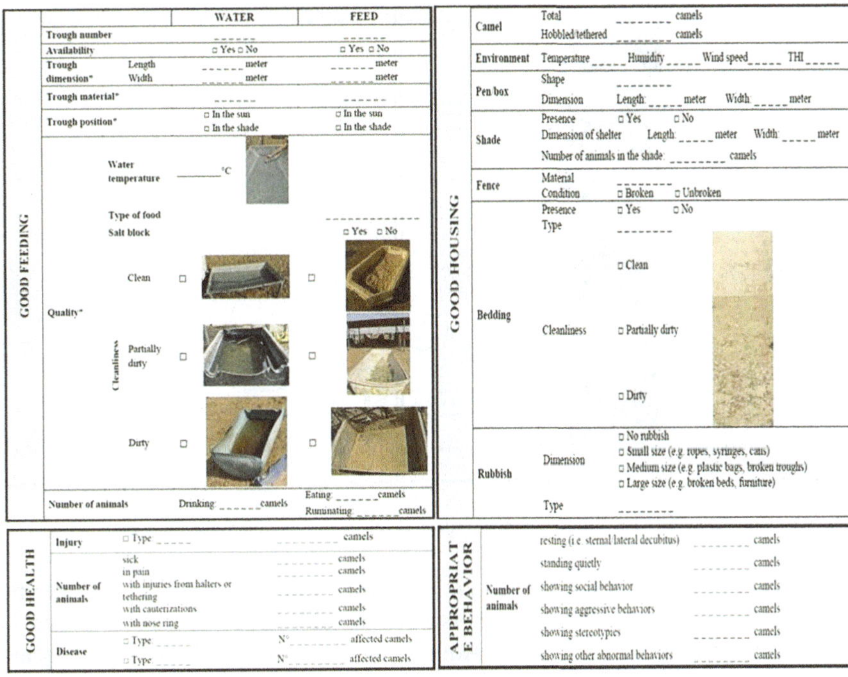

Fig. 3.5 Camel welfare recording sheet at herd level (from Padalino and Menchetti 2021)

inside to check some of the parameters, such as the temperature of the water. However, for this level, no interaction and approach with any camels are required. It may last about 30–60 min, depending on the size of the herd and of the pen.

For the animal levels, the requested number per herd should be chosen randomly. Then each of these animals should be observed from outside the pen and during the observation; the presence/absence of specific behaviour (positive social interaction, stereotypy, feeding ruminating, aggressive behaviour) is noted. The behavioural observation is followed by the approaching test, a clinical visual examination, and a bucket test. The behavioural responses to the approaching test are scored as positive, neutral, and negative. The clinical examination includes the assessment of Body Condition Score, presence and type of disease and injuries, the presence of swollen joints, lameness, discharges and cough, abnormal udder, or breathing patterns. The assessor should also score whether the observed camel is in pain, but an official pain score has not been validated in camels yet (see chapter on good health for details) (see Padalino and Menchetti 2021 for details) (Fig. 3.6). The animal level contains only ABMs (Fig. 3.7) and it takes about 10 min per animal.

The Padalino and Menchetti protocol was the first attempt to develop a tool for camel welfare assessment in line with the protocols used for other livestock species and the multidisciplinary concept of welfare. However, it only provides the application in semi-intensive and intensive systems, and it still requires refinements and a long validation process.

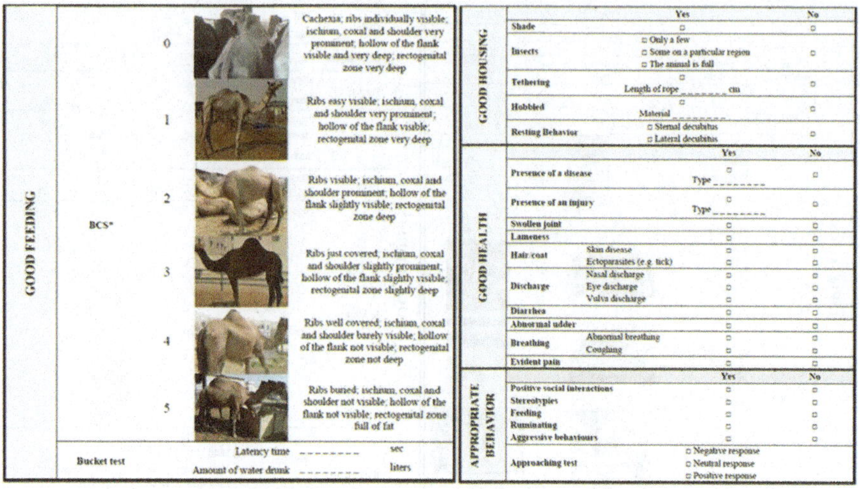

Fig. 3.6 Camel welfare recording sheet at animal level (from Padalino and Menchetti 2021)

In recent years, interest has been growing in investigating the effect of specific factors on camel welfare. Many authors have evaluated specific aspects related to their health (Padalino et al. 2021; Aqib et al. 2022; Muluneh et al. 2022) and behaviour (Fatnassi et al. 2014; Pastrana et al. 2021; Farsi et al. 2022) as well as the impact of feeding (Mohamed et al. 2009; Faraz et al. 2020; El Shoukary et al. 2021), watering (Bekele et al. 2011; Faraz et al. 2021), and housing systems (El Shoukary et al. 2020a; Zappaterra et al. 2021; Hussen and Al-Sukruwah 2022) on their welfare. Many of these studies were considered when Padalino and Menchetti were developing their protocol, while the new one can be useful to refine the protocol of dromedary camels kept in captivity. Dioli (2022) instead recently reviewed the husbandry practices utilized in pastoral livestock systems highlighting their major welfare concerns. He emphasized the need to contextualize the assessment of animal welfare in pastoral areas, taking into account the unique environmental, cultural, ecological, and economic settings. The author pointed out that the definition of animal welfare and what acceptable or unacceptable is should be viewed in perspective, taking into consideration the unique ecosystem humans and livestock create in a nomadic set-up (Dioli 2022). No protocol to assess the welfare of dromedary camels kept in extensive systems and under pastoralism has been developed yet, and the complexities of these issues need to be addressed in the future. Overall, it seems clear that research on welfare in dromedary camels is needed, and, in particular, it is crucial to validate new feasible ABMs, before being able to propose welfare standards for camels.

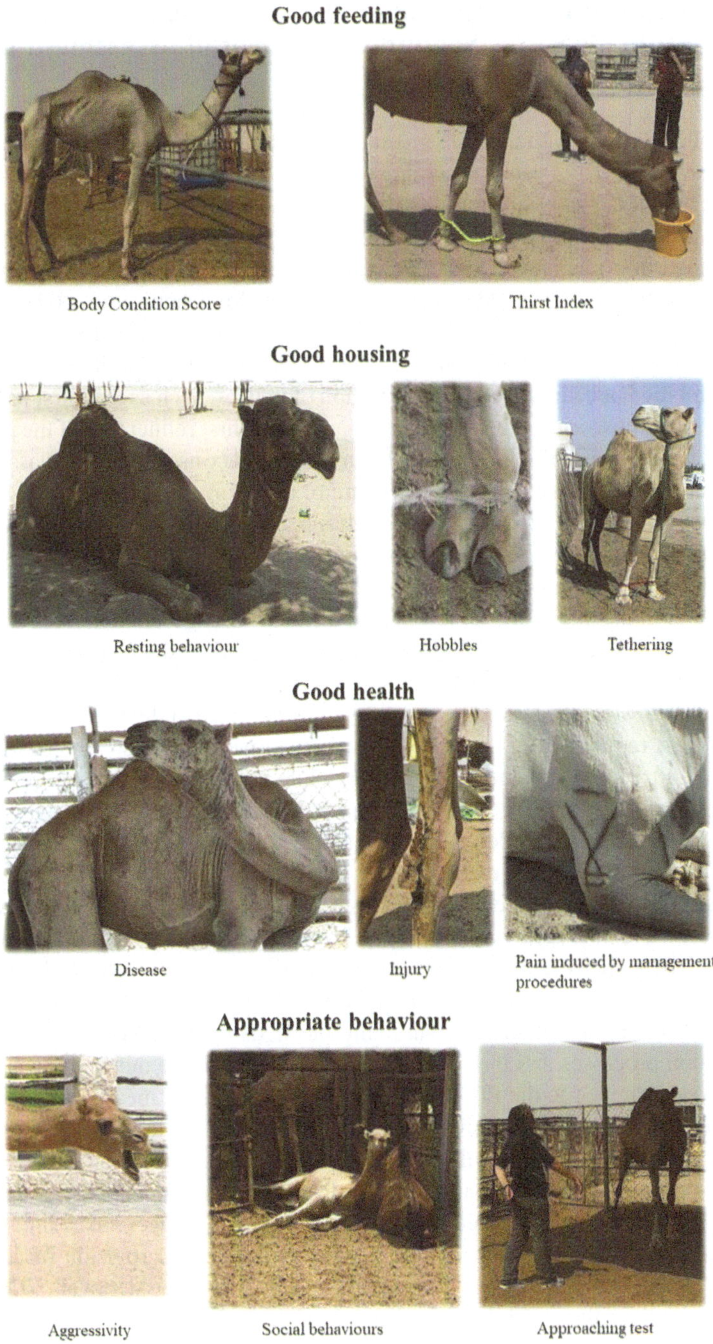

Fig. 3.7 Some of the animal-based measures included in the animal-level recording sheet used in the protocol by Padalino and Menchetti (2021) and classified according with the welfare principle

3.3 Application of the Camel Welfare Assessment Protocol: From Measures Scoring to the Overall Assessment of a Camel Farm

The application of a welfare assessment model implies the on-farm collection of the measures selected for each principle as well as their scoring and aggregation to produce an overall assessment of the farm (Botreau et al. 2007; Botreau et al. 2009). Padalino's method has so far only been applied to evaluate the welfare of dromedary camels kept at a market in Qatar considering each pen as a unit (farm) since they were managed differently (Menchetti et al. 2021b; Menchetti et al. 2021a). About 50 workers, 76 pens, and more than 500 camels were involved in this study. A total of 76 measures were selected from the protocol based on their feasibility while some were combined or categorized for the creation of indices, such as the thirst index, and their subsequent scoring. The application has led to a first refinement of the protocol, and the development of a method to obtain a total index useful for the classification of the pens according to the level of welfare of the animals kept in them (Menchetti et al. 2021b). Briefly, a four-step process has been proposed (Fig. 3.8).

During the first step, the collected measures were scored from 0 to 2, where 0 was considered good welfare and 2 unacceptable welfare. The scoring was developed by the authors using literature review and expert knowledge elicitation (Table 3.4). Clearly, the scoring should be further validated, refined, and adapted to a particular camel category.

In a second step, the scored measures were aggregated and converted into partial indices (PIs) ranging from 0 (the worst welfare situation) to 100 (the best welfare situation). PI can be obtained for each principle and assessment level (e.g., good feeding at caretaker level, good feeding at herd level, good feeding at animal level; good housing at caretaker level, good housing at herd level, and good housing at animal level; Fig 3.9, green rectangles). In the third step, the aggregate indices at the assessment level (LAIs; caretaker index, herd index, and animal index; yellow ovals in Fig 3.9) and the aggregate principle levels (PAIs; good feeding index, good housing index, good health index, and appropriate index, red ovals in Fig. 3.9) were calculated by summing the PIs. In the fourth step, the total welfare index was calculated. It can be obtained by summing the LAIs or the PAIs (Fig. 3.9). In the aggregation processes and the production of the total welfare index, a lower weight (20%) was attributed to the PIs of the caretaker level as these measures are based on information reported by the caretaker and not directly collected by the assessor.

The pens/farms where camels are kept can be classified based on the TWI or the combination of the scores obtained in each PAI. The most functional method is the latter. This system, also adopted by the Welfare Quality project, takes into account the multidimensional concept of welfare and facilitates the provision of advice to farmers on solving welfare problems (Botreau et al. 2009; Veissier et al. 2011). As a result, it classifies the pens/farm into four categories, namely, unacceptable, unsatisfactory, satisfactory, and excellent (Fig. 3.10).

At the camel market in Doha, most of the pens were classified as "unsatisfactory" (61.8%) and none as "excellent". Using this type of classification, the criticalities of

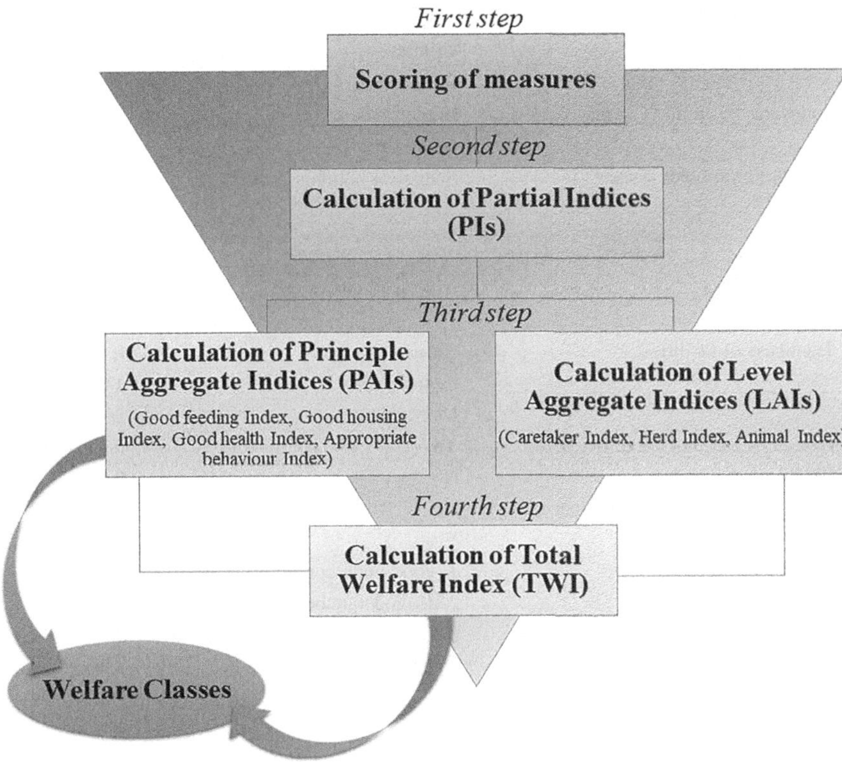

Fig. 3.8 The four-step process of scoring and aggregation of the collected measures to obtain an overall welfare index useful for the classification of dromedary camel pens/farms (Menchetti et al. 2021b). The process begins with the attribution of a score to each measure. Then, the scores are aggregated at level (LAIs) or at principle (PAIs) and subsequently used to calculate a single welfare index (i.e. total welfare index, TW). Pens/farms can be classified in welfare classes using the aggregate scores or the TWI. Adapted from Menchetti et al. (2021b)

Table 3.4 Scoring system developed for the measures included in the camel welfare protocol (from Menchetti et al. 2021b)

Measure	Criteria	Scores
Who carries out health assessment or medical treatment	A veterinarian	0
	A non-veterinarian	1
	Not conducted	2
Grade of caretaker's ability in identifying a camel in distress/pain	High–very high	0
	Moderate	1
	Low–some	2
Years of caretaker's experience	>10 years	0
	6–10 years	1
	0–5 years	2
Food/water distribution	Ad libitum	0

(continued)

Table 3.4 (continued)

Measure	Criteria	Scores
	Rationed	2
Food/water position[a]	In the shade	0
	In the sun	2
Continuous variables related to facilities[a,b]	Statistical binning (tertiles)	0 (best situation)
		1 (second tertile group)
		2 (worst situation)
Cleanliness of facilities[a]	Clean	0
	Partially dirty	1
	Dirty	2
Presence of salt block, shelter, shade, bedding	Yes	0
	No	2
Presence of rubbish, brocken fence, insects	No	0
	Yes	2
Body condition score	3 (good body condition)	0
	2, 4 (moderate body condition)	1
	0–1, 5 (poor body condition, lean or obese)	2
Thirst index	0	0
	1	1
	2–3	2
Presence of a disease, physical injuries, pain or behaviour indicating poor welfare[c]		
Animal level	No	0
	Yes	2
Herd level	Percentage of animals with the disease/injury/pain/behaviour	0 (0%) – 2 (100%)
Presence of behaviour indicating good welfare[d]		
Animal level	Yes	0
	No	2
Herd level	Percentage of animals showing the behaviour	0 (100%) – 2 (0%)
Tethering/hobbled	No	0
	Yes	2
Responses during the approaching test	Positive	0
	Neutral	1
	Negative	2

[a]When more than one trough was present in the pen, the score was attributed to a randomly chosen one
[b]Dimension and number of troughs, water temperature, space allowance, trough space, shaded space allowance
[c]Aggressive behaviours, stereotypies, and other abnormal behaviours
[d]Resting, standing quietly, positive social behaviours, feeding, rumination

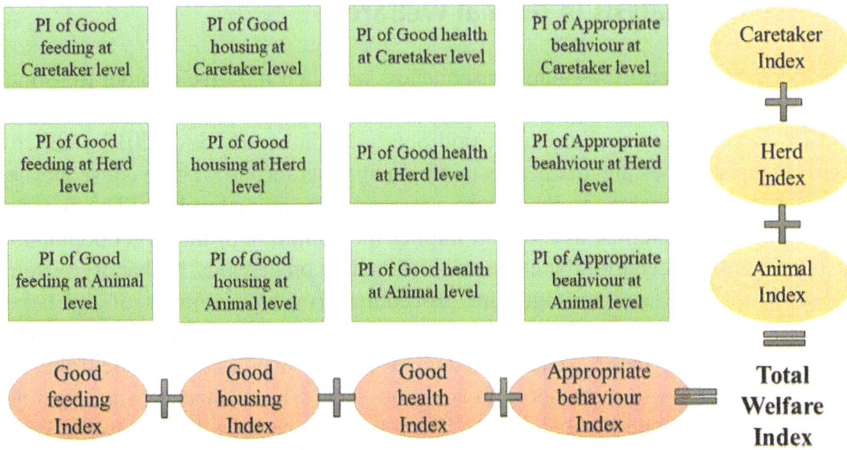

Fig. 3.9 The combination of the partial indices (PIs, the 12 green rectangles) into aggregate indices (4 at principle level, red circles, and 3 at assessment level, yellow ovals), and the combination of the aggregate indices into a total welfare index. A lower weight (20%) was attributed to the PIs of the Caretaker level. Adapted from Menchetti et al. (2021b)

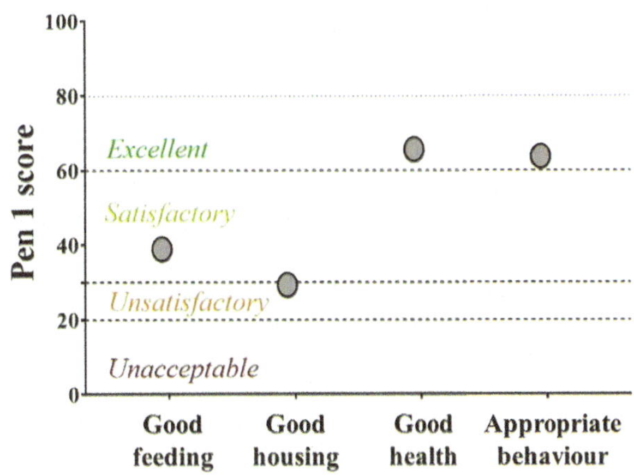

Fig. 3.10 Classification of the pen is based on the profile of the scores obtained by the four welfare principles. This classification system allows the identification of specific deficiencies in the welfare state and thus facilitates the provision of advice to farmers on solving the problems. The pen in the figure could be classified as "unsatisfactory". Adapted from Menchetti et al. (2021b)

the farm are easy to identify, as a poor score in good housing in the pen 1 reported in Fig. 3.10. Consequently, the authors were able to identify the welfare issues presented in each pen and give recommendations to improve camel feeding practices, housing, health practices, general management, and consequently camel welfare. Some of the recommendations have already been put in place in Doha, and the authors look forward to applying the protocol again to show the enhancement obtained and consequently the utility of the application of a standardized method to assess welfare in dromedary camels.

3.4 Risk Analysis in Animal Welfare

Another way to identify welfare consequences and their hazards, in order to formulate preventive and corrective measures, is to apply the risk analysis in animal welfare as suggested by European Food Safety Authority (EFSA) (2012a, 2022b). This model describes the link between influencing factors and welfare consequences (WCs) and defines some formal steps to assess risks for animal welfare. The first step of the formal welfare risk assessment implies the definition of the target population and the exposure scenario. It includes all information on genetics, husbandry system, nutrition, farming, and management procedures to which animals of the target population are subjected. The second step is the identification of the relevant WCs that may occur in these systems or due to the practice. The WCs describe the animal's responses to influencing factors and should be evaluated using feasible, sensitive, specific, and reliable ABMs. Then, relevant hazards, leading to the WCs, could be identified. The hazards are factors having the potential to improve or impair directly or indirectly the animal welfare in the target population. These factors usually include resource- and management-based measures (EBMs). The last step of the risk assessment is the risk characterization which consists of the likelihood estimation for WCs when exposed to a factor. The application of this formal model should produce recommendations to prevent and/or correct the hazards and to mitigate the WCs (EFSA 2012b, 2012a, 2022b).

This approach developed by EFSA was applied for the risk assessment for camel welfare by Menchetti et al. (2021a) (Fig. 3.11). This study aimed at assessing the risks for the welfare of dromedary camels reared in an intensive system, namely, at the Doha market. The authors selected a short list of ABMs covering all the principles of welfare (good feeding, good housing, good health, and appropriate behaviour), avoiding overlap but including both positive and negative indicators. The selected ABMs were as follows: body condition score, thirst index, presence of disease, injuries, stereotypies, and pain induced by management procedures, the proportion of camels showing restricted movements, resting and aggressive behaviours, and responses to the approaching test. Thus, the selected ABMs intended to assess not only the physiological, pathological, and behavioural responses of camels but also their needs and affective states as well as the human-animal relationship (EFSA 2012a, 2022b; Mellor et al. 2020). The authors then identified the factors influencing these ABMs, namely, caretaker experience, space allowance, bedding, food and water management, shaded space allowance, tethering, hobbles, and caretaker's experience. In the final step, statistical models were developed to define the strength of the association between ABMs and factors. Models showed that the risk of poor camel welfare was linked to limited space allowance, lack of shade, dirty bedding, rationed distribution of feeding and water, and short caretaker's experience (Menchetti et al. 2021a).

The risk assessment for animal welfare is not only a way to recognize welfare issues but also to propose practicable corrective actions and guidelines to protect the welfare of camels (Menchetti et al. 2021a). The study at the Doha market suggested that a minimum space allowance of 19 m^2/camel and adequate shaded areas should

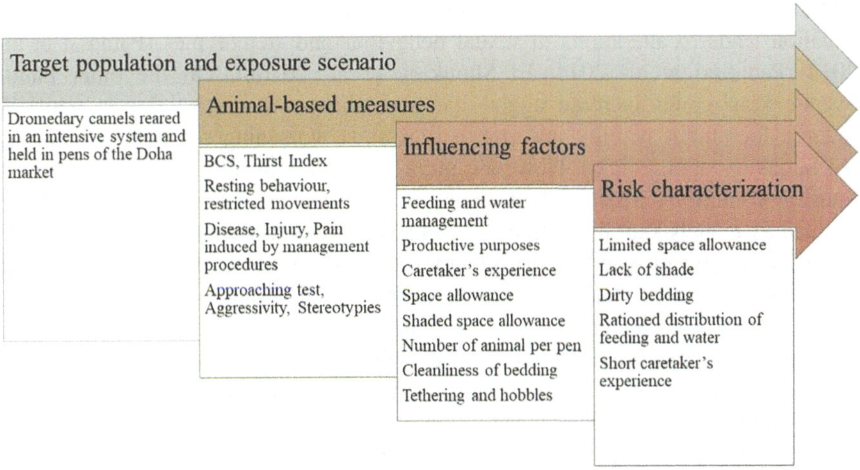

Fig. 3.11 Schematic representation of the application of the model for risk analysis in animal welfare proposed by EFSA at the camel market of Doha. BCS = Body Condition Score

be guaranteed in order to prevent heat stress and enhance dromedary camel welfare in this setting (Menchetti et al. 2021a).

3.5 Welfare Concerns and Gaps of Knowledge

Apart from the aforementioned issues, other several welfare concerns have been identified in dromedary camels in the last decades. Recent literature has indeed dealt with specific issues highlighting the negative association between management practices and camel welfare (Dioli 2022). However, there are still many aspects to be investigated and we have tried to highlight some gaps in knowledge for each principle of welfare.

Although the camel has extraordinary abilities to adapt to resource-poor environments, the principle of good feeding provides much food for thought. In general, studies on the physiological mechanisms that are involved in these abilities could offer insights not only for improving camel welfare (Delavaud et al. 2013; El Shoukary et al. 2021) but also for other animal species facing similar environmental challenges. Moreover, further studies on the effects of dietary supplements and different water regimes could help improve not only welfare but also their productivity, both in grazing systems and in modern intensive farms (Mohamed et al. 2009; Bekele et al. 2011; Faraz et al. 2020, 2021; Nagy et al. 2022) (see chapter on good feeding).

Recent studies related to the principle of good housing confirm that housing is a welfare concern mainly due to lack of freedom of movement, limited space allowance, and social contacts. Hussen and Al-Sukruwah (2022) have shown that restricted movement alters the camel immune system, including changes in blood immune cell composition and function, while several authors found that social isolation leads to alterations in sexual behaviour and stereotypies (Fatnassi et al. 2014; Padalino et al. 2014; El Shoukary et al. 2020b). On the other hand, overstocking could increase aggressive behaviours and cortisol levels while it reduces locomotor activity and rumination, thus worsening the body conditions (El Shoukary et al. 2020a). However, more studies are needed to identify the best housing conditions which will meet the behavioural and physiological needs of dromedary camels kept in captivity (see chapter on good housing).

The principle of good health includes several emerging themes. Camel calf mortality is a critical challenge in some areas, and inadequate colostrum, feeding, and watering management, as well as poor healthcare and the lack of practices reducing the spread of infections, have been identified as predisposing factors (Muluneh et al. 2022). The risk of mastitis, a problem affecting around 50% of the camel population, could be reduced by implementing appropriate hygiene practices (Aqib et al. 2022; Nagy et al. 2022). In general, the duty of care, cleanliness, vaccination and other preventive measures, and early and appropriate treatments could be recommended to implement the health of camels under both intensive and semi-intensive systems (El Harrak 2017; Padalino et al. 2021; Muluneh et al. 2022). Further research is, however, needed to understand the spread of camel infectious diseases, validate diagnostic tests, and develop control and prevention programmes, also in light of the recent pandemic events (El Harrak 2017; Padalino et al. 2021). Monitoring programmes could also limit the risk of the spread of infectious diseases common to other species of domestic animals such as poxvirus, brucellosis, and trypanosomosis. It should also be taken into account that, although camels are present in Europe in negligible numbers, some diseases may be also transmitted through the movement of germinal material (Desquesnes et al. 2008; Zema et al. 2022). The implementation of diagnostic procedures, moreover, could also improve camel productivity. For example, the use of diagnostic investigations for problems related to infertility, very common in camel farms, could increase birth rates and, thus, profitability (Zema et al. 2022) (see chapter on reproduction). The healthcare approach is still rather based on traditional ethnoveterinary practices. In some remote pastoral areas, it is often the only type of treatment available and its value cannot, therefore, be diminished (Antoine-Moussiaux et al. 2007; Dioli 2022). The use of plants without coercive methods for their administration, moreover, does not seem a serious welfare issue. However, other procedures, such as branding, firing, and thermocautery, are undoubtedly painful, cause injury to the animal, and are unnecessary (Dioli 2022). Moreover, the improper use of antibiotics, the development of drug resistance, and increased risk of drug residues in food-animal products could constitute real problems (Antoine-Moussiaux et al. 2007; Padalino et al. 2021). The criterion of the "Absence of pain and pain induced by management procedures" remains a major concern. In addition to some ethnoveterinary practices, coercive methods, improper and inhumane handling, have been often reported in the literature (Dioli 2022). In the authors' view, this is mainly due to the fact that the camels are often considered very aggressive and dangerous animals, so people are often afraid and overreact, often hitting the camels because they are not able to handle them using learning principles. The use of nose pegs, hobbles, sticks, short ropes, and other coercive and pain-induced management practices are still largely applied in the camel industry (Menchetti et al. 2021a; Dioli 2022). Recent data collected at the market in Cairo support this hypothesis (Animals' Angels 2022), highlighting the need of educating camel handlers on gentle and proper handling (see Chap. 8 for details). In general, the spread of information among scientists and stakeholder engagement can contribute to improving the living conditions of camels. The first workshop on camel welfare was recently organized by the World Organisation for Animal Health (WOHA) in the Emirates and more workshops are needed. Further efforts are required from the scientific community to investigate other issues related to good health principle. For instance, studies on the appropriate use of anaesthesia and analgesia to minimize suffering during surgical procedures and on how to implement camel healthcare are consequently needed. Moreover, since camels have a remarkable ability to bear the pain, new ABMs and specific pain scales should be developed to identify welfare concerns and refine the welfare assessment protocol (Padalino and Menchetti 2021) (see chapter on good health).

The poor knowledge of camel ethology is not limited to indicators of negative states such as pain and fear, so the principle of appropriate behaviour remains quite unexplored. The development of specific behavioural tests, greater knowledge of camels' preferences and motivations, as well as their ability to adapt to different farming systems, could contribute to the refinement of the welfare assessment protocol. For example, some authors (Padalino et al. 2021; Zappaterra et al. 2021) have recently shown that, despite their abilities to adapt to arid climates, camels have a preference for shade. Providing adequate shaded areas and avoiding the prolonged use of camels as draught animals could also prevent sunstroke episodes. More studies investigating the behavioural needs of camels are therefore crucial to define welfare standards (see chapter on appropriate behaviour). Moreover, camels are social animals with good communication and cognitive skills (Nagy et al. 2022), and thus ethological studies could contribute to developing and optimizing training protocols favouring, for example, more animal-friendly handling.

Smart technologies could be useful to enhance the knowledge and the monitoring of all the welfare principles listed above could be implemented, similar to what is happening to other livestock species. These technologies enhance both farmers' and animals' welfare, setting alarms in case of early health and production problems. A scoring system of welfare using precision livestock farming (PLF) tools has been proposed for livestock. A variety of technologies are available that can monitor the different welfare facets. However, they must be specific to the species and farming system, and the benefits, as well as the challenges to animal welfare, must be considered from time to time (Schillings et al. 2021). PLF is based on the identification of each single animal, which is then followed 24 h for 7 days a week, and when a shift from its routine is noticed, an alarm is sent to the farmer (Schillings et al. 2021). In dromedary camels, unfortunately, identification is still an issue. In many places, branding is still used as an identification method, with the pain and welfare consequences mentioned before. Microchips and other forms of electronic devices have been tested under different farming conditions (Caja et al. 2016). These devices should be implemented in the camel industry not only to increase traceability but also as a tool to allow PLF. In camels, PLF can be implemented to enhance camel health, production, and welfare, but studies are needed to adapt and validate these technologies for use with camels.

The legislation on camel welfare also seems to fall behind compared to other livestock species and properly defined welfare standards still seem a long way off. There exist no specific recommendations concerning the welfare of farmed dromedary camels within European legislation (Previti et al. 2016). As above mentioned, the method proposed by Padalino and Menchetti (2021) is applicable only for dromedary camels kept in semi-intensive and intensive farming scenarios, and the tool still requires much refinement and must undergo a thorough validation process. The fine tuning of specific protocols would moreover be necessary for assessing the welfare of dromedary camels bred for specific purposes, such as dairy farming and racing. Furthermore, it is important to highlight that there are still no regulations protecting the welfare of dromedary camels during transport and slaughter and more research is needed on these topics to provide evidence to the policy makers (see Chaps. 10 and 11). As it is commonly requested in Europe, a certificate of competencies should be enforced also for camel caretakers. Inappropriate handling and lack of education in workers have been indeed identified as risk factors in all animal species on-farm, during transport, and at slaughter (EFSA 2022c, 2022a). For the camels reared in the nomadic pastoral husbandry system, there is a necessity for specific protocols which also take into account the logistical characteristics of farms and the fact that animals may be less accustomed to the presence of humans. In conclusion, the application of protocols for the assessment of dromedary camel welfare from "farm to fork" and from birth to death as well as the development of specific standards and regulations protecting the camel welfare within several environmental and farming contexts is urgently needed.

3.6 Conclusions

Overall, the number of studies on dromedary camel welfare remains limited. The protocol to assess welfare in dromedary camels is in its infancy and also the scoring systems were based on limited pieces of knowledge. In the following chapters, each welfare principle will be presented in depth, reporting updated information regarding camels reared in different systems and countries. However, even if it seems clear that more research is required in order to suggest welfare standards for this species, the first step could be to spread the current scientific information within the camel industry to raise the level of knowledge of camel welfare.

References

Animals' Angels (2022) *Camel transport*. https://www.animals-angels.de/en/projects/animal-transports/camel-transports.html Accessed 9 Feb 2022

Antoine-Moussiaux N, Faye B, Vias GF (2007) Tuareg ethnoveterinary treatments of camel diseases in Agadez area (Niger). Trop Anim Health Prod 39(2):83–89. https://doi.org/10.1007/s11250-007-4404-1

Aqib AI et al (2022) Pathological insights into camel mastitis. Acta Trop 231:106415. https://doi.org/10.1016/j.actatropica.2022.106415

AWIN (2015a) Welfare assessment protocol for GOATS, pp 1–70. https://doi.org/10.12120/AWIN_GOΛTS_2015

AWIN (2015b) Welfare assessment protocol for HORSES, pp 1–80. https://doi.org/10.13130/AWIN_HORSES_2015

Bekele T, Lundeheim N, Dahlborn K (2011) Milk production and feeding behavior in the camel (Camelus dromedarius) during 4 watering regimens. J Dairy Sci 94(3):1310–1317. https://doi.org/10.3168/jds.2010-3654

Botreau R et al (2007) Aggregation of measures to produce an overall assessment of animal welfare. Part 1: a review of existing methods. Animal 1(8):1179–1187. https://doi.org/10.1017/S1751731107000535

Botreau R, Veissier I, Pern P (2009) Overall assessment of animal welfare: strategy adopted in welfare quality®. Anim Welf 18(4):363–370

Broom DM (1996) Animal welfare defined in terms of coping with the environment. Acta Agricultura Scandinavia Supplement 27:22–28

Caja G et al (2016) Comparison of visual and electronic devices for individual identification of dromedary camels under different farming conditions. J Anim Sci 94(8):3561–3571. https://doi.org/10.2527/jas.2016-0472

Dalla Costa E et al (2014) Equine on-farm welfare assessment: a review of animal-based indicators. Anim Welf 23(3):323–341. https://doi.org/10.7120/09627286.23.3.323

Delavaud C et al (2013) Plasma leptin, glucose and non-esterified fatty acid variations in dromedary camels exposed to prolonged periods of underfeeding or dehydration. Comp Biochem Physiol A Mol Integr Physiol 166(1):177–185. https://doi.org/10.1016/j.cbpa.2013.05.026

Desquesnes M et al (2008) First outbreak of Trypanosoma evansi in camels in metropolitan France. Vet Rec 162(23):750–752. https://doi.org/10.1136/vr.162.23.750

Dioli M (2022) Observation on dromedary (Camelus dromedarius) welfare and husbandry practices among nomadic pastoralists. Pastoralism 12(1):1–23. https://doi.org/10.1186/s13570-021-00221-5

EFSA (2012a) Guidance on risk assessment for animal welfare. EFSA J 10(1):2513. https://doi.org/10.2903/j.efsa.2012.2513

EFSA (2012b) Statement on the use of animal-based measures to assess the welfare of animals. Panel on Animal Health and Welfare (AHAW). EFSA Journal 10(6):2767. https://doi.org/10.2903/j.efsa.2012.2767

EFSA (2022a) *Animal welfare*. https://www.efsa.europa.eu/en/topics/topic/animal-welfare. Accessed 9 Feb 2022

EFSA (2022b) Methodological guidance for the development of animal welfare mandates in the context of the farm to fork strategy. EFSA J 20(7):7403

EFSA (2022c) Welfare of pigs. EFSA 20(8):7421

El Harrak M (2017) *Infectious diseases of camelids*. Abu Dhabi. Available at: Infectious diseases of %0Acamelids

El Shoukary RD, Osman A, Mohammed A (2020a) Effects of stocking density on some behavioral and some blood biochemical parameters in camel during the rut period. Egypt J Vet Sci 51(2):253–262. https://doi.org/10.21608/ejvs.2020.24526.1153

El Shoukary RD et al (2020b) Housing management of male dromedaries during the rut season: effects of social contact between males and movement control on sexual behavior, blood metabolites and hormonal balance. Animals 10(9):1–11. https://doi.org/10.3390/ani10091621

El Shoukary RD et al (2021) Time-budget and blood parameters of the dromedary camels (Camelus dromedarius) under different feeding management strategies. Adv Anim Vet Sci 9(6):895–903. https://doi.org/10.17582/journal.aavs/2021/9.6.895.903

Faraz A et al (2020) Comparative growth response related to hair mineral analysis in dromedary camel calves. Open Vet J 10(4):392–399. https://doi.org/10.4314/ovj.v10i4.6

Faraz A et al (2021) Effect of different watering regimes in summer season on water intake, feed intake, and milk production of marecha she-camel (Camelus dromedarius). Animals 11(5):1342. https://doi.org/10.3390/ani11051342

Farsi H et al (2022) Seasonal variations in locomotor activity rhythm and diurnal activity in the dromedary camel (Camelus dromedarius) under Mesic semi-natural conditions. Chronobiol Int 39(1):129–150. https://doi.org/10.1080/07420528.2021.1984936

Fatnassi M et al (2014) Effect of different management systems on rutting behavior and behavioral repertoire of housed Maghrebi male camels (Camelus dromedarius). Trop Anim Health Prod 46(5):861–867. https://doi.org/10.1007/s11250-014-0577-6

Fraser D (2009) Assessing animal welfare: different philosophies, different scientific approaches. Zoo Biol 28(6):507–518. https://doi.org/10.1002/zoo.20253

Fraser D et al (1997) A scientific conception of animal welfare that reflects ethical concerns. Anim Welf 6:187–205. http://docserver.ingentaconnect.com/deliver/connect/ufaw/09627286/v6n3/s1.pdf?expires=1548957253&id=0000&titleid=75000207&checksum=E85F310F26B118493EB73EB14755F7F0%0Ahttp://www.fao.org/fileadmin/user_upload/animalwelfare/Fraser1997.pdf

Hussen J, Al-Sukruwah MA (2022) The impact of the animal housing system on immune cell composition and function in the blood of dromedary camels. Animals 12(3):317. https://doi.org/10.3390/ani12030317

Mellor DJ (2016) Updating animal welfare thinking: moving beyond the "five freedoms" towards "a lifeworth living". Animals 6(3):21. https://doi.org/10.3390/ani6030021

Mellor DJ, Beausoleil NJ (2015) Extending the "five domains" model for animal welfare assessment to incorporate positive welfare states. Anim Welf 24(3):241–253. https://doi.org/10.7120/09627286.24.3.241

Mellor D, Stafford K Patterson-Kane E (2010) The sciences of animal welfare. *UFAW Animal Welfare Series*

Mellor DJ et al (2020) The 2020 five domains model: including human–animal interactions in assessments of animal welfare. Animals 10(10):1–24. https://doi.org/10.3390/ani10101870

Menchetti L et al (2020) Camel welfare: the first survey on camel caretakers' perspective. J Camelid Sci. Paper accepted

Menchetti L, Faye B Padalino B (2021a) New animal-based measures to assess welfare in dromedary camels, Tropical Animal Health and Production 53.6 (2021):533.

Menchetti L et al (2021b) Application of a protocol to assess camel welfare: scoring system of collected measures, aggregated assessment indices, and criteria to classify a pen. Animals 11(2): 1–24. https://doi.org/10.3390/ani11020494

Mohamed MI et al (2009) Feed intake, digestibility, rumen fermentation and growth performance of camels fed diets supplemented with a yeast culture or zinc bacitracin. Anim Feed Sci Technol 149(3–4):341–345. https://doi.org/10.1016/j.anifeedsci.2008.07.002

Muluneh B et al (2022) Constraints and incidence of camel calf morbidity and mortality in Borana rangeland Southern Ethiopia. J Arid Environ 206:104841. https://doi.org/10.1016/j.jaridenv.2022.104841

Nagy PP, Skidmore JA, Juhasz J (2022) Intensification of camel farming and milk production with special emphasis on animal health, welfare, and the biotechnology of reproduction. Anim Front 12(4):35–45. https://doi.org/10.1093/af/vfac043

Padalino B, Menchetti L (2021) The first tool for assessing welfare of camels. Front Vet Sci 7: 631876. https://doi.org/10.3389/fvets.2020.631876

Padalino B et al (2014) Could dromedary camels develop stereotypy? The first description of stereotypical behaviour in housed male dromedary camels and how it is affected by different management systems. PLoS One 9(2):e89093. https://doi.org/10.1371/journal.pone.0089093

Padalino B et al (2021) Dromedary camel health care practices reported by caretakers working at a permanent market. Emir J Food Agric 33(4):348–361. https://doi.org/10.9755/ejfa.2021.v33.i4.2688

Pastrana CI et al (2021) A tool for functional selection of leisure camels: behaviour breeding criteria may ensure long-term sustainability of a European unique breed. Res Vet Sci 140:142–152. https://doi.org/10.1016/j.rvsc.2021.08.007

Pond W, Bazer F (2011) Animal Welfare in Animal Agriculture. CRC Press, Boca Raton USA

Previti A, Guercio B, Passantino A (2016) Protection of farmed camels (Camelus Dromedarius): welfare problems and legislative perspective. Anim Sci J 87(2):183–189. https://doi.org/10.1111/asj.12446

Schillings J, Bennett R, Rose DC (2021) Exploring the potential of precision livestock farming technologies to help address farm animal welfare. Front Anim Sci 2. https://doi.org/10.3389/fanim.2021.639678

Szucs E et al (2012) Animal welfare in different human cultures, traditions and religious faiths. Asian Australas J Anim Sci 25(11):1499–1506. https://doi.org/10.5713/ajas.2012.r.02

Veissier I et al (2011) Highlighting ethical decisions underlying the scoring of animal welfare in the welfare quality® scheme. Anim Welf 20(1):89–101

Welfare Quality Network (2009) *Welfare Quality® project*. http://www.welfarequality.net/en-us/news/assessment-protocols/

World Organization for Animal Health (2022) *Terrestrial code*. https://www.woah.org/en/what-we-do/standards/codes-and-manuals/terrestrial-code-online-access/

Zappaterra M et al (2021) Do camels (Camelus dromedarius) need shaded areas? A case study of the camel market in Doha. Animals 11(2):1–16. https://doi.org/10.3390/ani11020480

Zema E et al (2022) European regulations on camel germplasm movement within the European Union: a current framework based on safety. Animals 12:2255. https://doi.org/10.3390/ani12172255

Good Feeding: Nutrition and Feeding of the Arabian Camel (*Camelus dromedarius*)

4

Rafat Al Jassim

Abstract

This chapter discusses the impacts of modern camel farming on the welfare principles of the nutrition and feeding of the Arabian camel (*Camelus dromedarius*). In order to better manage camels under an intensive husbandry system, we need to develop a better understanding of the camels' biological system and the environmental factors that impact on its health and well-being. The nutritional aspects dealt with in this chapter are related to digestion, absorption and metabolism, while the feeding aspects are mainly to do with the provision of nutrients. Therefore it is important to meet the behavioural needs alongside the provision of required nutrients. Confining a browsing animal, that has evolved to wander around and walk for long distances, to a yard and hand feeding it may result in a negative impacts. Also, treating the camel like a dairy cow or a racing horse without knowing the exact necessary nutrient requirements puts the camel under stress that may be overwhelming with welfare consequences. Forcing the camel to lose weight in order to be lighter for racing may force the camel into ketosis. On the other hand, feeding a camel a low-fibre diet and a large volume of starch-rich feed in one or two meals a day could lead to fermentative acidosis and laminitis and abnormal behaviours. Exposure to foreign items such as plastic bags, hay bale robs, or metal objects may have lethal consequences.

R. Al Jassim (✉)
Centre for Animal Science, Queensland Alliance for Agriculture and Food Innovation (QAAFI), The University of Queensland, St Lucia, Australia

African Sustainable Agriculture Research Institute (ASARI), Mohammed VI Polytechnic University (UM6P), Ben Guerir, Morocco
e-mail: r.aljassim@uq.edu.au; rafat.aljassim@um6p.ma

B. Padalino, B. Faye (eds.), *Dromedary Camel Behavior and Welfare*, Animal Welfare 24, https://doi.org/10.1007/978-3-031-48600-5_4

Keywords

Camel · Feeding behaviour · Nutrient requirements · Metabolic disorders · Welfare

4.1 Introduction

Camels evolved as browsing herbivores with a fermentation compartmental stomach that allows them to store ingested feed and break it down to its constituents. The fermentation processes are carried out by a vast number and diverse microbiota of protozoa, bacteria and anaerobic fungi. This microbiota evolved together with the host animal and benefited from the physiologic conditions its forestomach provides, so they developed a symbiotic relationship. The host camel provides the physiologic conditions, including temperature around 39 °C, neutral osmosis, buffered environment with pH between 6 and 7 and a continuous supply of nutrients. The microbiota breaks down structural components and releases organic acids such as the short chain volatile fatty acids (SCFA), synthesise microbial protein, detoxify antinutritional factors found in plants, synthesise essential nutrients and stimulate the immune system of the host animal. Microbial protein (MP) synthesis in the compartmental stomach of the camel was estimated to be 95 g MP/kg digestible organic matter intake (Guerouali et al. 2004). Studies into the microbial ecosystem of the camel have revealed the novelty of the bacterial system, with vast numbers and a diverse bacterial community, with different operational taxonomic units (OTUs) from those identified in cattle but they do the same function. Based on their functionality, they can be studied as fibrolytic, proteolytic, amylolytic, or more specific to other substrates (for more details see Al Jassim 2022).

4.1.1 The Digestive System of the Camel

The gastrointestinal tract of the camel has peculiar anatomical and physiological features that are specific to the Camelidae family and different from that of the Bovidae. However, the forestomachs of the camelids and ruminant animals share some important functional characteristics, due to the parallel evolution the two groups underwent during the Miocene epoch, about 25 million years ago (mya). The two species diverged from each other much earlier than this date, during the early and middle Eocene epoch, about 50 mya. The forestomach is of particular interest, from a nutritional point of view. It consists of three compartments in the camel (C1, C2 and C3, Fig. 4.1), while it is four compartments in cattle (rumen, reticulum, omasum and abomasum). Some refer to C1 and C2 as the rumen and C3 as the abomasum, with the omasum missing in the camel (Wang et al. 2000; Rabee et al. 2022; Rabee et al. 2023; Srivastava et al. 2023).

The characteristic feature of the external surface of the rumen shows the presence of cellulae (misnomer old name was water sacs). The external grooves are clear as

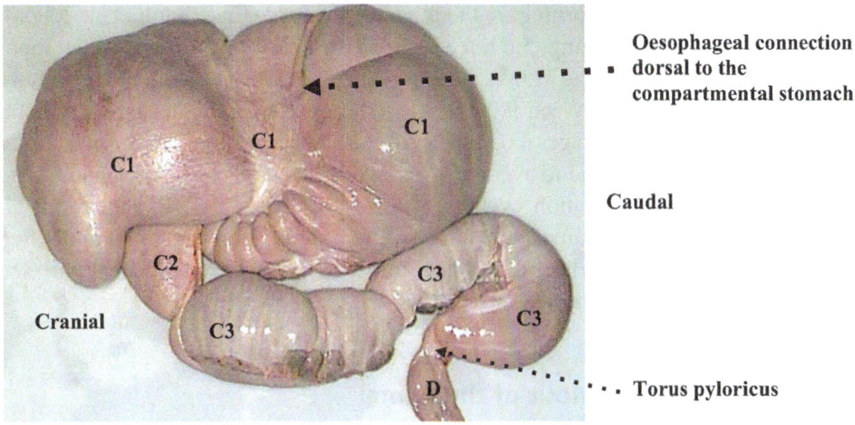

Fig. 4.1 The compartmental stomach of the Arabian camel (*Camelus dromedarius*) with the different parts marked: C1 = Compartment 1; C2 = Compartment 2; C3 = compartment 3

depressions on the C1 surface. These grooves formed due to the presence of thick muscular thickening inside the C1 wall. The first compartment (C1) occupies most of the left side of the abdominal cavity, pushing the viscera to the right of the body including the left kidney. Both C2 and C3 are located on the right of the midline, with C2 a little higher than the C3 which lies on the floor of the abdominal cavity (Engelhardt et al. 1988).

The C1 compartment in the camel is lined with stratified squamous epithelium like the ruminant's rumen except for the region of the cellulae. The unique cellulae are lined with simple columnar epithelium and have tubular glands. The second compartment (C2) is lined with stratified squamous epithelium, while C3 is like the abomasum in the ruminants and lined with simple columnar epithelium with glandular tissues (Engelhardt et al. 1988).

The digestive processes in the compartmental stomach of the camel are like those in the compartmental stomach of the cow, despite the anatomical, histological and physiological differences. The freshly ingested feed enters the compartmental stomach and adds to the digesta pool, which undergoes mixing and further reduction of particle size by rumination and microbial degradation. Mixing is carried out by two consecutive contractions, A and B (Kaske et al. 1989). It starts with a contraction of the C2 followed by a contraction of the caudal part of the C1 approximately 4 sec later, while the canal between C2 and C3 is relaxed. B-contractions start with the contraction of the cranial part of C1 followed by a contraction of C2 and the caudal part of C1 9 sec later. These contractions occur at higher frequency during eating time, up to 130 contractions per hour (A + B), compared with 80–100 contractions when the feed was removed. The A:B ratio also changed, being 1:6 during eating compared to 1:2–3 when feed was removed (Kaske et al. 1989).

Rumination was recorded in handfed camels on farms (Kaske et al. 1989; Engelhardt et al. 2006a). In both experiments, camels were offered their diets at 8:00 h. In the first experiment (Kaske et al. 1989), rumination started after midnight and lasted until the next morning. In the other experiment (Engelhardt et al. 2006a), camels spent 8.3 h ruminating, 5.6 h feeding and 10.1 h resting. Camels spent more time ruminating than eating (71.0 vs 61.3 min per kg DM hay). Rumination activity peaked in the morning between 9:00 and 11:00 and after midnight between 02:00 and 04:00 am, with an average of 67 boluses regurgitated per hour. Chewing activity was also recorded and found to average 45 s per bolus and 68 chews per minute. The pause between two rumination cycles lasted an average of 9 s (Engelhardt et al. 2006a). It is not known if camels will have different rumination behaviour under free browsing environments when they have free access to browse vegetation (see Chap. 7 on appropriate behaviour).

4.1.2 Feeding Behaviour of the Camel

As a browsing herbivore, the camel under browse feeding conditions spends long hours browsing on vegetation available in its ecosystem. The camel tends to select the parts of plants, from a variety of shrubs and trees, that are most nutritious and succulent. This behaviour is of particular interest regarding survival of the camel when vegetation is scarce and for the preservation of its fragile desert environment.

Studies from Australia investigating the feeding behaviour of feral camels and their impacts on the natural vegetation of the central desert of Australia showed that camels predominantly feed on trees and shrubs. However, after a generous rainfall, they alter their feeding behaviour and feed on ground storey vegetation (Phillips et al. 2001). In Ethiopia, in semi-arid area, Faye and Tisserand (1989) observed browsing time of a camel herd: at dry season, the animals started by grazing grass and small shrubs in the morning and in the evening while during the hottest time of the day (from 10 am to 4 pm approximately) all the herd spent time under forage trees (mainly *Acacia* sp.); at rainy season, the part of time for eating grass increased. In addition to being a browsing and selective eater, the camel has also been described as a conservative eater in its choice of plants. Camels tend to feed on the plants they know and have eaten before. When moved into a new paddock with a different mix of vegetation, they feed on plant species they knew from previous paddocks (Phillips et al. 2001). Seasonal changes in feeding behaviour were observed and recorded for feral camels in Australia. Acacia species are the preferred diet for camels, with *Acacia estrophiolata* (ironwood), *Acacia victoriae* (acacia bush), *Acacia aneura* (mulga), *Acacia georginae* (Georgina Gidgee), *Acacia kempeana* (wanderrie wattle) and *Acacia ligulata* (dune wattle) as the most preferred species. Other tree species, including *Atalaya hemiglauca* (whitewood) and *Grevillea striata* (beefwood), are also among the most preferred topfeed, while *Atriplex* species (*Atriplex enchylaena* and *Atriplex salsola*) are the preferred forbs. Several other species were also monitored and recorded during the study of Phillips et al. (2001). The change in feeding behaviour during the wet season and following substantial rainfall is of particular interest and suggests the need for alternative grazing management. This is particularly important when other species of animals such as cattle are co-grazed with camels because the change in camel feeding behaviour may impact on the availability of forage for cattle. It was observed that camels feed on the freshest part of the

Fig. 4.2 Wire mesh roller for hay feeding

preferred trees and shrubs: young branches, leaves and flowers (Phillips et al. 2001). Feeding on young shoots, leaves and flowers ensures that camels obtain a relatively stable supply of quality diet that are not affected by season or stage of maturity. This is also advantageous to the camel and explains how camels have managed to thrive in the harsh and hostile conditions of the Australian desert where other domestic animals such as cattle and sheep are unable to.

In contrast, camels that are confined to a yard will be unable to perform species-typical behaviour and acquire necessary skills and making them vulnerable to conditions imposed on them by the farm management. It is well established that both genotype and the environment influence behaviour (Breed and Sanchez 2010), but it is not known if the feeding behaviour of the camel is instinctive. Instinct suggests that "a behaviour is performed without thought and cannot be modified by learning" (Breed and Sanchez 2010). However, the fact that it has taken the camels a few weeks before adapting to the new plant species when moved to a new paddock (Phillips et al. 2001) may reflect an association between an instinctive response and a learning behaviour. The learning behaviour is essentially developed through exposure and practice.

In the intensive dairy farming system, the calves are separated from their mothers and raised in separate group-barns with other calves (Nagy et al. 2022). They not only miss the opportunity to interact with mature animals, particularly their mothers, and learn from them, but they are also forced against their nature and changed from browsing animals to handfed, intensively managed animals. The lack of mobility and the easy availability of the feeds with high energy and protein concentration in intensive farm leads also the camel to spend few times for eating with higher risks of metabolic disorders and even obesity.

To avoid such bad consequences, in Canary Islands, a system of hay distribution in rolling hay feeder has been experimented (Fig. 4.2).

The hay is covered with a wire mesh that allows small amounts of hay being obtained at each bite, which mimics the browsing behaviour of the camel. This system seems to work well for young camels as it encourages foraging and interaction with items and other camels foraging at the same time. It can be also convenient for lactating camel, especially if they don't spend time browsing.

4.1.3 Physiological Adaptation to Arid and Semi-Arid Environments

As indicated earlier, camels are browsing herbivores evolved with the ability to forage on a range of desert vegetation and select parts of the plants that are most nutritious. Special physiological adaptation enables camels to benefit the most from their diet. Fluid retention time in the camel's forestomach is shorter compared with cattle, sheep and goats (Lechner-Doll et al. 1990). However, retention time for both fractions, the liquid and solid of digesta, increase during the dry season, but the increase in the camel is smaller compared with cattle and sheep. This is mainly due to the higher quality of diet selected by camels and goats during the dry season (Lechner-Doll et al. 1990). The liquid fraction of the digesta spends less time in the forestomach than the solid particles (~14 vs 49 h) and retention time for longer particles is longer than for small particles (Heller et al. 1986). Fermentation processes are always associated with loss and, therefore, the faster passage rate of the soluble fraction, the more benefits obtained from the readily available carbohydrates and proteins that are associated with the liquid phase of the digesta. At the same time, some of the soluble plant toxins require a longer retention time to be completely degraded and detoxified by forestomach microorganisms. These toxins may escape microbial attack, enter the intestines and get absorbed into the body. A good example for such implication is the toxin indospicine, which is a free amino acid found in the leguminous plant *Indigofera* spp. such as *I. spicata* (Tan et al. 2016, 2017). At the same time, the longer retention time of solid particles enables the forestomach microbiota to maximise the extraction of high-quality nutrients from the rigid structural components of feed. The longer retention time of solid particles in the forestomach of the camel may also explain the lower feed intake in the camel, because emptying time influences gut fill and feed intake.

Mean retention time (MRT) of large (20 mm) and small (2 mm) particles in the forestomach of camels (*Camelus dromedarius*) was investigated by Lechner-Doll (1990) during the wet and dry seasons in Kenya and compared with that in cattle, sheep and goats. MRT for large particles in the forestomach of camels was double that of the small particles (52.7 vs 26.5 h) and about 5 times that of the fluid (10.6 h). A longer retention time was observed during the dry season compared with the wet season. The increase of particle MRT in the dry season, as a percentage of the values in the wet season, was lowest in camel (18%) and highest in sheep (46%), with cattle and goats between 27% and 22%, respectively. The increase in MRT in the forestomach of the camel was associated with a 5% increase in the digestibility of the slowly digestible low-quality feed (Rutagwenda et al. 1990). Using a similar experimental protocol, Dittmann et al. (2015) have estimated the MRT for the fluid and solid particles of different sizes (2, 10 and 20 mm) in the Bactrian camels (*Camelus bactrianus*), the llama and vicugna and compared their results with data of Lechner-Doll et al. (1990) of the dromedary camel. MRT of the liquid has averaged 34 h and for the solid particles were 47, 66 and 67 for the 2, 10 and 20 mm, respectively. So, even if camels are able to cope with long fasting, they can suffer from prolonged hunger and the associated negative emotions. They should therefore have access to roughage all the time to meet the principle of good feeding and be in a positive welfare status. Inappropriate feeding regime will lead to behavioural changes, including the development of stereotypy (see chapter on appropriate behaviour) and to health issues.

4.1.4 Metabolic Disorders of Intensively Managed Camels: Acidosis

Camels under intensive management, such as that of dairy operations or race camels, are fed large volumes of grain-based concentrate diets that are low in fibre and rich in starch. The fermentability of starch in the forestomach of the camel is rapid, leading to the production of large quantities of volatile fatty acids (VFAs) and lactic acid. Such change in the fermentation pattern will overwhelm the absorptive and buffering capacities of the forestomach, leading to the accumulation of these organic acids which leads to a decline in forestomach pH. Normal pH is around 6.5 when camels are fed a roughage-based diet, but it drops to less than 5.5 when a grain supplement is added to their diet (Ghali et al. 2019). Accumulation of VFAs will drop the pH down and encourage the proliferation of acid-tolerant bacteria, such as the predominant *Streptococcus bovis* which is a key lactic acid-producing bacterium in the forestomach of the camel (Ghali et al. 2011). This bacterium is the main causative agent of fermentative acidosis in ruminant animals, in horses and in the camel.

In contrast to high-starch diets, browsing and feeding on high-fibre diets lead to a balanced gut environment, particularly in the forestomach, mainly due to the slow fermentation processes of fibre and the efficient removal of produced VFAs (Ghali et al. 2019). The efficient and continuous removal of organic acids prevents their accumulation and therefore maintains a buffered environment. Mastication and rumination of fibre-rich feeds lead to the production of a copious volume of saliva which helps in the buffering and facilitates mixing and efficient digestion. The economic and physiologic impacts of clinical and sub-clinical acidosis in the camel are not determined, but incidences of severe diarrhoea and laminitis in camels are common. The damage to the lining of the forestomach and beyond, due to the accumulation of excessive amounts of acids, is not known. However, if the impact of acid accumulation is like that occurring in other species of animals, such as the horse (Andrews et al. 2008), it may compromise the integrity of the gut lining or cause leaky bowel, which have severe adverse effects on the physiological function of the gut and the health of the camel.

4.1.5 Nutritional Management of Race Camels

Field observations from farm visits reveal that race camels are forced to lose excessive amounts of fat to flatten the hump and produce lighter camels. The amounts of fibre-rich feeds offered to race camels are also reduced in order to reduce gut fill. Alternatively, race camels are fed concentrate mixes and special dietary ingredients such as ghee, honey, eggs and dates.

Although camels are known to be less susceptible to the development of ketosis when deprived from feed (Wensvoort et al. 2001), it is not known if the severity and duration of feeding below maintenance will overwhelm the camel's regulatory system and compromise such ability and could be considered linked to the welfare consequences of chronic hunger.

The plasma concentration of ketone bodies in camels is much lower than that in ruminant animals such as sheep (Faye and Bengoumi 2018). This is mainly due to the lower *ß-hydroxybutyrate dehydrogenase* enzyme activity in the epithelium of the camel's forestomach (Chandrasena et al. 1979).

In a comparative study, the ability of the various tissues to oxidise butyrate to ketone bodies and CO_2 in the camel was compared with those from sheep and goats (Emmanuel 1980). The forestomach epithelium and the liver tissues of the camel converted negligible amounts of butyrate to ketone bodies, while the kidney of the camel metabolised more butyrate than in sheep and goats. This biochemical phenomenon has implications on energy precursor usage during a shortage of feed supply and may result in a higher demand for amino acids and proteins for the gluconeogenic pathway, leading to more glucose produced by gluconeogenesis. This may also explain the higher plasma and urine urea concentration that increase during feed deprivation (Dahlborn et al. 1992). This adds to the complexity of the biological system and the interdependency on alternative sources and the activation of other pathways when the supply substrate for normal pathways is compromised. Ketone bodies are acidic, and efficient removal of them is necessary to maintain normal blood pH and normal body function. Wensvoort et al. (2001) described the camel system as having the "ability to control lipolytic and gluconeogenic activities to prevent or postpone the state of ketosis".

Limited access to fibre diets increases the incidence of ingestion of foreign objects such as plastic bags, bale robs and metals. It is important to shift race camels back to normal diets and offer them free access to roughages when the racing season is finished.

In order to avoid excessive energy intake and undesirable overweight, the roughage diet should be chosen carefully and may have to dilute the high-quality roughage such as lucerne hay with a low or medium quality roughage, so that camels can have greater access to a fibre-rich diet while maintain the same nutrient intakes. Feeding strategies as reported above (Fig. 4.1) or as applied in racing horses to reduce oral stereotypy (Mazzola et al. 2016) and enhance their welfare could be suggested for racing camels, but they need further studies.

The current practices of offering camels what it has not evolved to feed on and digest need to be evaluated and evidence be sought from research to support such practices.

4.1.6 Glucose Homeostasis in Camels

Glucose homeostasis in the camel is different from that in the ruminant animals and equids. Basal plasma glucose level is higher in camels compared with sheep and ponies (Elmahdi et al. 1997, Emmanuel 1981), and the rate of glucose elimination following intravenous administration was lower in camels than in sheep and ponies. Comparative study of insulin responsiveness comparing camel to sheep, ponies and pigs (Kaske et al. 2001) has shown that insulin responsiveness was much lower in camels than other species. This can be partly explained by the lower insulin sensitivity in camels and possible higher gluconeogenesis as compared to other species of animals (Emmanuel, 1981). Deprivation of camels of feeds for 5 days did not change plasma glucose level, serum concentration of triglycerides and beta hydroxybutyrate (Abdoun et al. 2011).

4.1.7 Nutrient Requirements

Limited information is available on nutrient requirements for camels. Energy requirements were determined by calorimetric and energy balance experiments (Guerouali and Wardeh 1998) and, earlier, by regression analysis (Farid, 1995). This work was recently reviewed by Al Jassim (2019) who also made further extrapolation for energy and protein requirements for maintenance and gain based on reliable experimental estimates. At this stage all these estimates need to be verified under different feeding conditions and systems. Regarding other nutrients, few experiments have aimed at determining the requirements for selenium (Se) and vitamin E.

Because of the absence of standard feeding tables for camels, researchers continue to follow traditional feeding practices or apply standard feeding tables for cattle to camels or use levels that they consider optimal for camels. For example, they consider requirements to be lower than those of domestic animals, such as cattle, and accordingly an intake of 1.5 to 2% body weight of roughage is considered to satisfy the requirements for maintenance and weight gain of up to 1 kg per day (El Badawi 2018). This corresponds to 9–12 kg DMI for a 600 kg camel. At the same time, it was reported that a 600 kg dairy camel excretes an average of 16 kg faecal DM per day (Abdel-Rahman et al. 2020) while a drought camel excretes 11 kg. These figures may correspond to an intake of 40 kg DM of diet with 60% DM digestibility by a dairy camel and about 27 kg DM by a drought camel, which is 6.7% and 4.5% body weight. In a growth experiment comparing camels with steers, camels consumed 1.75% of BW while the steers consumed 2.5% BW. For the experimental 376 kg BW camels, the DMI was 6.6 kg (El-Badawi and Yacout 1999). Recent work by

Laameche et al. (2019), using diets with different levels of concentrate mixes and roughage, reported a DMI of 1.3 to 1.96 kg DM per 100 kg BW. Increasing the amounts of concentrate decreased roughage intake, while an increase in the roughage intake led to a decrease in total DMI. Intakes from concentrate increased from 2.39 to 6.0 kg per day, while intakes from hay dropped from 3.05 to 1.93 kg per day. These intakes correspond to an estimated ME intake of 60.4 to 96.1 MJ of ME per day, which satisfies maintenance and milk yield requirements according to Al Jassim (2019).

4.1.8 Salt and Water Deprivation Tolerance

Adaptation of the Arabian camel to arid desert conditions implies infrequent access to limited water supply and feeding on plants that are often rich in antinutritional secondary compounds and concentrated in salt. The camel is known to tolerate high levels of dietary salt and browsing camels prefer natural forbs that contain high salt, such as the *Atriplex* species. Salt content is not the only factor that makes such plants palatable to camels; the nutritive value of these plants is also a factor. However, tolerance should not be confused with requirements, and the camel adopts different physiological strategies to cope with high salt especially if that is accompanied by deprivation of water. Early research has emphasised the role of the kidney and its response to high salt and water shortage. Such conditions increase sodium excretion, produce highly concentrated urine and conserve water.

Recent research has investigated the role of the liver and the intestines. Zhang et al. (2020) suggested the involvement of long non-coding-RNA (lncRNA) and micro-RNA (miRNA) in the modulation of the transcription of transporters involved in absorption and excretions of sodium. In their work, they have focused on differential alternative splicing (AS) and gene expression in the liver and ileum. It was found that a number of genes were upregulated while others were downregulated in camels under salt stress and water deprivation stress. Among the downregulated genes in the ileum of camels under water deprivation and salt stresses is the *AQP5*, which encodes aquaporin 5, a water channel protein located on biological cell membranes that facilitate the passage of water and small molecules through the lipid bilayer. It was suggested that such downgrading is necessary to maintain osmotic homeostasis and prevent dehydration of cells (Zhang et al. 2020). The other gene that was downregulated in the ileum of salt- and water-deprived camels is *MUC6*. This gene encodes mucin 6, a glycoprotein which is an important member of the mucin family, linked to Na^+/Ca^{2+} pump. Downregulation of *MUC6* prevents excessive Na^+ import by the Na^+/Ca^{2+} exchanger. The reference to these two genes, *AQP5* and *MUC6*, is just an example to provide an insight into the complexity of the biological system and the adaptation of more than a couple of organs in the Na homeostasis and hydration of the body under salt and water deprivation stresses. The alteration on the expression of some genes could see as a coping strategy or as a first sign of impaired welfare as reported in other species (Tomas Marques 2017).

4.1.9 Water Intake

Limited information is available on water requirements and intakes by camels under different conditions. Water is a precious commodity which is fundamental for the sustainability of the agricultural systems and life. Climate changes and desertification is becoming increasingly challenging and important to consider when deciding on what animals to raise and what crops to grow. It is often overwhelming when reviewing the quantities of water required to produce 1 kg beef or 1 kg wheat or lucerne hay under current water shortages worldwide. In order to produce 1 kg of beef you need 15,000 to 70,000 litres of water and about 900 to 2000 litres to produce 1 kg of wheat. The list of agricultural produce is long, and the water requirements raise concerns. The requirement for water to produce daily diet per capita in the Middle East and North Africa is estimated to be 2940 litres (Schreier 2002). These estimates pose a challenge to agriculturists and decision-makers, due to the severe water shortage in the region and the costly impact of that shortage on rural communities and on the sustainability and development of agricultural systems.

The camel is often described as the animal of choice under drought and desert conditions and has been praised for its ability to thrive and produce under these conditions. However, little information is available on its water requirements and drinking behavioural needs. Previous sections have dealt with feed requirements. This is not comprehensive, but it is informative enough to appreciate the difference between camels and other domestic animals.

Camels, like other livestock, obtain water from three sources: the feed they eat, metabolic water that results from the oxidation of organic materials (i.e. carbohydrates, fat and proteins) and drinking water. When camels are penned, deprived from browsing and fed hay and concentrate diets that supply very little water (~10%), their requirements for drinking water increase significantly. Other factors including the breed of the camel, the physiological status, activity or productivity, food intake and food quality and climatic conditions must be considered. Water salinity is an important factor and camels may have an advantage over other species of animals because of their ability to concentrate their urine. In general, camels like other animals must consume a greater volume if the water is saline, because the animal will use some of the water it drinks to excrete the salts that are dissolved in the drinking water (Dryden 2008).

Camels weighing approximately 655 kg consumed an average of 22.3 ± 2.31 litre/day when they had access to ad libitum water and hay feed. However, after 11 days of water deprivation, they consumed 97.3 ± 24.1 litre/day within 1 hr. of allowing them free access to water again, which is about 4.5 times their daily intake before water deprivation (Engelhardt et al. 2006b). Water deprivation for 11 days reduced feed intake to a very low level of 10% of the normal intake.

Water intake from fresh forage plus free water was 26.95, 22.56 and 26.21 for lactating camels watered daily, every 4 days and every 6 days, respectively (Faraz et al. 2021). At the end of the 6-day water deprivation period, camels compensated the water loss by drinking about 4.6 times the control. Water deprivation had some effect on milk yield, but milk contents were not affected as camels produced the same amount of fat and slightly lower protein and lactose.

On the other hand, feed deprivation for 4 days had little effect on the fluid balance in camels. Camels maintained plasma volume and glucose concentration and lowered their body temperature (Dahlborn et al. 1992). Deprivation from food did not affect water intake on day 0 when the first meal was omitted, but it dropped to only 27% of the pre-deprivation level during days 1 and 2 and to 33% during day 3. On day 4 the camels did not drink appreciable amounts of water, and when refed, they refused to drink any water until after they consumed all the feed they were offered. They then consumed slightly more than their pre-feed deprivation level of water intake (113%).

As a welfare issue, clean fresh water should be provided to animals in care, preferably ad libitum (the camel can regulate rapidly his water requirements), particularly when those animals are confined to fenced areas and do not have access to natural water sources such as rivers and dams. An average daily water intake from three trials carried out in different countries and under different conditions ranged between 2.9 and 4.6 L/100 kg BW (Dahlborn et al. 1992; Engelhardt et al. 2006b; Faraz et al. 2021). In relation to feed dry matter intake (DMI), water intake was approximately 2.63 L/kg DMI (Faraz et al. 2021) and 2.69 L/kg DMI (Engelhardt et al. 2006b). The ability of camels to withstand and adapt to water deprivation should not be confused with requirements and behavioural needs, and the water requirements for ensuring the absence of prolonged thirst in camels in different physiological status, age, season and feeding conditions should be measured experimentally.

4.1.10 Intensification and the Spread of Parasitic Diseases

Camels conserve water and therefore excrete drier faeces compared with other species of animals. As browsing animals, camels spread their faeces over the sandy desert, which is also dry, therefore the dung desiccates rapidly making the environment less favourable for parasites, if present, to survive and be transmitted. Our experience in Australia is that camels in the desert are almost free of the common internal parasite we find in production animals such as cattle and sheep. It is only when camels are confined to a yard and in farms close to the coastline and in wet areas that internal as well as external parasites become a health issue.

Camel faeces carry a range of seeds from the plants the camel feeds on and spreads them in the desert. The survival of seeds through the camel's gastrointestinal tract and later in the faeces is vital for the regeneration of desert flora (Trabelsi et al. 2017). In a study in the northern Saharan desert of Algeria, seeds were collected from 48 faecal samples dispersed by camels in the desert. The seeds were cultivated in pots and, upon germination, 712 seedlings were successfully obtained. They belonged to 15 plant types and 13 species of annuals and perennials belonging to 9 botanical families. This finding emphasises the importance of camels to the sustainability of the desert ecosystem. Removing camels from the environment may protect some trees that are highly palatable and are highly preferred by camels, but the benefits to the environment would be greater if camels are managed correctly by allowing them to browse in certain seasons.

4.2 Conclusion

The Arabian camel (*Camelus dromedarius*) is well adapted to the arid and semi-arid conditions of the Arabian and north African deserts. The main features of the desert climate are the scarce vegetation and water, particularly during the summer season, the high ambient summer temperature that can reach 55 °C, and the impact of urbanisation on the natural habitat of camels and the cameleers. Camels survive these most inhospitable desert environments by a range of adaptational physiological mechanisms both at the systemic and cellular level. Now that we have access to molecular tools, we can explore with depth these mechanisms at the cellular levels and explain the involvement of the different organs in maintaining normal function while conditions are adverse. As a browsing herbivore, the camel must have access to rangelands with the type of vegetation it prefers and has evolved to utilise. This point has been reiterated because of its importance. The management of the camel herd should be an integral part of the overall management of the rangeland where camels are found, to preserve plant cover and prevent soil erosion. The quantity and quality of available vegetation must be monitored and nutritionally assessed regularly to determine their adequacy and suitability for camels and to avoid overstocking. In case of any shortage, a fibre-rich diet must be provided. A special consideration is to be given to the quality of drinking water, especially if bore water is in use.

These final concluding remarks are a reminder of what good nutrition and feeding is all about. It is to provide the camels with the nutrients they require from diets that are most suitable for their digestive system. The diet must be palatable to be consumed, digestible to be broken down in the digestive tract, absorbable so the digestion end products will be absorbed and finally metabolisable to be assimilated into animal products or generate work. Overall, it is important to highlight that being able to cope and survive in a harsh condition do not mean that camels are in a good welfare state while kept in these conditions. To ensure the welfare principle of good feeding, therefore, the camels should be kept in absence of hunger and thirst, making sure that not only the nutritional requirements are met but also the behavioural and psychological needs are met.

References

Abdel-Rahman EH, Alzahrani FH, Sulieman AE (2020) Potential of camel dung as promising organic manure in Saudi Arabia. Adv Life Sci 7(4):202–207. www.als-journal.com

Abdoun K, Alameen A, Elmagbol W, Makkawi T, Al-Haidary A (2011) Assessment of the nutritional status in feed deprived one-humped camels. J Camel Pract Res 18(2):209–212

Al Jassim R (2019) Metabolisable energy and protein requirements of the Arabian camel (Camelus dromedarius). J Camelid Sci 12:33–45

Al Jassim R (2022) Foregut microbiology of the Arabian camel (*Camelus dromedarius*). Anim Front 14(4):46–51. https://doi.org/10.1093/af/vfac049

Andrews FM, Buchanan BR, Elliott SB, Al Jassim RAM, McGowan CM, Saxton AM (2008) In vitro effects of hydrochloric and lactic acids on bioelectric properties of equine gastric squamous mucosa. Equine Vet J 40(4):301–305

Breed M, Sanchez L (2010) Both environment and genetic makeup influence behaviour. Nat Educ Knowl 3(10):68

Chandrasena LG, Emmanuel B, Hamar DW, Howard BR (1979) A comparative study of ketone body metabolism between the camel (*Camelus dromedarius*) and sheep (*Ovis aries*). Comp Biochem Physiol B 65:109–112

Dahlborn K, Benlamilih S, Zine-Filali R, Gueroulali A, Hossaini-Hilali J, Oukessou M (1992) Food deprivation and refeeding in the camel (*Camelus dromedarius*). Am J Physiol 262(6):R1000–R1005. https://doi-org.ezproxy.library.uq.edu.au/10.1152/ajpregu.1992.262.6.R1000

Dittmann MT, Runge U, Ortmann S, Lang RL, Moser D, Galeffi C, Schwarm A, Kreuzer M, Clauss M (2015) Digesta retention patterns of solute and different-sized particles in camelids compared with ruminants and other foregut fermenters. J Comp Physiol B 185:559–573. https://doi.org/10.1007/s00360-015-0904-x

Dryden GML (2008) Water use and requirements, chapter 8. In: Dryden GML (ed) Animal nutrition science. CABI Publications, Wallingford, Oxfordshire, U.K., pp 115–130. https://doi.org/10.1079/9781845934125.0115

El Badawi AY (2018) The present situation of animal protein in Egypt and the role of camels in providing cheap and healthy meat for people in poor greenery lands. Int J Avian Wildl Biol 3(4): 319–322

El-Badawi AY, Yacout MHM (1999) Comparative study on growth performance of camel (*Camelus dromedarius*) calves and cattle steers in the feed-lot system. Egypt J Nutr Feed 2: 319–330

Elmahdi B, Sallmann HP, Fuhrmann H, von Engelhardt W, Kaske M (1997) Comparative aspects of glucose tolerance in camels, sheep, and ponies. Comp Biochem Physiol A Physiol 118 (1):147–151. https://doi.org/10.1016/s0300-9629(96)00449-5. PMID: 9243815

Emmanuel B (1980) Oxidation of butyrate to ketone bodies and CO2 in the rumen epithelium, liver, kidney, heart and lung of camel (*Camelus dromedarius*), sheep (*Ovis aries*) and goats (*Carpa hircus*). Comp Biochem Physiol B 65:699–704

Emmanuel B (1981) Glycolysis in camel (*Camelus dromedarius*) and sheep (*Ovis aries*). Comp Biochem Physiol B 68:547–550

Engelhardt WV, Haarmeyer P, Kaske M, Lechner-Doll M (2006a) Chewing activities and oesophageal motility during feed intake, rumination and eructation in camels. Comp Biochem Physiol B 176:117–124

Engelhardt WV, Haarmeyer P, Lechner-Doll M (2006b) Feed intake, forestomach fluid volume, dilution rate and mean retention of fluid in the forestomach during water deprivation and rehydration in camels (*Camelus* sp.). Comp Biochem Physiol A 143:504–507

Engelhardt WV, Lechner-Doll M, Heller R, Schwartz HJ, Rutagwenda T, Schultka W (1988) Physiology of the forestomach in camelids with particular reference to adaptation to extreme dietary conditions. A comparative approach. Anim Res Dev 28:56–70

Faraz A, Khan NU, Passantino A, Pugliese M, Eyduran E, Pastrana CI, Ismail A, Tauqir NA, Waheed A, Nabeel MS (2021) Effect of different watering regimes in summer season on water intake, feed intake, and milk production of marecha she-camel (Camelus dromedarius). Animals 11:1342. https://doi.org/10.3390/ani11051342

Farid MFA (1995) Nutrient requirements of dromedary camels: protein and energy requirements for maintenance. J Arid Environ 30:207–218

Faye B, Bengoumi M (2018) Camel clinical biochemistry and hematology. Springer Publ, New-York, p 346

Faye B Tisserand JL (1989) Problèmes de la détermination de la valeur alimentaire des fourrages prélevés par le dromadaire. Séminaire sur la nutrition et l'alimentation du dromadaire, Ouargla, Algérie. *Options méditerranéennes* Séries séminaires n°2, p 61–65

Ghali MB, Alhadrami GA, Al Jassim R (2019) Protozoa population and types in the forestomach of the Arabian camel (*Camelus dromedarius*) associated with diet changes. Short Communication. J Camelid Sci 12:46–51

Ghali MB, Scott PT, Alhadrami GA, G. A. and Al Jassim, R. A. M. (2011) Identification and characterisation of the predominant lactic acid-producing and lactic acid-utilising bacteria in the foregut of the feral camel (*Camelus dromedarius*) in Australia. Anim Prod Sci 51:597–604

Guerouali A, El Gass Y, Balcells J, Belenguer A, Nolan J (2004) Urinary excretion of purine derivatives as an index of microbial protein synthesis in the camel (*Camelus dromedarius*). Br J Nutr 92:225–232

Guerouali A, Wardeh MF (1998) Assessing nutrient requirements and limits to production of the camel under its simulated natural environment. In: Proceedings of the Third Annual Meeting for Animal Production under Arid Conditions. United Arab Emirates University Publishing Unit 36: 36–51

Heller R, Lechner M, Weyreter H, Engelhardt WV (1986) Forestomach fluid volume and retention of fluid and particles in the gastrointestinal tract of the camel (*Camelus* dromedarius). J Veterinary Med Ser A 33:396–399

Kaske M, Elmahdi B, Engelhardt WV, Sallmann H-P (2001) Insulin responsiveness of sheep, ponies, miniature pigs and camels: results of hyperinsulinemic clamps using porcine insulin. J Comp Physiol B 171:549–556. https://doi.org/10.1007/s003600100205

Kaske M, Osman T, Lechner-Doll M, Larsson M, Engelhardt WV (1989) Circadian changes of forestomach motility and of rumination in camels. Asian-Austral J Anim Sci 2(3):301–302

Laameche F, Chehma A, Faye B (2019) Effect of diet composition on dry matter intake of dairy she-camels. Trop Anim Health Prod 51:2513–2519

Lechner-Doll M, Rutagwenda T, Schwartz HJ, Schultka W, Engelhardt WV (1990) Seasonal changes of ingesta mean retention time and forestomach fluid volume in indigenous camels, cattle, sheep and goats grazing a thornbush savannah pasture in Kenya. J Agric Sci Cambridge 115:409–420

Mazzola S, Palestrini C, Cannas S, Fè E, Bagnato GL, Vigo D, Frank D, Minero M (2016) Efficacy of a feed dispenser for horses in decreasing cribbing behaviour. Vet Med Int 2016:1–6. https://doi.org/10.1155/2016/4698602

Nagy PP, Skidmore JA, Juhasz J (2022) Intensification of camel farming and milk production with special emphasis on animal health, welfare, and the biotechnology of reproduction. Anim Front 12(4):35–45. https://doi.org/10.1093/af/vfac043

Phillips A, Heucke J, Dörges B O'Reilly G (2001) Co-grazing cattle and camels. A report for the Rural Industries Research and Development Corporation. www.rirdc.gov.au/reports/Index.htm

Rabee AE, Abd El Rahman T, Lamara M (2023) Changes in the bacterial community colonizing extracted and non-extracted tannin-rich plants in the rumen of dromedary camels. PLoS One 18(3):e0282889. https://doi.org/10.1371/journal.pone.0282889

Rabee AE, Sayed Alahl AA, Lamara M, Ishaq SL (2022) Fibrolytic rumen bacteria of camel and sheep and their applications in the bioconversion of barley straw to soluble sugars for biofuel production. PLoS One 17(1):e0262304. https://doi.org/10.1371/journal.pone.0262304

Rutagwenda T, Lechner-Doll M, Schwartz HJ, Schultka W, von Engelhardt W (1990) Dietary preference and degradability of forage on a semiarid thornbush savannah by indigenous ruminants, camels and donkeys. Anim Feed Sci Technol 31:179–192

Schreier H (2002) Water and agriculture: harvesting the water before harvesting the crop. Workshop and think tank on water and the future of life on earth. Simon Fraser University Morris J. Wosh Centre for Dialogue. May 22–23, 2002, invited presentation. Proceedings; chapter 17 continuing studies in science at Simon Fraser University

Srivastava S, Bombaywala S, Jakhesara SJ, Patil NV, Joshi CG, Purohit HJ, Dafale NA (2023) Potential of camel rumen derived *Bacillus subtilis* and *Bacillus velezensis* strains for application in plant biomass hydrolysis. Mol Gen Genomics 298:361–374. https://doi.org/10.1007/s00438-022-01987-y

Tan E, Al Jassim R, Cawdell-Smith J, Ossedryver S, D'Arcy B, Fletcher M (2016) Accumulation, persistence and effects of indospicine residues in camels fed Indigofera plant. J Agric Food Chem 64:6622–6629

Tan ETT, Al Jassim R, D'Arcy BR, Fletcher MT (2017) *In vitro* biodegradation of hepatotoxic indospicine in *Indigofera spicata* and its degradation derivatives by camel foregut and cattle rumen fluids. J Agric Food Chem 65(34):7528–7534

Tomas Marques AF (2017) Molecular biomarkers of welfare in poultry. PhD Thesis, University of Milano. https://air.unimi.it/handle/2434/485613#

Trabelsi H, Chehma A, Al Jassim R, Senoussi A (2017) Camel as seed disperser in the northern Sahara rangelands of Algeria. Int J Biosci 10(4):58–65

Wang J-L, Lan G, Wang G-X, Li H-Y, Xie Z-M (2000) Anatomical subdivisions of the stomach of the Bactrian camel (*Camelus bactrianus*). J Morphol 245:161–167

Wensvoort J, Kyle DJ, Ørskov ER, Bourke DA (2001) Biochemical adaptation of camelids during periods where feed is withheld. Rangifer 21(1):45–48

Zhang D, Pan J, Zhou H, Cao Y (2020) Evidence from ileum and liver transcriptomes of resistance to high-salt and water-deprivation condition in camels. Zool Lett 6(8):1–13. https://doi.org/10.1186/s40851-020-00159-3

5 Good Housing: Camels and Their Interaction with the Environment

Asim Faraz, Syeda Maryam Hussain, Carlos Iglesias Pastrana, and Martina Zappaterra

Abstract

Camel farming is undergoing a profound transformation. To date, camels can be reared in many different ways, ranging from traditional nomadic breeding to more technological housing systems, with milking parlours and intensive farming systems similar to those used in dairy cows. The different types of housing systems used for camel farming will be described in this chapter. Improper housing systems and facilities can cause severe welfare issues in camels. The chapter emphasizes how a poor environment may affect camel welfare and provides suggestions for designing facilities in compliance with the welfare principle named "good housing". In particular, good housing is ensured when camels are reared in an environment and building meeting the welfare criteria of "comfort around resting", "ease of movement", and "thermal comfort". Providing camels with adequate space allowance, clean beddings, shelters and shaded areas, and building facilities following the camels' behavioural needs for movement and sociability are key points for ensuring farming systems respecting the principle of good housing. Overcrowding, lack of movement, and insufficient social interactions are among the main welfare issues associated with improper housing

A. Faraz
Department of Livestock and Poultry Production, Bahauddin Zakariya University, Multan, Pakistan

S. M. Hussain (✉)
Department of Livestock Production and Management, Pir Mehr Ali Shah Arid Agriculture University, Rawalpindi, Pakistan
e-mail: syedamaryam@uaar.edu.pk

C. Iglesias Pastrana
Faculty of Veterinary Sciences, Department of Genetics, University of Cordoba, Cordoba, Spain

Department of Agricultural and Food Sciences, University of Bologna, Bologna, Italy

M. Zappaterra
Department of Agricultural and Food Sciences, University of Bologna, Bologna, Italy

B. Padalino, B. Faye (eds.), *Dromedary Camel Behavior and Welfare*, Animal Welfare 24, https://doi.org/10.1007/978-3-031-48600-5_5

systems. In addition, concerning the "thermal comfort" criterium, more studies are needed to provide more precise guidance on the temperature and humidity ranges to keep camels within their thermal comfort zone.

Keywords

Housing · Restriction of movements · Thermal stress · Discomfort

5.1 Introduction: Camel's Welfare Requirements in Relation to the Good Housing Principle

Animal husbandry covers various and diverse topics, having a direct or oblique impact on the final product of the animal. Breeding, feeding, housing, health status, and disease management all affect the growth and production of animals, with large effects also on human health and economic growth. Perfectly in line with this vision that interconnects man and animal, dromedary camels are multipurpose animals having had an important role in the lives of humans. Their role has been particularly important in arid regions due to their ability to survive in harsh situations with excessive solar radiation and preserve themselves on coarse fodder and salty and thorny trees (Faraz et al. 2019). However, even if they are particularly able to cope with the heat, they still need a good environment around them to be kept with good welfare standards. The animal welfare principle of good housing corresponds to the question "Are the animals properly housed?" (Botreau et al. 2007). In order to assess whether the animal is reared in an environment that allows it a life worth living and expresses its full productive potential, three criteria have been included in the good housing principle: "Comfort around resting", "Ease of movement", and "Thermal comfort" (Botreau et al. 2007; AWIN 2015; Welfare Quality 2009).

5.1.1 Comfort Around Resting

The "comfort around resting" criterium answers the question "Are the animals able to comfortably rest?". This criterium can be evaluated by assessing the space allowance per animal and the type and cleanliness of bedding. Since the latter factors also depend upon managerial decisions, camel caretaker experience and thoughts are also of importance. Generally, camels display a strong attachment to resting places and prefer the quietest locations (Schulte and Klingel 1991). Similar to other ruminants, dromedary camels mainly assume sternal recumbency as the resting position, with the head up or lying down according to the different stages of sleep, and sometimes also lateral recumbency (El Allali et al. 2022). Camels assume a lateral or sternal recumbency position with their head and neck stretched out on the floor when in a sleep-like behaviour (El Allali et al. 2022). In intensive farming systems, however, overcrowding does not allow camels to rest stretching their necks and assuming the head-lying-down position. Thus, comfort around resting criterium is not met in overcrowded pens, nor in cases with dirty or uncomfortable beddings. Garbage can indeed limit the space for resting and also for walking.

5.1.2 Ease of Movement

The "ease of movement" criterium answers the question "Are the animals able to freely move?" and thus refers to the animal's need for a good space allowance for freedom of movement. In the protocol for assessing animal welfare (Padalino and Manchetti 2021) the "ease of movement" can indeed be evaluated by assessing the space allowance per animal and monitoring the presence of tethered and restrained animals. Since the latter factors also depend upon managerial decisions, camel caretaker experience and thoughts are also of importance. Camels are normally calm and docile and in feral situations, live in herds moving over extensive regions of land (Beerda et al. 1999). Generally, wild camels graze for 8–12 h/day and spend the same time ruminating. As they normally live in areas with scarce vegetation spread over large distances, dromedary camels graze during the daytime and rest at night (Gauthier-Pilters and Dagg 1981). Dromedary camels are however in some cases restrained in small places or tied with short ropes and hobbled (Padalino et al. 2015). Retraining and tethering are welfare issues as they prevent the animals from expressing their natural behaviours, such as grazing and walking, and expressing their social behaviours.

They like to play or rub their legs and neck against each other or on trees and roll in sandy places; therefore these types of enrichments (e.g. brush) must be provided when these animals are kept in pens and intensive systems (Glauthier-Pilters and Dagg 1981). Like other species (Faye and Barnouin 1985), limited place, restraining, and social isolation are the cause of continual stress in camels causing vices (Padalino et al. 2014). In the long vision, movement restriction affects metabolism, while exercise and the freedom of movement improve camel's health and metabolism (El Shoukary et al. 2020b). Therefore, the design of the barn also plays an important role in creating favourable conditions for camels kept in enclosed systems.

5.1.2.1 Space Allowance

The smallest space dimension to have enough freedom and for exhibiting natural behaviour is one of the most important factors to be evaluated in confinement (Petherick and Phillips 2009). The World Organisation for Animal Health's annual "Terrestrial Code" (OIE 2021) addresses the most typical issues such as the excessive stocking densities at beef and dairy cow, broiler, and pig production systems and offering suggestions on how to reduce these issues (OIE 2021). Respecting the necessary space per head in livestock scenarios does not always come in applications, however, making it easier to create precise quantitative criteria. Instead, it is advised that interested parties refer to pertinent national or international documents pertaining to accepted farming methods for the welfare of animals used for food production.

The scientific literature aimed at the precise definition of the spatial requirements per animal in camel farming systems continues to be limited. Only three referring documents on minimum space allowance for dromedary camels exist. They all share the premises that camels have to be offered enough area for freedom of movement (horizontally and vertically), to lie down, to escape when necessary, from other congeners that may exhibit territorial or aggressive behaviour, but also to permit their efficient and safe handling.

The oldest document is just a magisterial guideline on camel husbandry (Moretti 2008) that recommends adopting the standards of the "Exhibited Animals Protection Act" (Wales 1986). Thus, a space of 100 m^2 with a 6 m minimum width is stated to be required for one camel, being extended by 50 m^2 for every additional congener included in the same yard.

The other two studies are both experimental-based and include a former analysis of the correlations between minimum space permitted per animal and the frequency of expressed behaviours (maintenance, posture, aggression, and stereotypies), the body condition score, and some biochemical parameters (thyroid function and cortisol). El Shoukary and Osman (2020) researched on the effects of group size on the behavioural repertoire of dromedary males during the rutting season and conclude that a minimum space of 15 m^2 per mature male produces no harmful consequences or distress. On the other hand, Menchetti et al. (2021b) aimed at identifying outcome-based measures and propose a categorical division for this good housing-related welfare indicator. In that case, the authors define an area of 19.1–40.0 m^2 per animal as regular or acceptable, while values below and above this range are catalogued as limited and ample spaces, respectively. Notwithstanding, this categorical classification requires external validation in different camel breeding regimes since the thresholds defined were obtained by statistical binning as no species-related contrasted bibliography was available (Menchetti et al. 2021a).

Hence, to favour the large-scale replication of these proposals to reach a consensus on the minimum space requirements per housed dromedary camel, some animal body dimension measurements and behavioural expressions need to be considered as potential influencing factors. The total area occupation is known to vary according to the animal morphometrics, the total time that an animal passes in a certain position (standing, feeding, or recumbency), and the number of times that an animal changes its position during the day (Pastorelli et al. 2006). These characteristics will intrinsically differ depending on the duration of the confinement, the age and sex cohorts within a herd (Pastrana et al. 2021), the environmental surrounding conditions, and the camel breeds and types' differential behaviour (Iglesias et al. 2020a, b; Alhajeri et al. 2021).

In particular, a complex interaction does exist between the thermal environment, the locomotor activity, and the space allowance. The ability of camels to thermoregulate to maintain heat balance would be largely governed by the stocking density and the materials used for the fabrication of the space enclosure (Petherick and Phillips 2009). Similarly, the frequency of movement is known to vary as a function of the static and dynamic friction caused by flooring surface attributes, the number of animals placed in the pen, the airflow rate, the relative humidity, and the temperature

(Phillips and Morris 2000; Keane et al. 2017). Both the capacity of animals to maintain a consistent internal environment and moderate physical activity are key elements for the long-term welfare of the animals.

On that account, it becomes crucial to appoint a proper design of the installations have to contemplate the minimum space per animal in the enclosed pen equally including the chance for animals to freely access and stay at shelter areas. So, camels would be able to spend more time in recumbency and ruminating in the shaded space, which can widely safeguard camel welfare and improve their growth and production potential, especially in extremely hot environments (Zappaterra et al. 2021). Besides, the physical accommodation of camels should include a relatively dry area where animals can lie down comfortably and hygienically. In addition, providing the camels with the chance for grazing in nearby free-ranging areas (Dereje and Udén 2005) will enhance their physical and psychological health.

In conclusion, ongoing studies on camel good housing are encouraged to be based on animal morphometrics (static zoometry), basic and species-typical motion behaviour (dynamic zoometry) (Mellor and Beausoleil, 2015; Sugiono et al. 2018), social organization (Pastorelli et al. 2006), and materials engineering, to guide the ergonomic design of facilities for this animal species. Artificial intelligence-based methods could constitute powerful data modelling tools capable to infer the inter-relationship between the studied animal welfare-influencing factors and would help to reduce the costs associated with manual measurements (Sugiono et al. 2018). Some factors that should also be taken into consideration and that must lead to the need of extra space are large groups of animals or bull camels; land with untrampeable areas that can not be exploited by the animals. Other specific cases requiring specially designed spaces are given below.

5.1.3 Thermal Comfort: The Thermal Needs of Dromedary Camels

5.1.3.1 An Introduction to Thermo-regulation, Thermal Comfort Zone, and Thermoneutral Zone

Thermoregulation is a complex mechanism made of several processes (physiological, endocrinological, physical, and behavioural) used by animals to regulate and maintain internal body temperature in an acceptable range to keep metabolic functions (Osilla et al. 2022). Animals can be classified into two major categories based on their way to thermoregulate: *poikilotherms* or "cold-blooded" animals (which cannot generate their body heat and thus conform to the ambient temperature) and *homeotherms*, or "warm-blooded" animals. The latter can produce heat and demand to maintain their body temperature within a specific range of temperature (Akin 2011; Osilla et al. 2022). In homeotherms, core body temperature normally ranges between 36 and 42 °C (97 and 107 °F) (Ivanov 2006; Akin 2011; Ivanov 2006).

Homeotherm animals need to maintain a constant core body temperature, with shifts from the normal core temperature that in most of the homeotherm animal species should not exceed some tenths of Celsius degree (Ivanov 2006). In mammal homeotherms, this range is even narrower, with maximum core body temperature in the range of 37–39 °C (about 98–102 °F). These values are considered in most mammals the upper-temperature-temperature limit of life and seem to depend on the thermal resistance of proteins and cell membranes, which become de-natured and disrupted when the temperature rises above 40 °C (Ivanov 2006; Bowler and Manning 1994). For this reason, homeotherm organisms employ several physiological, endocrinological, and behavioural responses to thermoregulate their body temperature. Different thermoregulation responses are carried out depending on how far from the biophysical requirements are the ambient, skin, and body core temperatures.

Thermal balance is indeed dependent on a combination of these temperatures, and depending on the biological requirements of each homeotherm organism, different ambient temperature ranges are tolerated. A stable core temperature can be maintained only when heat production and heat loss are balanced. Thermal homeokinesis is a steady state where the internal body temperature of a homeotherm animal is kept constant at the normal core temperature level with little additional energy expenditure. Thermal homeokinesis is kept when ambient temperatures, and more generally speaking environmental parameters, are within the range of the thermal comfort zone (TCZ).

Thermal comfort zone (TCZ) was first hypothesized based on the human perception of the thermal environment (Kingma et al. 2014) but is now applied in other animals as well. The TCZ is the range of environmental parameters (mainly temperature) where the energetic and physical efforts of thermoregulation are minimal and within which an animal expresses satisfaction with the thermal environment and does not need to change its behaviour to cope with the environment (Kingma et al. 2014; Schlader et al. 2011; Robbins 2021; EFSA AHAW Panel 2022). Outside the TCZ, the animal starts to experience thermal discomfort, which drives thermal-related behaviours (e.g. huddling, posture adjustments, searching for shaded places, etc.) that anticipate autonomic thermoregulatory mechanisms. The TCZ is comprised of a wider range of ambient temperatures, namely, the thermoneutral zone.

The thermoneutral zone classical definition was "the range of ambient temperature at which temperature regulation is achieved only by control of sensible (dry) heat loss, i.e. without regulatory changes in metabolic heat production or evaporative heat loss" (Lups 2001). Thus, inside the range of TNZ, the thermoregulation functions consist of levelling internal and external temperature excitations that continuously arise due to minimal muscle activity (i.e. walking) or small changes in the environmental parameters (Ivanov 2006). TNZ boundaries are represented by upper critical temperature (UCT) and lower critical temperature (LCT). Temperatures (absolute or perceived) above the UCT lead to severe heat stress in homeotherm animals. At this point, physiological, endocrinological, and behavioural responses are activated to counteract the increase in core body temperature. The energy expenditure to activate these responses increases; thermoregulation is mainly sought with increased water evaporation from the surface of a body (thermal sweating) or the mucosa of respiratory ways (thermal panting) (Ivanov 2006). These responses are often accompanied by other endocrinological and physiological processes that have the scope of maintaining a stable core body temperature. When heat stress is prolonged over time, animals experience health issues, infertility, decreased growth and production, decreased immune system efficiency, and cellular and mitochondrial oxidative damage (Belhadj Slimen et al. 2016). In the most serious cases, when the body's ability to thermoregulate becomes disrupted, it can result in heat stroke (hyperthermia); extreme and prolonged hyperthermia causes organ failure and death (Osilla et al. 2022).

On the other hand, temperatures (absolute or perceived) below the LCT lead to cold stress in homeotherm animals. Below LCT, metabolic heat production increases, as the animal attempts to keep body core temperature in an acceptable range using shivering (irregular frequent muscle contractions), vasoconstriction, and in some animal species also by activating brown adipose tissue catabolism (Grigg et al. 2004). When environmental temperature and humidity values are below the LCT and the physiological responses activated by the body are not able to maintain or restore an acceptable core temperature, the animal enters hypothermia. Hypothermia is associated with several organ failures and cardiovascular dysrhythmias such as ventricular fibrillation and pulmonary oedema. The central nervous system's electrical activity is also noticeably diminished (Kurz 2008). Either hypothermia or hyperthermia has therefore deleterious effects on the various body systems, leading to ischaemia and multiple organ failure (Osilla et al. 2022).

5.1.3.2 Dromedary Camels and Thermal Comfort

Dromedary camels are homeotherm animals but compared with other homeotherm animals evolved to cope with extreme environments. These animals have some key features at the anatomical, physiological, and molecular levels that allow them to cope with extreme environments and high environmental temperatures (reviewed in Hoter et al. 2019). Among them, camel arteries and veins evolved to mitigate the temperature of blood reaching the brain, thus protecting the animal from potential brain damage. This mechanism was referred to as "selective brain cooling" (Ouajd and Kamel 2009). Furthermore, camel nostrils have a muscular nature and thus can be controlled by the camel, which can fully open them when trying to cool the internal temperature or close them in case of sandstorm events (Gebreyohanes and Assen 2017). To increase the body surface exposed to conduction (heat exchange with a solid floor) or convection (heat exchange with a fluid, e.g. air), in the recumbent position, the camel sternum flattens in a "plate-like" conformation (Ouajd and Kamel 2009). From a physiological point of view, camel kidneys are also unique, as they can filter blood and excrete highly concentrated urine, avoiding as much as possible to lose water (Siebert and Macfarlane 1971). Camelids red blood cells have an elliptical shape that was hypothesized to be an evolutionary strategy developed to allow red blood cells to circulate in the blood vessels of dehydrated animals (Warda et al. 2014) and have a peculiar cell membrane composition that protects camels' blood cells from osmotic problems deriving from dehydration (Warda et al. 2014; Warda and Zeisig 2000). Camels' blood platelet and cell membranes are also able to resist high temperatures of 43–45 °C, which in humans and other animals cause marked structural and functional alterations (Al Ghumlas et al. 2008).

Together with the exceptional mechanisms developed during evolution, camels are homeotherm animals that developed "adaptive heterothermy", a water-conserving mechanism in which, as reported by Willmer et al., body temperature can widely fluctuate, daily, with heat storage during the day and heat dissipation during the night (Willmer et al. 2009). In the absence of heat stress, the daily fluctuations in camels' core body temperature are about 2 °C, but a dehydrated camel in a hot environment may display fluctuations in its body temperature to the extent of 6 °C (Schmidt-Nielsen et al. 1967). Under these circumstances, before dawn, camels' body temperature may fall to below 35 °C and increase up to 41 °C after the heat load occurs during the day (Schmidt-Nielsen et al. 1967). The adaptive heterothermy in dromedary camels showed differences across the diurnal light–dark cycle, and these changes were found to be modulated by daily heat, water restriction, and also the associated reduction of food intake (Bouâouda et al. 2014). During thermal comfort conditions, camels' core temperature ranges between 36 and 37 °C based on individual variations.

At the time of writing this book, the number of studies aimed at estimating TNZ intervals in dromedary camels is very small. However, based on the little evidence reported to date, it is possible to point attention to some environmental temperature ranges. Based on the research carried out on ten dromedary camels subjected to different temperatures, TNZ in dromedary camels would range between 10 °C (indicated as LCT) and 40 °C (being the UCT) (Samara and Al-Haidary 2014). However, differences may exist among camels belonging to different breeds, having different ages, hydration statuses, and body condition scores, or between males and females (Brown-Brandl 2009; Al-Haidary et al. 2013).

Among the breeds tested by Alhaidary et al., Alzargeh and Almajaheem breeds seemed to be the ones that were able to best cope with the high temperatures during the summer period in Saudi Arabia (environmental temperature = 40 °C; relative humidity = 10%; black globe temperature = 50 °C), as they showed a smaller increase in respiratory rates and superficial and core temperatures when compared with Alsafrah and Almaghatir breeds (Al-Haidary et al. 2013). However, it should be considered that these temperatures and those of TNZ indicated by Samara and Al-Haidary were obtained at low relative humidity values (Samara and Al-Haidary 2014). Relative humidity is a major environmental factor, as it can influence the temperature perceived by the animals. The lower the relative humidity, the higher the UCT. Temperature and humidity values can be used to calculate the temperature-humidity index (THI), a single value representing the combined effects of air temperature and humidity associated with the level of thermal stress in animals. Habte et al. estimated the THI using the equation reported by Kendall et al. (Kendall et al. 2008) and found that dromedary camels showed an exponential increase in body core temperature and skin temperature when THI started increasing above 71. A THI of 71 means that the environmental temperature was about 24 °C with 50% of relative humidity (Habte et al. 2021). The body core temperature that increased up to a peak of about 38 °C at 3 p.m. in the dry season suggested that camels at 31 °C with a relative humidity of 34% (Kendall THI of 77) were failing to maintain their heat balance and were thus outside the TCZ, probably approaching the upper boundary of TNZ (i.e. UCT). However, even during the hot and dry season reported by Habte et al., camels did not show significant decreases in milk production, suggesting that they were probably not facing severe heat stress. A Kendall THI of 81 (environmental temperature = 40 °C; relative humidity = 6.9%) was instead found to change blood metabolites and haematological indices, suggesting that the camels were facing severe heat stress (Abdoun et al. 2012).

Dehydration plays a central role in the way camels can cope with heat stress, and thus water availability should also be considered when evaluating whether a camel is under heat stress or not (Schroter et al. 1987; Al-Haidary 2005). For example, hydrated camels start increasing their respiratory rate when their body core temperature rises above 37 °C, while severely dehydrated camels increase their respiratory rate when their body core temperature exceeds 38 °C, to avoid evaporation and water loss (Schroter et al. 1987). Therefore, the increase in the respiratory rate cannot be read in light of the existing literature in other production animals, as when dehydrated dromedary camels do not increase the respiratory rate despite being under heat stress. In addition, camels do not express panting, thus making it more difficult to identify the presence of animals in severe heat stress. It is therefore extremely important to design pens and paddocks with housing conditions that can prevent camels from entering a state of heat stress, which would result in reduced welfare, growth losses, consumption of body fat, and decreased milk production.

On the other side, cold stress should also be taken into consideration. At the time of writing this book, there are not many studies concerning the minimum temperatures below which camels begin to implement changes in behaviour and physiology. The suggested LCT for dromedary camels is 50 °F (10 °C) (Samara and Al-Haidary 2014). Contrary to what one might think, dromedary camels are however still able to survive even severe temperature drops, as happens during night hours in deserts such as the Sahara, where temperatures are on average of 24 °F (about –4 °C) during the nights. During the coldest season, dromedary camels may meet thermal comfort by thickening their furs and using their adipose tissue for thermoregulation. Dromedary camels reared in geographic areas characterized by severe climatic conditions in winter should however be reared during the cold months in environmentally controlled facilities or in pens with shelters allowing thermal insulation.

5.1.4 The Good Housing Principle

Summarizing this first section, the good housing principle can be assessed by means of the combination of the three criteria "comfort around resting", "ease of movement", and "thermal comfort" (Fig. 5.1). As reported in recent scientific publications (Padalino and Menchetti 2021), these three criteria can be evaluated at the caretaker level, at the herd level, and at the animal level (see Fig. 5.1 and Chap. 3).

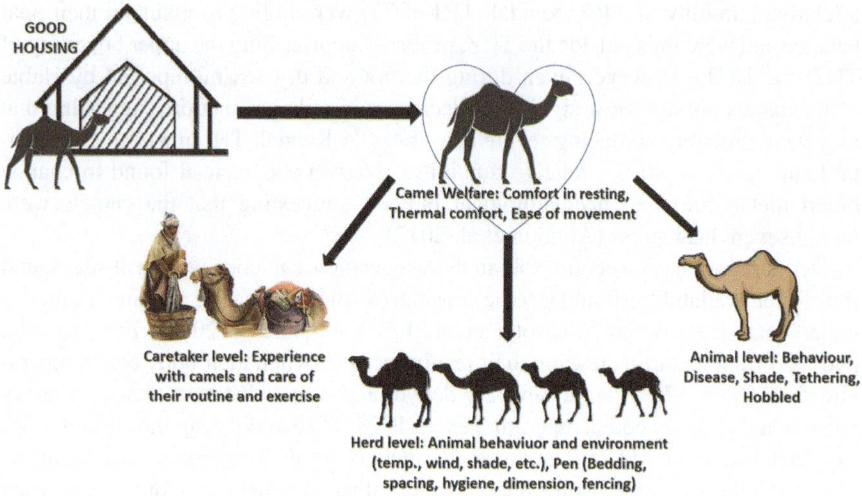

Fig. 5.1 Camel welfare indicators for good housing (adapted from Padalino and Menchetti 2021)

5.2 Dromedary Camel's Housing Systems

The three main types of housing systems are the following:

1. **Free-grazing systems and large-spaced fenced areas:** camels kept in free-grazing systems graze outside (continuous grazing of one area over a long period of time or intensive rotational grazing on small areas over short periods of time). In some cases, this management system consists of a fence, with shelters open on all sides or without shelters but only with the provision of natural shade. Camels are usually herded during the daytime on communal grazing lands and kept during the night in traditional kraals made of thorny bushes and tree branches around homesteads or settlements as protection from predators and thieves/raiders. It is not a common practice to keep camels with other species in a single kraal. In some areas, camels may also be left to roam around during the night (Mirkena et al. 2018).
2. **Semi-intensive housing systems:** camels are set free during the daytime and collected at the evening to be herded in fences and/or in enclosed facilities during night.
3. **Intensive housing systems:** Camels are kept, raised, and bred in confined, partially or completely enclosed, spaces. These systems allow the monitoring of farm microclimate and the feeding regime.

The choice among these three types of management systems mainly depends on the purpose of the camels (i.e. milk; meat; wool; work; breeding) and the geographical area they are bred. For example, dromedary camels are managed in free-grazing systems in pastoral areas, and in desert areas, where these animals have a major role for men living in desertic areas. Camel housing is however changing due to the increased cultivation, resulting in a significant decrease in free grazing areas, and consequently increased intensification (Faye 2016). The food search, the invasive urbanization, and sedentary life are causing people to abandon camel holding and flip to simpler approaches to living. Conventional camel raising is therefore living an intensification process and needs to be modernized and revolutionized as an economical enterprise. However, due to its important ecological role in desert ecosystems, camel rearing must not be abandoned for regular utilization of deserted areas (Gauthier-Pilters 1979). There is an instant relation between the cultural habits of man and camel raising. With the fast change in lifestyle and preferences, the nomadic community is also changing drastically. Nomadic societies rear camels utilizing both free-grazing and partially enclosed farming systems, depending on the season. In traditional camel raising, the whole community is living together in the cold season spending most of the time aiding the mating and calving of the camel. During cold seasons, therefore, nomadic societies also use stall systems, with fencing can be erected that would restrain the animals within a specific period, and shelters (Evans and Powys 1979). Thus, the different alternatives for housing camels have normally been considered under nomadic life, but under commercial farming systems, intensification is the one that fits better, as it is quite inexpensive and allows for a more controlled farming environment. As an example, milk-producing dromedary camels are today mainly kept in intensive farming systems, and they need as much care as dairy cows. Ventilated sheds, an adequate amount of water, and quality fodder are a must for camels producing 10–15 l of milk daily.

Concerning the types of housing, there are different types of pens and shelters or enclosures. To date, camels are however still kept following some traditional husbandry systems, such as in open areas (Fig. 5.2), kept loose in groups inside fences (Fig. 5.3), or in individual fences (Fig. 5.4).

Other types of traditional housing systems comprise also camels kept in open areas, with fences, natural shade and shelters made with natural materials, and centralized mangers (Fig. 5.5).

These types of open-area pens are often used also in markets. In a camel market in Qatar, the pens were generally square-shaped with natural shaded areas (trees or bushes) or with open shelters covering about 20% of the pen area. All pens had at least one feeding point; however, some may have not. The number of camels being kept also varied; therefore, space allowance, feeding, and water space were also very variable (Zappaterra et al. 2021). The location variability in rearing conditions in the examined market was in line with the different camel farms (Al-Ahmadi et al. 2020; Traoré et al. 2014). The types of camel farming structures were investigated in Saudi Arabia by Abdallah and Faye (2013). The authors identified several categories of camel farms which substantially differed for the camels' purpose, feeding, and health management, which were quite different along with the lifestyle of farmers. The farm locations and infrastructure varied to a great extent from traditional farms inside the barren region, with a nomadic lifestyle, to farms managed via owners residing within the town with modern-day business purposes. Other types of housing systems

Fig. 5.2 Camels kept in an open area (adapted with publication permission from Dioli 2013)

Fig. 5.3 Loose housing type (adapted with Publication permission from Dioli 2013)

comprise also concrete buildings, with individual or group boxes and open areas where camels could move, graze, and express their social behaviours. An example can be found in Figs. 5.6 and 5.7 from a camel experimental farm of the Camel Breeding and Research Station Rakh Mahni.

Fig. 5.4 Loose camels kept in individual fences (adapted with Publication permission from Dioli 2013)

Fig. 5.5 Camels kept in open areas with a centralized manger and some shaded parts, with natural shades or covered areas (adapted with publication permission from Dioli 2013)

Fig. 5.6 An aerial view of camel house and pens at Camel Breeding and Research Station Rakh Mahni (photo by Dr. Asim Faraz)

Fig. 5.7 Camel individual pens and a common central manger at Camel Breeding and Research Station Rakh Mahni (photo by Dr. Asim Faraz)

5.3 Housing Systems in Camel Breeding

Based on the knowledge of the chapter's authors, the types of structures adopted in camel breeding are of various types and will be described below.

5.3.1 Thatched-Roofed Open-Type Kucha Shelter

The open-type shed is provided with 1.5 m high barbed wire fencing. The open-type kucha shelter is constructed according to locally available and eco-friendly agricultural materials (Figs. 5.8 and 5.9) (Akbar et al. 2012), such as kemp (*Leptadenia pyrotechnica*), which can be used for the construction of the roof. It may have covered as well as open areas. One side of the roof is supported by a wall and the other side by pillars (36 cm). Kucha manger may be constructed with mud plastering below a covered area (Figs. 5.10 and 5.11). The enclosure is made up of balli/bamboo 1.83 m in height. The front side of the enclosure has a sliding gate. The height of the roof is 4.5 m (back) and 3.6 m (front) with sufficient slope. The covered area is 10 m (length) × 4.3 m (breadth). The floor is Kucha with loose dunes. The total area of this shelter has a length of 12 m and a breadth of 10 m. Experimental animals requiring individual feeding are kept under a roofed shed separately with a provision of 30–35 m^2 space and manger of 75 cm × 75 cm × 40 cm internal dimension (Fig. 5.12). Pregnant and recently calved animals should also be kept in a separate open-type shed. Sick animals should be housed in a separate housing to limit the spreading of disease. The soiled sand of the floor is regularly replaced with new sand (Chakrabarti 2006).

Manger Length 10 m, breadth 76 cm, inside depth 46 cm, the height of the front wall from ground 107 cm.

Fig. 5.8 Asbestos-roofed open-type shelter

Fig. 5.9 Thatched-roofed open-type kucha shelter

Fig. 5.10 Camels kept in groups in a shelter with a manger in the centre

5.3.2 Asbestos Roof Close-Type Concrete Shelter

It may have covered as well as open area. The covered area is covered by an asbestos roof with concrete walls on three sides. The floor is made with loose dunes. The

Fig. 5.11 Camels with head-to-head mangers in an open shed (photo by Dr. Zahid Kamran/Dr. Naeem Tahir)

Fig. 5.12 Asbestos-roofed open-type shelter with an animal tied on the individual manger

manger is of concrete/pucca type. The open area is enclosed with balli/bamboo having a gate at the front side. The gate is of a sliding type. The total area (covered and open), the height of the roof, and the dimension of the manger and enclosure are almost similar to the thatched-roofed open-roofed kucha shelter. A single camel may be kept in an enclosure measuring at least 5 × 10 m, while one male and two females may be kept in an area of at least 10 × 20 m.

5.3.3 Concrete Housing (Solid Construction)

An example of this type of camel housing may be found at Camel Breeding and Research Station (CBRS) Rakh Mahni, Desert Thal, Punjab Pakistan (Figs. 5.6 and 5.7). The boxes are made of solid construction made with cement, concrete, and bricks. The camels are housed in separate boxes with open areas where they can walk, graze, and express their social behaviours. The mangers are also made of solid construction. The pen is made in dimensions of 7.5 m in length, 3.9 m in width, and 6.1 m in height for one adult camel. However, the open area is provided in the dimensions as 41 m in length and 1.5 m in width for ten adult camels, while the feeding mangers inside the pen have a dimension of 76 cm height from ground level, 61 cm width, a depth of 46 cm, and a length of 3.9 m. The mangers in the open areas have a length of 2.7 m, a height of 1.07 m from ground level, a total width of 61 cm, and a depth of 46 cm (Figs. 5.13, 5.14, 5.15, and 5.16).

Fig. 5.13 Sick pen for camels at Camel Breeding and Research Station Rakh Mahni (photo by Dr. Asim Faraz)

Fig. 5.14 Individual camel pen (internal view) at Camel Breeding and Research Station Rakh Mahni (photo by Dr. Asim Faraz)

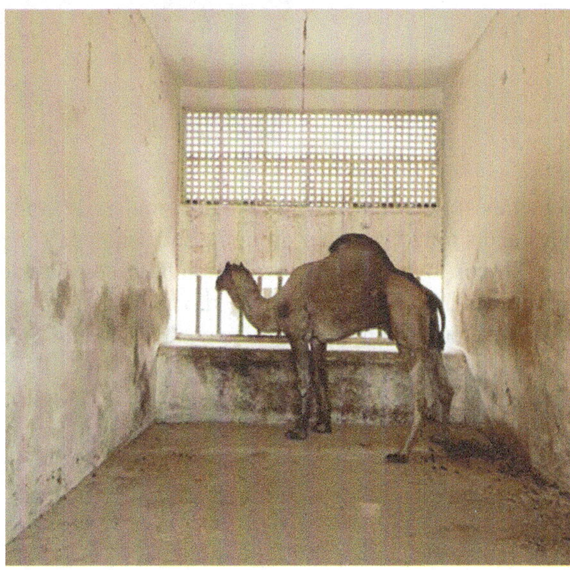

Fig. 5.15 Camel housed at Camel Breeding and Research Station Rakh Mahni (photo by Dr. Asim Faraz)

5.3.4 Under a Tree Shed or Loose Housing

An open-type paddock is provided, and it is enclosed by wire fencing, supported by an iron angle. Some trees (such as kikar trees, *Prosopis juliflora*) can be used to provide shade for camels. The floor is kucha with loose dunes. Kucha manger can be round-shaped and constructed with mud plastering just beneath the tree shade. The used mangers are almost equal to the mangers used in the other types of housing.

Fig. 5.16 Camel feeding mangers of the box-like type without a lid (photo by Dr. Asim Faraz)

Sometimes, electric wires are used in commercial farms. The pen can be made with thick and strong bushes cut from the surrounding. The pen should be swept at least once a week to avoid the accumulation of faeces. Over-exposure to cold breezes can cause respiratory disorders, and a strongly built pen protects calves from predators while regular cleaning helps in the control of ticks.

Manger A manger about 90–120 cm high is constructed. A square pillar about 40 cm in height is erected, and over it, a platform of about 70 cm × 70 cm is constructed with a wall of 15 cm thickness built on all four sides of it from 30 to 45 cm in height. Figure 5.16 shows a box-like manger without a lid.

5.3.5 Other Types of Camel Housing

Cattle yards can also be used for feral camel by doing some alterations in infrastructure (Sharp 2012). These are the following:

- The top of race partitions should be elevated to 1.8 m.
- The peak of bows over race and gate slides needs to be increased to 2.4 m.
- Metallic loading races must be blanketed with dust to lessen the hollow sound and to save the steel cleats from the camel's tender toes

There may be other types of mangers. The cheapest ones are constructed by:

- A dip in the ground of 50–70 cm diameter and 50 cm deep.
- The manger is up of kacha bricks or mud and cow dung or pucca bricks.

5.3.6 Calves Shed

Freshly parturated she-camels must be kept with their calves in an open free stall with soft, dry, and clean bedding material (wheat straw, sawdust, etc.). This type of shed normally facilitates keeping animals of various age groups under a single shed with a common manger for all. Also, the calves' paddocks can be located near their dams to allow contact but restricting suckling at any time (Fig. 5.17). There can be many compartments under a single roof with some sort of segregation most probably pipe usage for defining various stages of productivity or age. To keep calves with their mother is with animal welfare and good health perspective for both mother and calf. Parturition is a stressful process and can result in negative behaviours of the she-camel, which may in some cases reject the calf after birth. Hence this type of housing will enable and encourage the natural mother-calf relationship and boost calf immunity by proper milking.

Fig. 5.17 Freshly parturated she-camels and calves shed (photo by Dr. Bernard Faye)

5.3.7 Breeding Bulls Shed

The simple rule is that larger distances make larger animals happy; this is true, especially in the case of breeding camels. In smaller areas, the males divert to fighting or aggressive behaviours (McGee Bennett 2014). However, the housing system must also ensure the presence of shade, water, and mineral (e.g. salt blocks) for all members of a male group. Overall, 8 × 10 m per breeding bull space is generally required, but in case of more space allowance can cause an aggressive and bad-tempered response. The manger dimension within the pen is 1-m width × 1.5-m depth × 2-m height. All the camels must be provided within spacious paddocks with shades without any requirement for cooling technology. Generally, a good idea is to provide some extra space/pen to remove any chances of defence, competition, and limited resources issues. The pen size, composition, and structure can manipulate the animal's behaviour and health. Generally, males prevent their females from bachelor's males by standing or walking between them and driving them away. That's why the shed must be segregated based on age or sex to prevent this aggression or stereotypical behaviours.

As per Schulte and Klingel (1991), the distance between young males must be maintained at a distance from females. The bulls may chase heifers of ages up to 5 years in their side by pens. Whenever they came too close to the herd, they were attacked and chased up to 50 m or further away. The camel-breeding bull behaviour is quite interesting and more dramatic as posturing, clucking, rearing, and open-mouth breathing. In a very small pen, no matter what they do, animals are not provided with enough space which ends up with fights and stressful conditions both for dominant and subordinate males.

5.3.8 Lactating She-Camels Shed

Approximately 50 m^2 are required for one lactating she-camel, of which 15 m^2 should be shaded (Nagy et al. 2022). Water must be available at all times along with sufficient feeding space for all camels in the paddock if they want to feed at the same time. Approximately 80–100 cm of feeding space/camels is required to avoid camel competition and fighting (Fig. 5.18).

For mass treatment and handling of the animals, treatment areas (or the so-called catching areas) are installed in most paddocks that allow herding, selecting, separating, and individually treating the camels. Compared to pastoral systems where camels are allowed to roam freely, intensive production significantly restricts the movement of animals. To mitigate the suspected negative effect of this confinement, some experimental and modern farms have developed exercise facilities, namely, walking tracks, where dromedaries have daily controlled exercise for approximately 1 h. In some cases, a horse walker has also been installed for the male dromedaries allowing them to exercise separately.

Most countries as Saudi Arabia, Qatar, and the United Arab Emirates, where camel milk is produced at a large scale and is culturally utilized more, have very

Fig. 5.18 A Camel farm in France (photo by Dr. Bernard Faye)

advanced and technologically equipped camel farms with efficient and ergonomic milking parlours. A milking parlour can have a range of technologies from simple milk recording per camel to milk testing on spot for various components. But under any form, a milking parlour must be safe and comfortable for both animals and people. Milkers should have easy access to the udder, and the flooring of the parlour should be non-slippery to avoid injury to the animals. It is typical defensive behaviour of camels to sit down if there is no other way to escape from a difficult situation, so the design of the parlour should support effective handling in such an emergency. The walkways should also be planned well to allow the fast and safe movement of the various groups to and from the milking parlour.

The paddocks of the calves should be located adjacent to that of their dams to allow contact but prevent free suckling of the calves. Under confinement camels are normally allowed in special walking tracks for walks in larger open areas or exercise facilities generally for at least 1 h daily.

5.3.8.1 Milking Parlour

According to Faye (2020), even though most camels are kept under pastoral, extensive, or semi-intensive systems, well-planned intensification can make a significant improvement in the safety and traceability of camel milk production chain. A significant investment is however required to build a proper milking (Nagy et al. 2022). Milking parlour must ensure the animal health and animal welfare, as a well-designed, comfortable, and efficient milking parlour is crucial for intensive camel dairying. The main aim of the milking parlours should be the camel and their

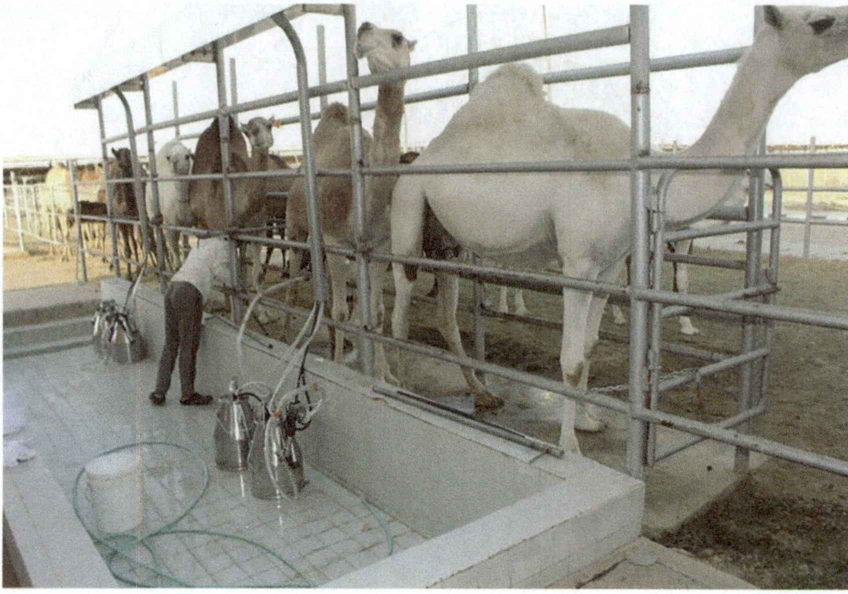

Fig. 5.19 Double cluster portable milking machine hooked upto the camels udders in open farm (photo by Dr. Bernard Faye)

milkers' comfort with having easy access to animal's udder, non-slippery floors for animals, humans, and equipment (Figs. 5.21, 5.22, and 5.23). The camel behaviour is very distinct when it comes to animal defence, and she prefers to sit if not be able to escape. Thus a milking parlour must be of broader sideways and walkways for allowing a safer and quicker movement of animals or handlers in any emergency/risk. Both traditional and modernized milking systems have their own pros and cons, which are described based on the experience of the co-authors (Fig. 5.19).

Traditional milking performed with one or more operators requires much less initial infrastructure expenditure (Fig. 5.20). If the human-animal relationship is good, she-camels may be less reluctant to be milked and therefore may get used to milking more quickly. However, traditional milking can only be performed on a maximum of one-to-two hundred she-camels, and as the size of the herd increases, it becomes more difficult to milk by hand. In addition, modernized milking farms make it possible both to handle larger numbers of animals (up to several thousand depending on the size and number of milking parlours) and to have higher and healthier milk yields. In fact, modern facilities are usually accompanied by increased attention to hygiene and the maintenance of the cold chain (Fig. 5.21). Modern milking parlours can have different designs, such as double-sided (Figs. 5.21 and 5.22) or herringbone milking parlours (Fig. 5.23).

Fig. 5.20 Fully automatic camel milking parlour with well-trained professionals for milking (photo by Dr. Bernard Faye)

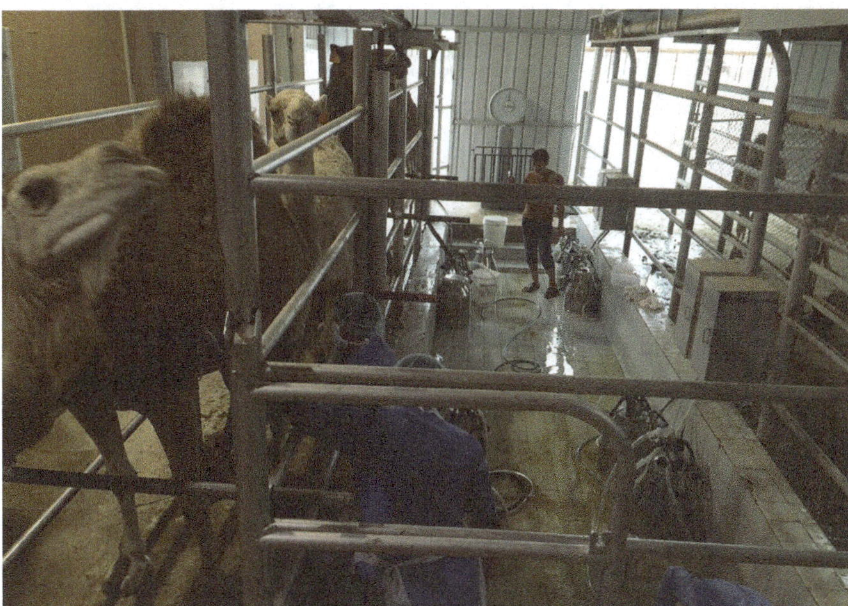

Fig. 5.21 A double-sided milking parlour of dromedaries with portable milking unit (photo by Dr. Bernard Faye)

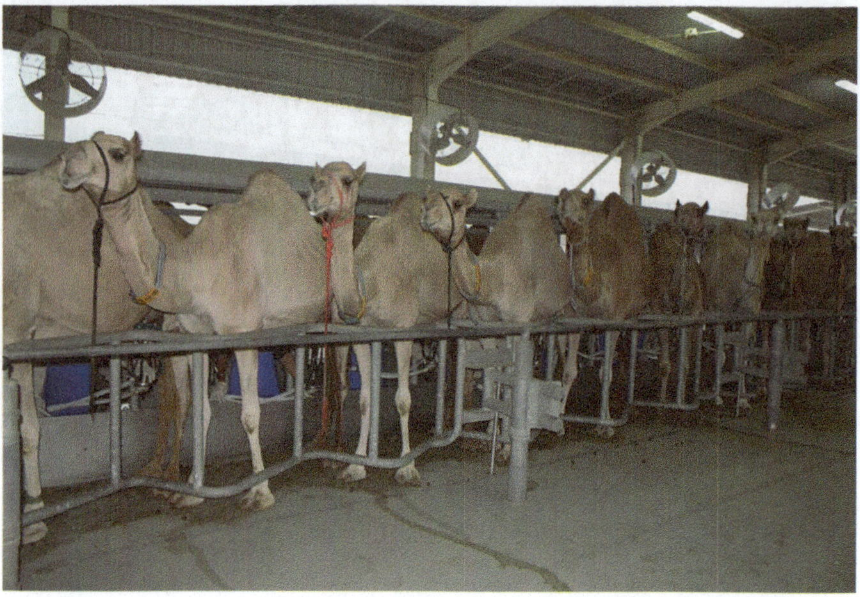

Fig. 5.22 A large-scale herringbone design milking parlour (photo by Dr. Bernard Faye)

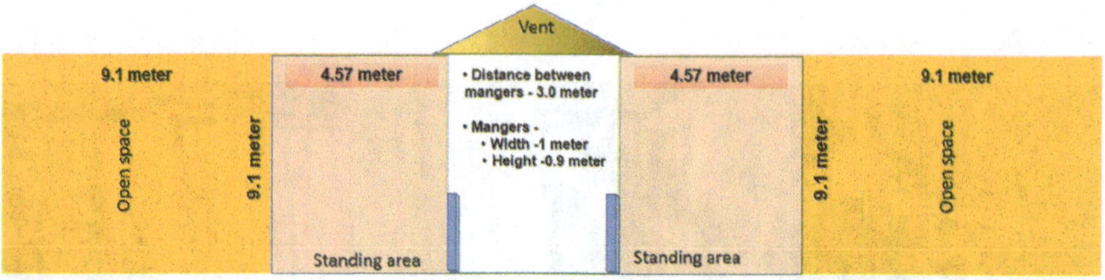

Fig. 5.23 Sketch of two camels shed dimensions taken at the Faculty of Veterinary and Animal Sciences, The Islamia University of Bahawalpur (photo by Dr. Zahid Kamran and Dr. Naeem Tahir)

5.4 The Design of Camel Housing for Ensuring a Good Welfare

Efficient camel housing must be designed to meet the behavioural needs of the camels and must prevent the spread of infectious disease. General functions of Housing are as follows:

- Protect the camels from adverse weather.
- Provide the animal with enough possibility to be having ample, fresh, and clean feed.
- Provide a comfortable space to reduce the chance of injuries, falls, and accidents due to drains, openings, slippery floors, etc.

- Provide a place for unique handling and coping with and good observation and care of animals.

Important points to be considered for good housing conditions in perspective of camel welfare are the following:

- How many camels are reared at the farm?
- Are there other species apart from camels?
- Which is the main purpose of rearing camels?
- Is any exercise facility or grazing routine available for camels?
- What is the type of house? Climatically suitable or not?

Unfortunately, studies describing the effects of different housing systems on camel welfare are scant; the authors have therefore described some important points based on their own experience.

5.4.1 Construction

Open-air shelters must be strong enough to resist the camels striking the fences. Concrete or synthetic floor surfaces should be kept minimum and non-slippery. The favourable floor types are floor coverings with grass or sand.

5.4.2 Fencing

Fences should be either wire mesh, clearly visible high tensile wire fitted with droppers, or made by any other wood or metal fence, up to a height of 1.9 m high. In the case of wire mesh, the mesh size must not be interrupted or create a risk of entangling the animals' head or legs. Fences should be checked regularly to ensure the lack of any sort of damage. The use of barbed wire is not recommended.

Enclosures must be large enough to:

- Allow animals to roam freely and meet their locomotory needs.
- Allow animals to interact and express social behaviours.
- Avoid air stagnation (well-ventilated).
- Provide protection from adverse weather conditions and predators.

5.4.3 Lighting and Ventilation

Camels should have access to natural light and fresh air/proper ventilation.

5.4.4 Drainage

The drainage of the enclosure must be quick, easy, sloppy, and smooth to remove all excess water. Drains should be designed to avoid camel injury, and any open drains, other than those carrying surface water, should be outside the shed. Any faecal material must be disposed of in an environmentally sound way.

5.4.5 Cleanliness

The indoor and outdoor areas of the camel shed must be clean and free from debris on a daily basis. The surfaces must be easy to clean, dry, and disinfect.

5.4.6 Prevention of Escape

Gates to the sheds must be able to lock whenever required, especially if the area is nearby to a public place or road. Unnecessary access to the public can cause the animal to be annoyed, and disrupted, and cause injuries, panic, and accidents with an effect on production and reproduction. This can be protected by building a secondary wall/fence on the accessible sides of the enclosure with some signs for warning.

5.4.7 Yard Design

Floors of yards, sheds, pens, and loading ramps need to have surfaces that minimize slipping. The yards should now not be moist, slippery, or boggy. It's vital to create a raised amount of sand in yards that are subjected to wet or boggy situations. Protecting yards ought to be designed without protruding items for avoiding any unwanted and unpleasant accidents or injuries. They have to be massive sufficient to allow all animals to lie down. If the yards are for holding for longer than some days, they must be sufficient to enable good enough exercising (Husbandry guidelines of Arabian camels 2008). Figure 5.23 can give a good vision of the animal house dimensions.

5.4.8 Enclosure Design

General principles and inclusions to exhibit design:

- The shelter should be open to sunlight with shaded areas provided throughout all daylight hours.
- A shelter must be effective against all climatic extremes. In desert areas, especially from the worst heat, wind, and sandstorms.

- The shelter may not allow animal's access to any harmful plants/materials as camels tend to indiscriminate browsers and may consume toxic plants.
- The enclosure must be constructed to limit or stop the access of:
  - Predators (e.g. snakes, carnivores), pets, rats, and wild animals.
  - The animals must not be able to escape.
- The enclosure must be environment friendly, safe, easy for maintenance and from the unnecessary public, and with adequate warning/safety signs.
- If access is large enough to allow entry of a drive-able vehicle, then the entry should be through successive gates to minimize animal escape.
- The enclosure must include enrichment tools. Below is a list of environmental enrichments based on the experience of the co-authors of this chapter and which can be used in camel enclosures:
  - Rubbing/scratching posts (e.g. telegraph poles, rock work, mechanical brushes similar to those used in dairy cow farms).
  - Substrate variety (e.g. sand, grass, dirt).
  - Lying down/resting areas (e.g. mulch pits, dirt patches, mud wallows).
  - Natural water opportunities (lakes).
  - Feeding stations at different distances.
  - Objects suspended from various heights (browse).
  - Raised keeper platforms used for conditioning and training.
  - Live local native/habitat native trees (Keekar trees or large palm trees).
  - Artificial trees with browse pots.
  - Mixed animal species exhibit social stimulation (but may be problematic for disease spreading among different animal species).
  - Misting/watering systems for alleviating heat stress.
  - Visual barriers to public and neighbouring animals.
  - Areas for scatter feeds.
  - The large clear area for free run/play away from objects.

5.4.9 Feeding/Watering Areas

To avoid the accumulation of faeces around water/feeding points:

- They must have an easy cleaning surface around feeding stations (e.g. concrete).
- They should be portable water/feeding stations in order to be easier to clean from dirt, feed residues, and substrates dried out (e.g. sand, mulch).

5.4.10 Restraint Area Design

A restraint area must always be available for several reasons including beauty contests, health, quarantine or reproduction, animal's restraining for medication, or before transport. If the restraint area is intended for handling a small group of camels,

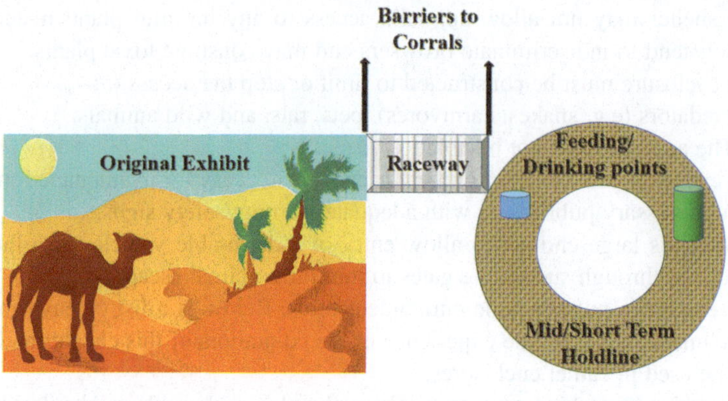

Fig. 5.24 Ideal camel holding area design (adapted from Dioli 2013. Modified version)

it should be large enough to allow the camel full movement in all directions with other camels, including turning, standing, or stretching.

Pre-departure restraint areas are intended for temporary/short/midterm location of camels and should be located nearby the original exhibit and allow the camels to be easily moved between them.

Additional housing or enclosure maintenance could be used to create a raceway from the exhibit into a smaller enclosure while cleaning or maintenance or for overnight security as a sleeping den (Fig. 5.24).

It should also have adequate protection from the weather (e.g. roof or shade cloth). It must also include a feeding and water station.

5.4.11 Enclosures

It is recommended to position shelters/night den in a particular direction, varying as per countries' geographical locations. Ideally, the face opening of the shed may be north, for protection against the winds of south, east, and west.

It is also recommended that in summer the camel must be sheltered from the sun and in winter it can get the winter sun. Therefore, the position, style, and design of the enclosure and surrounding areas will vary. Facilities must be present in the area for handling and comfort of the animals without unnecessary threat to animals or handlers. Well-trained camels require standard yards with adequate heights, efficient airflow, and protection from sunlight, rain, and predators. Similarly, yards should be properly tiered with dry regions to permit camels to sit down and rest whenever required.

5.4.12 Weather Protection

There is considerable evidence that shade prevents the deleterious effects of heat stress on health and production in domestic livestock (West 2003) and that cattle, horses, and buffaloes prefer shaded areas (Schütz et al. 2014; Holcomb et al. 2014; Almeida et al. 2019; Heitman et al. 1962; Mayorga et al. 2019). This evidence has been useful for implementing a code of practices for farm animal welfare (Primary Industries Standing Committee 2006; Animal Health Australia 2016), and shade is currently recommended by the World Organization for Health Terrestrial Animal Health Code (OIE 2018).

As a first recommendation, dromedary camels need shaded areas in the pens and paddock. As evidenced by the scientific publications reported in the section on camel thermal needs, camels may suffer from heat stress too, even though they evolved to survive extreme drought and the environmental conditions in arid and semi-arid regions. As observed by Zappaterra et al. (2021), camels prefer shaded areas when kept in a hot environment (environmental temperature ranging from 40 to 47 °C and relative humidity values between 16 and 40%). Indeed, as observed in that study, camels tended to crowd in the shade when present and subjects in the pen where shaded areas were absent spent less time ruminating and stayed for a longer period in a standing position rather than in recumbency (Zappaterra et al. 2021). It is therefore essential to provide the camels with shelters, which allow the animals to choose when to stand in the sun and when to take shelter. Shelters should be designed considering several points, which will be discussed here below.

5.4.12.1 Provision of Adequate Shaded Area

The shaded area should be large enough to allow all the animals in the pen to rest in lateral recumbency at the same time. Feeder(s) and drinking troughs should also be placed in the shaded area so that both water and food remain in the shade and protected as much as possible from direct sunlight, rain, and sand.

5.4.12.2 Use of Adequate Materials

The materials used to build the shelters must be adequate for the micro-climatic zone where the camels are bred. Roofs can be sloping or of the flat type depending on the annual precipitations and the flock size. Roofing materials should be strong and durable. In arid and semi-arid countries, metals (such as galvanized iron sheets) should be avoided due to the high conductivity of those materials. Instead, white-washing (with lime and white cement in equal amounts) and white painting on the outer surface of the roof are highly recommended as these strategies provide optimum comfort to the livestock housed inside the shelters. Today, there are several solutions on the market that increase the reflectance of roofs, which are technically named "cool roofs". These include various white paints that can significantly mitigate the heat produced by solar radiance, preventing overheating of roofs and thus eliminating the infrared thermal radiation from the ceiling. These strategies are being exploited over the world to counter overheating in both urban areas and dwellings but could be of great importance also for livestock buildings (Santunione et al. 2017).

5.4.12.3 Provision of an Adequate Roof Height

The roof height is also of great importance for ventilation inside the shaded area and for reducing the effect of roof temperature on the temperatures and radiant heat balance inside the shelter. The scientific literature lacks studies that specifically address the definition of the most adequate roof height for dromedary camels kept in arid and semi-arid environments. Based on what has been identified for cattle; however, the best roof height should be 10 m (33 ft) (Berman and Horovitz 2012). In cases where this is not possible, the authors suggest to keep at least a height of 4 m (13 ft), to reduce the heat inside the shelter and permit air circulation. This minimum height of 4-m is suggested because camels can exceed 2 m in height, and therefore there is a need to provide at least a minimum distance of 1.5 m between the roof and the camel's hump (the tallest point of the animal).

5.4.13 Enclosure Substrate

Based on co-authors' experience, enclosure substrates must be chosen following these suggestions:

- For the majority of the substrate in the enclosure, you should avoid anything abrasive or irritating to the camels.
- Camels are ungulates (hooved animals), and thus it is necessary to provide some abrasive substrate to curb the excessive hoof growth and damage (e.g. sand, textured cement or concrete, crushed gravels).
- Sand is the best substrate for the majority of the shelters under various zones as it is a part of their natural habitat, dried off quickly, and is not slippery.
- Feeding stations must be above ground and not fixed.
- For substrate under feeding stations, concrete would be more appropriate. In case of over-consumption of sand by camel and not getting enough fibre, it may lead to a health issue of sand impaction of the gut.

5.4.14 Bedding Material

Based on co-authors' experience, bedding materials must be chosen following these suggestions:

- Bedding material must be provided in a shelter or night den and should be easy to clean and dry from faeces.
- Appropriate substrate for bedding would be mulch, sand, soil, straw, hay, and any abrasive substances such as concrete, gravel, or asphalt that will cause discomfort

to a camel. Also, the cleanliness of faecal matter without removing large amounts of substrate each time will be not possible.

5.5 Welfare Concerns and Gap of Knowledge

Camels kept in overcrowded pens have limited possibility to walk freely and express natural behaviour and often show more aggressive behaviour and camel-camel relationships (Wang et al. 2016; Fu et al. 2016). Overcrowding is indeed the major welfare concern in intensive camel farming systems. Research has shown that in pens where the space allowance per animal is lesser than 19 m^2/animal, camels showed a worse thirst index and high frequency of aggressive behaviours (Menchetti et al. 2021b). Secondly, animals in overcrowded pens have less space to walk freely, thus experiencing unstable social relationships and inhibiting natural behaviours (Wang et al. 2016). The impulsive dominance hierarchies are caused by unnatural behaviours, and competitive ones can increase it (Fu et al. 2016; Faye 2020). Furthermore, overcrowding can enhance aggression and stereotypies along with hindering animals' health status and reproductive performance. The negative effects of restricted animal areas are obvious in feeding behaviour, abnormal behaviours, and frame lesions marked clearly in various farm animal species (Salak-Johnson et al. 2012; Fu et al. 2016; Raspa et al. 2020). El Shoukary et al. (2020a) showed that overstocking resulted not only in decreased lying and rumination time but also in increased serum cortisol concentrations, feed competition, aggressive behaviour, and production losses.

Concerning shade provision, behavioural modifications linked with the presence of shaded areas are proven in detail under various conditions in numerous species (cattle, sheep, buffaloes, and horses) (Schütz et al. 2010; Holcomb et al. 2015; Giro et al. 2019; De et al. 2020; Mishra 2021). The findings suggested that a shaded location of at least 7 m^2 per animal and comfortable bedding had useful outcomes on numerous components of the camel's welfare (Welfare Quality 2009; AWIN 2015). Furthermore, if the shaded space is adequate for rearing density, camels will show less competitive behaviour and will rest more, which is considered a positive welfare indicator (Mellor 2016).

The cleanliness of bedding showed to be another key point for ensuring a good welfare of the camels. Sick camels are indeed more likely to be found in boxes with dirt and garbage (Schwartz and Dioli 1992). Body condition score is also positively associated with the cleanliness of the bedding as camels living in pens with dirt beddings had lower BCS when compared with the other camels. The high BCS tends to be negatively associated with the presence of a health issue and constrained feeding or space.

Some welfare concerns have also been raised for bulls. Male dromedary camels housed in single stalls had higher cortisol levels and showed more stereotypic behaviour than the animals in open paddocks and interacted with females (Padalino et al. 2014). The freedom of movement and the olfactory contact with females permitted the expression of sexual behaviours and increased the testosterone levels in male camels (Fatnassi et al. 2014).

Despite the evident increase in the socio-economic interests on camel-intensive rearing and production (Khan et al. 2003) and the parallel evolution of camel science (Pastrana et al. 2020), empirical data are lacking in regard to this species' welfare and its procuration in such a regime. Contextually, since law and science are dependent on each other (Freeland 2012), the current limited academic progress in terms of camel well-being-related measurable is translated into the disability of the legal community in its practice of specific regulations' enforcement. Few studies have been carried out to assess the effects of housing systems on camels (Padalino et al. 2014; Fatnassi et al. 2014; Aubè et al. 2017; Menchetti et al. 2021a, b). However, these parameters must be seen as a starting point, and further studies are much needed to suggest minimum space allowances for the different categories of camels. In other respects, there is a lack of studies that can provide a suggested minimum size of shade per head. Given also the nature of the camel as an animal able to adapt to the desert, very few researches have defined the limits of thermal comfort for this animal. All too often, the camel is regarded as an animal that can be reared in extreme climatic conditions, without providing it with shelter from high temperatures and direct sunlight. However, there is a clear distinction between the extreme climatic conditions these animals can withstand and the conditions that allow them to live (Norton et al., 2014) with a good level of welfare.

5.6 Conclusions

Housing systems that guarantee social interaction and physical activity and thermal comfort are the most suitable housing systems allowing camels to exhibit their behavioural needs and to ensure health, production, and reproduction. Social interaction is of particular importance for maintaining the animals' physical and psychological health. Allowing locomotion activity improves camel metabolic status and also decreases their captivity stress. Fenced and exercised male dromedaries had significantly higher performance, health, and reproductive efficiency, while she-camels can have increased milk production and rear well their calves. Hence, while lack of adequate shaded areas and overcrowding may cause poor welfare outcomes and should be avoided, proving appropriate housing will lead to better animal health, performance, and welfare.

References

Abdallah HR, Faye B (2013) Typology of camel farming system in Saudi Arabia. Typology of camel farming system in Saudi Arabia. Emir J Food Agric 25(4):250–260

Abdoun K, Samara E, Okab AB, Al-Haidary AA (2012) A comparative study on seasonal variation in body temperature and blood composition of camels and sheeps. J Anim Vet Adv 11(6): 769–773

Akbar MA, Bhakat C, Yadav JS (2012) Camel rearing in India. Indian Farming 62(4):20–24, 33

Akin JA (2011) Homeostatic processes for thermoregulation. Nat Educ Knowledge 3(10):7
Al-Ahmadi K, Alahmadi M, Al-Zahrani A (2020) Spatial association between primary Middle East respiratory syndrome coronavirus infection and exposure to dromedary camels in Saudi Arabia. Zoonoses Public Health 67:382–390
Al Ghumlas AK, Abdel Gader AGM, Hussein MF, Al Haidary A, White JG (2008) Effects of heat on camel platelet structure and function—a comparative study with humans. Platelets 19(3): 163–171. https://doi.org/10.1080/09537100701882061
Al-Haidary AA (2005) Effect of dehydration on core body temperature of young Arabian camels (Camelus dromedarius). J King Saudi Univ 18(1):1–7
Al-Haidary AA, Samara E, Okab A, Abdoun K (2013) Thermophysiological responses and heat tolerance of Saudi camel breeds. Int J Chem Environ Biol Sci (IJCEBS) 1(March):173–176
Alhajeri BH, Alhaddad H, Alaqeely R, Alaskar H, Dashti Z, Maraqa T (2021) Camel breed morphometrics: current methods and possibilities. Trans R Soc S Aust 1–22
Almeida JCFD, Joset WCL, Noronha RDPP, Barbosa AVC, Lourenço Júnior JDB, Silva JARD, Almeida JCFD, Joset WCL, Noronha RDPP, Barbosa AVC et al (2019) Behavior of buffalo heifers reared in shaded and unshaded pastures during the dry season on Marajó Island, Pará, Brazil. Acta Sci Anim Sci 41. https://doi.org/10.4025/actascianimsci.v41i1.43088
Animal Health Australia (AHA) (2016) Australian Animal Welfare Standards and Guidelines—Sheep. http://www.animalwelfarestandards.net.au/files/2016/01/Sheep-Standards-and-Guidelines-for-Endorsed-Jan-2016-250116.pdf
Aubè L, Fatnassi M, Monaco D, Khorchani T, Lacalandra GM, Hammadi M, Padalino B (2017) Daily rhythms of behavioral and hormonal patterns in male dromedary camels housed in boxes. Peer J 5:3074. https://doi.org/10.7717/peerj.3074
AWIN (2015) AWIN Welfare Assessment Protocol for Horses, 1–80
Beerda B, Schilder MB, Van Hooff JA, De Vries HW, Mol JA (1999) Chronic stress in dogs subjected to social and spatial restriction. I. Behavioral responses. Physiol Behav 66:233–242
Belhadj Slimen I, Najar T, Ghram A, Abdrrabba M (2016) Heat stress effects on livestock: molecular, cellular and metabolic aspects, a review. J Anim Physiol Anim Nutr 100(3): 401–412. https://doi.org/10.1111/jpn.12379
Berman A, Horovitz T (2012) Radiant heat loss, an unexploited path for heat stress reduction in shaded cattle. J Dairy Sci 95(6):3021–3031. https://doi.org/10.3168/jds.2011-4844
Botreau R, Veissier I, Butterworth A et al (2007) Definition of criteria for overall assessment of animal welfare. Anim Welf 16:225–228
Bouâouda H, Achâaban MR, Ouassat M, Oukassou M, Piro M, Challet E, El Allali K, Pévet P (2014) Daily regulation of body temperature rhythm in the camel (Camelus dromedarius) exposed to experimental desert conditions. Physiol Rep 2(9). https://doi.org/10.14814/phy2.12151
Bowler K, Manning R (1994) Membranes as the critical targets in cellular heat injury and resistance adaptation. Temp Adapt Biol Membr 185–203
Brown-Brandl TM (2009) Overview of the progress in reducing environmental effects on cattle. In: Proceedings American Dairy Science Association 18th discover conference, 2–5 Nov 2009, Nashville, IN
Chakrabarti A (2006) Handbook of animal husbandry sciences. 2nd revised edition. Kalyani, New Delhi. B149805061. ISBN 81-272-2632-7
De K, Sharma S, Kumawat PK, Kumar D, Mohapatra A, Sahoo A (2020) Tree shade improves the comfort of sheep during extreme summer. J Vet Behav 40:103–107
Dereje M, Udén P (2005) The browsing dromedary camel: I. Behaviour, plant preference and quality of forage selected. Anim Feed Sci Technol 121(3–4):297–308
Dioli M (2013) Pictorial guide to traditional management husbandry and diseases of the one-humped camel
EFSA AHAW Panel (2022) Scientific opinion on the welfare of pigs during transport. EFSA J 20(9):7445. https://doi.org/10.2903/j.efsa.2022.7445

El Allali K, Beniaich Y, Farsi H, M Hani MEM, Jabal MS, Piro M, Achaâban MR, Ouassat M, Challet E, Besson M, Mounach J, Pévet P, Satté A (2022) Sleep pattern in the dromedary camel: a behavioral and polysomnography study. Sleep 45(8):zsac101. https://doi.org/10.1093/sleep/zsac101. PMID: 35512227

El Shoukary R, Osman A (2020) Mitigates the negative behavioral effects of rut season in camels: -group size. SVU Int J Vet Sci 3(1):39–50

El Shoukary RD, Osman A, Mohammed A (2020a) Effects of stocking density on some behavioral and some blood biochemical parameters in camel during the rut period. Egypt J Vet Sci 51:253–262

El Shoukary RD, Nasreldin N, Osman AS, Hashem NM, Saadeldin IM, Swelum AA (2020b) Housing management of male dromedaries during the rut season: effects of social contact between males and movement control on sexual behavior, blood metabolites and hormonal balance. Animals 10(9):1–11

Evans JO, Powys JG (1979) Camel husbandry to increase the productivity of rangeland camels. In: IFS symposium, Sudan, pp 241–250

Faraz A, Waheed A, Mirza RH, Ishaq HM (2019) The camel—a short communication on classification and attributes. J Fish Livest Prod 7(1):289

Fatnassi M, Padalino B, Monaco D, Aubé L, Khorchani T, Lacalandra GM, Hammadi M (2014) Effect of different management systems on rutting behavior and behavioral repertoire of housed Maghrebi male camels (Camelus dromedarius). Trop Anim Health Prod 46:861–867. https://doi.org/10.1007/s11250-014-0577-6

Faye B (2016) The camel, new challenges for a sustainable development. Trop Anim Health Prod 48:689–692. https://doi.org/10.1007/s11250-016-0995-8

Faye B (2020) How many large camelids are in the world? A synthetic analysis of the world camel demographic changes. Pastoralism 10:25

Faye B, Barnouin J (1985) Objectivation de la propreté des vaches laitières et des stabulations. L'indice de propreté. Bull Tech Cent Rech Zootech Vétérinaires Theix 59:61–67

Freeland DMH (2012) Speaking science to law. Georgetown Int Environ Law Rev 25:289

Fu L, Li H, Liang T, Zhou B, Chu Q, Schinckel AP, Yang X, Zhao R, Li P, Huang R (2016) Stocking density affects welfare indicators of growing pigs of different group sizes after regrouping. Appl Anim Behav Sci 174:42–50

Gauthier-Pilters H (1979) Some ecological aspects of the camel in Western Sahara. In: IFS symposium, Sudan, pp 387–398

Gauthier-Pilters H, and Dagg A. (1981). The camel. Its evolution, ecology, behaviour, and relationship to man. University of Chicago Press, Chicago.

Gebreyohanes MG, Assen AM (2017) Adaptation mechanisms of camels (Camelus dromedarius) for desert environment: a review. J Vet Sci Technol 8(6):1–5

Giro A, Pezzopane JRM, Barioni Junior W, Pedroso ADF, Lemes AP, Botta D, Romanello N, Barreto ADN, Garcia AR (2019) Behavior and body surface temperature of beef cattle in integrated crop-livestock systems with or without tree shading. Sci Total Environ 684:587–596

Grigg GC, Beard LA, Augee ML (2004) The evolution of endothermy and its diversity in mammals and birds. Physiol Biochem Zool PBZ 77(6):982–997. https://doi.org/10.1086/425188

Habte M, Mitiku E, Melesse M, Dereje A, Abiyot L, Birhanu A (2021) The influence of weather conditions on body temperature, milk composition and yields of the free-ranging dromedary camels in southeastern rangelands of Ethiopia. Edited by Pedro González-Redondo. Cogent Food Agric 7(1):1930932. https://doi.org/10.1080/23311932.2021.1930932

Heitman H, Hahn L, Bond TE, Kelly CF (1962) The effects of modified summer environment on swine behaviour. Anim Behav 10:15–19. https://doi.org/10.1016/0003-3472(62)90125-2

Holcomb KE, Tucker CB, Stull CL (2014) Preference of domestic horses for shade in a hot, sunny environment. J Anim Sci 92:1708–1717. https://doi.org/10.2527/jas.2013-7386

Holcomb KE, Tucker CB, Stull CL (2015) Shade use by small groups of domestic horses in a hot, sunny environment. J Anim Sci 93:5455–5464

Hoter A, Sandra R, Naim HY (2019) Cellular and molecular adaptation of Arabian camel to heat stress. Front Genet 10. https://doi.org/10.3389/fgene.2019.00588
Husbandry Guidelines for Arabian Camel, Camelus dromedarius (2008) Western Sydney Institute of TAFE, Richmond. Captive Animals
Iglesias C, Navas F, Ciani E, Capote CJ, Delgado BJV (2020a) Effect of research impact on emerging camel husbandry, welfare and social-related awareness. Animals 10:780. https://doi.org/10.3390/ani10050780
Iglesias C, Navas F, Ciani E, Arbulu AA, González A, Marín C (2020b) Zoometric characterization and body condition score in Canarian camel breed. Archivos de zootecnia 69(265):102–107
Ivanov KP (2006) The development of the concepts of homeothermy and thermoregulation. J Therm Biol Second Int Meet Physiol Pharmacol Temp Regul 31(1):24–29. https://doi.org/10.1016/j.jtherbio.2005.12.005
Keane M, McGee M, O'Riordan E, Kelly A, Earley B (2017) Effect of space allowance and floor type on performance, welfare and physiological measurements of finishing beef heifers. Animals 11(12):2285–2294
Kendall PE, Tucker CB, Dalley DE, Clark DA, Webster JR (2008) Milking frequency affects the circadian body temperature rhythm in dairy cows. Livest Sci 117(2):130–138. https://doi.org/10.1016/j.livsci.2007.12.009
Khan BB, Arshad I, Riaz M (2003) Production and management of camels. University of Agriculture, Faisalabad, Department of Livestock Management
Kingma BRM, Frijns AJH, Schellen L, van Marken Lichtenbelt WD (2014) Beyond the classic thermoneutral zone. Temperature 1(2):142–149. https://doi.org/10.4161/temp.29702
Kurz A (2008) Physiology of thermoregulation. Best Pract Res Clin Anaesthesiol 22(4):627–644. https://doi.org/10.1016/j.bpa.2008.06.004
Lups TC (2001) Glossary of terms for thermal physiology. Revised by the Commission for Thermal Physiology of the International Union of Physiological Sciences (IUPS Thermal Commission). Jpn J Physiol 51(2):245–280
Mayorga EJ, Renaudeau D, Ramirez BC, Ross JW, Baumgard LH (2019) Heat stress adaptations in pigs. Anim Front 9:54–61. https://doi.org/10.1093/af/vfy035
McGee Bennett M (2014) Understanding male behavior. In the Alpaca. Alpacas Magazine, 30–34
Mellor DJ (2016) Updating animal welfare thinking: moving beyond the "five freedoms" towards "a life worth living". Animals 6:21
Mellor DJ, Beausoleil NJ (2015) Extending the "Five Domains" model for animal welfare assessment to incorporate positive welfare states. Anim Welf 24:241–253
Menchetti L, Zappaterra M, Costa LN, Padalino B (2021a) Application of a protocol to assess camel welfare: Scoring system of collected measures, aggregated assessment indices, and criteria to classify a pen. Animals (Basel) 11:494
Menchetti L, Faye B, Padalino B (2021b) New animal-based measures to assess welfare in dromedary camels. Trop Anim Health Prod 53:533
Mirkena T, Elias W, Nega T, Getachew Gari G, Abebe and S. Newman. (2018) Camel production systems in Ethiopia: a review of the literature with notes on MERSCoV risk factors. Pastoralism 8:30. https://doi.org/10.1186/s13570-018-0135-3
Mishra SR (2021) Thermo-regulatory responses in riverine buffaloes against heat stress: An updated review. J Therm Biol 96:102844
Moretti J (2008) Husbandry Guidelines for Arabian camel. Western Sydney Institute of TAFE, Richmond
Nagy PP, Julian AS, Judit J (2022) Intensification of camel farming and milk production with special emphasis on animal health, welfare, and the biotechnology of reproduction. Anim Front 12(4):35–45. https://doi.org/10.1093/af/vfac043
Norton PL, Gold JR, Russell KE, Schulz KL, Porter BF (2014) Camelid heat stress: 15 cases (2003–2011). Can Vet J 55:992–996

OIE—World Organisation for Animal Health (2018) Chapter 7.9. Animal welfare and beef cattle production systems. In: Terrestrial Animal Health Code. https://www.oie.int/fileadmin/Home/eng/Health_standards/tahc/2018/en_sommaire.htm
OIE—World Organisation for Animal Health (2021) Terrestrial Animal Health Code, 28th ed. ISBN 978-92-95115-40-8
Osilla EV, Marsidi JL, Sharma S (2022) Physiology, temperature regulation. StatPearls, Treasure Island, FL. https://www.ncbi.nlm.nih.gov/books/NBK507838/
Ouajd S, Kamel B (2009) Physiological particularities of dromedary (Camelus dromedarius) and experimental implications. Scand J Lab Anim Sci 36(1):12
Padalino B, Menchetti L (2021) The first protocol for assessing welfare of camels. Front Vet Sci 7: 631876
Padalino B, Aubé L, Fatnassi M, Monaco D, Khorchani T, Hammadi M, Lacalandra GM (2014) Could dromedary camels develop stereotypy? The first description of stereotypical behaviour in housed male dromedary camels and how it is affected by different management systems. PLoS One 9:e89093. https://doi.org/10.1371/journal.pone.0089093
Padalino B, Monaco D, Lacalandra M (2015) Male camel behavior and breeding management strategies: how to handle a camel bull during the breeding season? Emir J Food Agric 27:338–349
Pastorelli G, Musella M, Zaninelli M, Tangorra F, Corino C (2006) Static spatial requirements of growing-finishing and heavy pigs. Livest Sci 105(1-3):260–264
Pastrana CI, González FJN, Ciani E, Capote CJB, Bermejo JVD (2020) Effect of research impact on emerging camel husbandry, welfare and social-related awareness. Animals 10(5):780
Pastrana CI, González FJN, Ciani E, Arbulu AA, Bermejo JVD (2021) The youngest, the heaviest and/or the darkest? Selection potentialities and determinants of leadership in Canarian dromedary camels. Animals 11(10):2886
Petherick JC, Phillips CJ (2009) Space allowances for confined livestock and their determination from allometric principles. Appl Anim Behav Sci 117(1-2):1–12
Phillips C, Morris I (2000) The locomotion of dairy cows on concrete floors that are dry, wet, or covered with a slurry of excreta. J Dairy Sci 83(8):1767–1772
Primary Industries Standing Committee (2006) Model code of practice for the welfare of animals: the camel, 2nd ed. PISC Report 86; Csiro, Collingwood, Australia. https://www.publish.csiro.au/ebook/download/pdf/5204
Raspa F, Tarantola M, Bergero D, Bellino C, Mastrazzo CM, Visconti A, Valvassori E, Vervuert I, Valle E (2020) Stocking density affects welfare indicators in horses reared for meat production. Animals 10:1–23
Robbins LA (2021) Into the comfort zone: understanding swine thermal preference. Thesis, Purdue University Graduate School. https://doi.org/10.25394/PGS.14137937.v1
Salak-Johnson JL, DeDecker AE, Horsman MJ, Rodriguez-Zas SL (2012) Space allowance for gestating sows in pens: Behavior and immunity. J Anim Sci 90:3232–3244
Samara E, Al-Haidary AA (2014) Estimating the thermoneutral zone of dromedary camels using surface heat transfer modelling
Santunione G, Libbra A, Muscio A (2017) Cool roofs with high solar reflectance for the welfare of dairy farming animals. J Phys Conf Ser 796(January):012028. https://doi.org/10.1088/1742-6596/796/1/012028
Schlader ZJ, Simmons SE, Stannard SR, Mündel T (2011) The independent roles of temperature and thermal perception in the control of human thermoregulatory behavior. Physiol Behav 103(2):217–224. https://doi.org/10.1016/j.physbeh.2011.02.002
Schmidt-Nielsen K, Crawford EC, Newsome AE, Rawson KS, Hammel HT (1967) Metabolic rate of camels: effect of body temperature and dehydration. Am J Physiol Legacy Content 212(2): 341–346. https://doi.org/10.1152/ajplegacy.1967.212.2.341
Schroter RC, Robertshaw D, Baker MA, Shoemaker VH, Holmes R, Schmidt-Nielsen K (1987) Respiration in heat stressed camels. Respir Physiol 70(1):97–112. https://doi.org/10.1016/S0034-5687(87)80035-X

Schulte N, Klingel H (1991) Herd structure, leadership, dominance and site attachment of the camel, Camelus dromedarius. Behavior 118(1/2):103–114
Schütz KE, Rogers AR, Poulouin YA, Cox NR, Tucker CB (2010) The amount of shade influences the behavior and physiology of dairy cattle. J Dairy Sci 93:125–133
Schütz KE, Cox NR, Tucker CB (2014) A field study of the behavioral and physiological effects of varying amounts of shade for lactating cows at pasture. J Dairy Sci 97:3599–3605. https://doi.org/10.3168/jds.2013-7649
Schwartz HJ, Dioli M (1992) The one-humped camel in Eastern Africa, livestock production science. Verlag Josef Margraf Scientific Books, Weikersheim, Germany
Sharp T (2012) Mustering of feral camels. Standard Operating Procedure. PestSmart-Centre for Invasive Species Solutions. Govt. of Australia. https://pestsmart.org.au/toolkit-resource/mustering-of-feral-camels
Siebert BD, Macfarlane WV (1971) Water turnover and renal function of dromedaries in the desert. Physiol Zool 44(4):225–240. https://doi.org/10.1086/physzool.44.4.30152494
Sugiono S, Soenoko R, Lukodono RP (2018) Static/dynamic zoometry concept to design cattle facilities using back propagation neural network (BPNN). Artif Intell Emerg Trends Appl. https://doi.org/10.5722/intechopen.75136
Traoré B, Moula N, Toure A, Ouologuem B, Leroy P, Antoine-Moussiaux N (2014) Characterisation of camel breeding practices in the Ansongo Region, Mali. Trop Anim Health Prod 46:1303–1312
Wales MfAaWNS (1986) Exhibited Animals Protection Act (No 123)
Wang FX, Shao DF, Li SL, Wang YJ, Azarfar A, Cao ZJ (2016) Effects of stocking density on behavior, productivity, and comfort indices of lactating dairy cows. J Dairy Sci 99:3709–3717
Warda M, Zeisig R (2000) Phospholipid- and fatty acid-composition in the erythrocyte membrane of the one-humped camel [Camelus dromedarius] and its influence on vesicle properties prepared from these lipids. Dtsch Tierarztl Wochenschr 107(9):368–373
Warda M, Prince A, Kim HK, Khafaga N, Scholkamy T, Linhardt RJ, Jin H (2014) Proteomics of old world camelid (Camelus dromedarius): better understanding the interplay between homeostasis and desert environment. J Adv Res 5(2):219–242. https://doi.org/10.1016/j.jare.2013.03.004
Welfare Quality® (2009) Welfare Quality® Assessment protocol for cattle. Welfare Quality® Consortium, Lelystad
West JW (2003) Effects of heat-stress on production in dairy cattle. J Dairy Sci 86:2131–2144. https://doi.org/10.3168/jds.S0022-0302(03)73803-X
Willmer P, Stone G, Johnston I (2009) Environmental physiology of animals. Wiley, Oxford
Zappaterra M, Menchetti L, Costa LN, Padalino B (2021) Do camels (Camelus dromedarius) need shaded areas? A case study of the camel market in Doha. Animals 11:480

Good Health: Recognition and Prevention of Disease and Pain in Dromedary Camels

6

Mohammed Hocine Benaissa and Carlos Iglesias Pastrana

Abstract

Good health is one welfare principle. However, minor species such as camels continue to be generally disregarded for prevention and clinical management of disease, injury and pain. Hence, detrimental effects on camel health, production and welfare emerge when diseases, injuries and noxious stimuli cannot be properly managed. Such negative repercussions are susceptible to becoming increasingly patent in a contemporaneous scenario in which dromedary camels are being translated from traditional extensive to more intensive production regimes. Notwithstanding, the relatively high presence of camels in extensive nomadic systems that practice ancient husbandry and medicine has also many impediments that prejudice the basic principles of animal welfare assurance. Hence, the present chapter aims to provide an overview of the most common injuries and diseases compromising dromedary camel health, welfare and productivity to serve as a practical guide for owners, researchers and stakeholders in this field. Concerning pain assessment, given the fact that no sensitive scales do exist for pain recognition and scoring in camels through physiological and behavioural responses, a camel composite pain scale was developed based on the literature. Additionally, a theoretical body on risk factors for poor health, the animal-dependent variables potentially modulating camel reactivity to pain/discomfort and the limitations of traditional farming and medicine for good health promotion are presented. To sum up, a list of recommendations on how to enhance physical and psychological health in dromedary camels is derived from the outlined content.

M. H. Benaissa
Scientific and Technical Research Centre for Arid Areas (CRSTRA), Nezla, Touggourt, Algeria

C. Iglesias Pastrana (✉)
Faculty of Veterinary Sciences, Department of Genetics, University of Cordoba, Cordoba, Spain

B. Padalino, B. Faye (eds.), *Dromedary Camel Behavior and Welfare*, Animal Welfare 24, https://doi.org/10.1007/978-3-031-48600-5_6

Keywords

Dromedary camels · Health disorders · Pain · Risk factors · Diagnosis · Treatment · Prevention

6.1 Health Disorders and Welfare

Good health is a principle of animal welfare. Injuries and diseases affect animals' feelings as well as their capacity for future good life. Diagnosis and treatment of diseased animals are crucial in all livestock systems to meet contemporary animal welfare standards (Moltumo et al. 2020).

In the past few decades, global attention given to camel pathologies has slowly increased (Koenig 2007). Indeed, until now there has been a very high potential for underreporting of camel diseases due to a lack of knowledge among veterinary health professionals regarding specific camel diseases (Higgins 1986; Eckstein et al. 2022). In extensive areas of camel countries, the weakness of the veterinary network is contributing to insufficient care of the animals except through ethnoveterinary medicine having either more or less questionable efficiency or based sometimes on coercive methods, painful for the animal (Antoine-Moussiaux et al. 2007). In reverse, in intensive camel farming systems which are emerging over the last 20 years to meet the increased demand for camel products, despite better veterinary management, an increased incidence of health problems is observed such as metabolic disorders (Faye et al. 1995; Nagy et al. 2022) and a decline in longevity and fertility (Tibary et al. 2006). Moreover, animals are more prone to suffer from contagious infectious and parasitic diseases due to high stocking density (Jones et al. 2013; Khalafalla and Hussein 2021).

In spite of the fact that camel science and its impact have noticeably increased in a contemporaneous scene, a parallel evolution of specific welfare laws is lacking (Iglesias Pastrana et al. 2020). Therefore, this chapter is not limited to the presentation of the prevalence and incidence of the most common injuries and diseases in camels but is including also a set of behavioural features and clinical signs that could help in the identification of pain status in camels, as well as a list of recommendations on how to enhance physical and psychological health in these animals. This conglomerate of information is intended to serve as a practical guide for breeders, clinicians, academicians, business stakeholders and animal welfare advocacies to boost the reorganization of camel breeding systems and policy reforms to undertake specific mandatory regulations for the promotion of camel wellbeing.

The aim of the present chapter is not to give a comprehensive list of health disorders in camels but rather to suggest the links between health problems and welfare. For a complete presentation of camel diseases, we can refer to some books devoted to this subject (e.g. Wernery et al. 2014). From a welfare point of view, we could distinguish the diseases having a low impact on welfare (e.g. subclinical diseases) and those provoking suffering. At the herd level, assessment of health management by indicators such as mortality, morbidity and udder health is crucial to meet the standard of animal welfare (Padalino and Menchetti 2021).

Usually, animal disorders are categorized into infectious, parasitic, metabolic, nutritional diseases and physical lesions (accidents notably). The pain generated by these different categories can be highly variables. Diseases causing minimally painful clinical signs and possessing a short duration have a relatively small effect on animal welfare, whereas diseases with a long duration or causing more painful clinical signs will have a greater impact at both the individual and herd levels. In veterinary medicine, treatment methods are assessed using a variety of indicators such as behavioural responses and physiopathological manifestations, which are known to be the most reliable signals to be related to pain intensity (Coetzee 2011).

Another categorization of diseases is those affecting all types of farming systems (with eventually devastating economic impact as transboundary diseases—TADs) and those having only local impact. For example, in recent decades, some outbreaks of transboundary animal diseases (TADs) were reported in different parts of the world (Faye 2019). The TAD outbreaks have included the following: (a) trypanosomosis (surra) in Iran (Zakian et al. 2017), Europe (Gutierrez et al. 2010) and Algeria (Benaissa et al. 2020b; Boushaki et al. 2019); (b) brucellosis in Sudan (Musa et al. 2008); (c) camelpox in Sudan (Khalafalla and Abdelazim 2017); (d) peste des petits ruminants (PPR) in Iran (Zakian et al. 2016); (e) contagious ecthyma in and Sudan (Azwai et al. 1995); and (f) Rift Valley fever in the Arabian Peninsula (Balkhy and Memish 2003). In that frame, the major zoonotic diseases in dromedary camels are Middle East respiratory syndrome (MERS), brucellosis, *Echinococcus granulosus* and Rift Valley fever (Zhu et al. 2019). Known camel pathogens and their diagnostic methods are briefly summarized in Table 6.1. For those diseases, often very contagious, only national or international prevention plans can be proposed to avoid their dissemination among camel farms from different countries and farming systems. The camel owners have a low ability to limit the suffering of their animals. In that sense, welfare policy is based on TAD surveillance, collective prevention policy, vaccination campaigns if any and massive antiparasitic prophylaxis (see, e.g. Surra, Diall et al. 2022)

Regarding diseases with local impact, most of them are multifactorial diseases and are depending on individual farm management. These health disorders can be managed not only by individual treatment based on visible symptomatology but also by the identification of the main risk factors and their elimination or alleviation. In such cases, the eco/pathological approach could be useful (Faye et al., 1999). However, diseases with an identified pathogen (mono-factorial diseases) can also occur in camel farms and affect one or several animals in the herd, requiring individual and specific treatment. Those diseases can have a systemic effect with general clinical signs expressing a global discomfort of the animals (weariness, loss of appetite, fever) or affect one specific organ leading to functional disorder (reproduction, respiration, digestion, etc.).

Table 6.1 List of the most common pathogens reported in dromedary camels over the past 20 years

Pathogen	Taxonomic classification	Pathogen characterization	Diagnosis	Reference(s)
Acinetobacter spp.	Bacteria	Gram negative	Culture, PCR	Adugna et al. (2013)
Anaplasma phagocytophilum	Bacteria	Gram negative	Serology, PCR	Bahrami et al. (2018)
Bartonella spp.	Bacteria	Gram negative	PCR, culture, serology	Ghaemi et al. (2019)
Blastocystis	Protozoa	N/A	PCR, fecal	Sazmand et al. (2019)
Blastomyces	Fungi	N/A	Culture, histopathology	Al-Ani and Roberson (2005)
Borrelia burgdorferi	Bacteria	Gram negative	Serology, PCR	Said et al. (2016)
Botulinum toxin	Bacteria	Gram positive	PCR, ELISA, CFT	Hussein (2021)
Bovine herpes virus-1 (BHV-1)	Virus	N/A	Ag-ELISA	Intisar et al. (2009), Benaissa et al. (2020c)
Clostridium difficile	Bacteria	Gram positive	PCR, culture	Rahimi et al. (2014a, b)
Coxiella burnetii	Bacteria	Gram negative	PCR, serology	Benaissa et al. (2017)
Cryptosporidium	Protozoa	N/A	Fecal, ELISA, R-T PCR, PCR	Wang et al. (2021)
Dermatophytes (*Trichophyton* and *Microsporum*)	Fungi	N/A	Culture, ELISA, PCR	Almuzaini et al. (2016)
Echinococcus	Helminth	N/A	Imaging/ surgery, serology/PCR-RFLP	Zhu et al. (2019)
Ehrlichia spp.	Bacteria	Gram negative	PCR, serology, blood smear	Younan et al. (2021)
Escherichia coli	Bacteria	Gram negative	Culture, PCR	El Wathig and Faye (2016)
Fasciola hepatica	Helminth	N/A	Fecal, serology	Ouchene-Khelifi et al. (2018)
Giardia spp.	Protozoa	N/A	Fecal, PCR	Sazmand et al. (2019)
Hepatitis E virus	Virus	SS positive-sense RNA	Serology, PCR	Woo et al. (2014)
Influenza A(H1N1) and A(H3N2)	Virus	SS negative-sense RNA	Serology, PCR	Chu et al. (2020)
Leptospira spp.	Bacteria	Gram negative	Serology, culture, PCR	Doosti et al. (2012)

(continued)

Table 6.1 (continued)

Pathogen	Taxonomic classification	Pathogen characterization	Diagnosis	Reference(s)
Listeria monocytogenes	Bacteria	Gram positive	Serology, culture, PCR	Rahimi et al. (2014a, b)
Middle East respiratory syndrome coronavirus (MERS-CoV)	Virus	N/A	Serology, PCR	Khalafalla and Hussein (2021)
Mycobacterium caprae and *Mycobacterium bovis*	Bacteria	Acid fast	(CIT) test, Serology, histopathology, PCR	Narnaware et al. (2015), Infantes-Lorenzo et al. (2020)
Onchocerca cervicalis	Helminth	N/A	Histopathology	Mirzaei et al. (2018)
Parainfluenza-3 virus (PI-3)	Virus	N/A	Ag-ELISA, RT-PCR	Intisar et al. (2009)
Rabies	Virus	SS negative-sense RNA	IHC, histopathology	Ahmed et al. (2020)
Respiratory syncytial virus (RSV)	Virus	N/A	Ag-ELISA, RT-PCR	Saeed et al. (2015)
Rickettsia spp.	Bacteria	Gram negative	PCR, serology, cytology	Sazmand et al. (2019)
Staphylococcus spp.	Bacteria	Gram positive	Culture, PCR	Alebie et al. (2021)
Streptococcus equi subsp. *zooepidemicus*	Bacteria	Gram positive	Culture, PCR	Younan et al. (2005)

6.2 Some Systemic Diseases

Systemic diseases are leading to general symptoms causing important discomfort to the animal but not necessarily specific pain. The causes of those systemic diseases are highly variable. Haemoparasites notably are very common in camel. The most common is camel trypanosomosis caused by *Trypanosoma evansi* (Benaissa et al. 2020b). The tabanids and stable flies are important vectors of *T. evansi*. Since the early twentieth century, camel trypanosomosis prevalence rate has been elevated in North Africa and the Middle East. Large outbreaks occurred in Algeria in 2005–2006 (Boushaki et al. 2019; Benaissa et al. 2020b), Saudi Arabia (El Wathig and Faye 2013) and metropolitan France (Desquesnes et al. 2008) because a major resurgence of haemoparasites occurred during the humid season in multiple countries. Affected camels are very meagre and become very weak.

Anaplasmosis is also a frequent haemoparasitosis. Serological and molecular evidence of infection of dromedary camels with *Anaplasma* spp. were reported in Saudi Arabia (Alshahrani et al. 2020) and Iran (Sharifiyazdi et al. 2017). Camels may serve as reservoirs of *A. platys* (Lorusso et al. 2016). More recently, a much higher prevalence of piroplasmosis and anaplasmosis was recorded among Somalian camels (Abdalla et al. 2017). Blood pathogens detected in camels include also *Theileria equi*, *Theileria mutans*, *Theileria ovis*, *Babesia caballi* and *Babesia behnkei* (Bahrami et al. 2017). All these parasites could provoke global weakness of the camel. However, it is difficult without laboratory analysis to distinguish those haemoparasitoses from some protozoan diseases as toxoplasmosis (caused by *Toxoplasma gondii*), a worldwide significant public health disease. The camel seroprevalence of *T. gondii* widely varies among regions and countries (Tonouhewa et al. 2017). Some bacterial diseases can be also systemic as Q-fever. High Q-fever (*Coxiella burnetii*) seroprevalence has been reported in camels from many countries in Africa, Asia and the Middle East (Devaux et al. 2020).

Some of these diseases can be transmitted by ticks. Tick infestation, when the abundance of these external parasites is high, can provoke important discomfort in the animals and even anaemia due to blood withdraw. In camels, vector-borne diseases transmitted by ticks are frequent, causing considerable economic losses due to the costs of medical treatments, reductions in meat and milk production, abortions and infertility (Selmi et al. 2022). Moreover, some pathogens express serious zoonotic risks (Bellabidi et al. 2020). Finally, it is admitted that global climate change may exacerbate the threat of both spatial and temporal spread of these vector-borne pathogens (Ogden and Lindsay 2016).

However, some systemic diseases can provoke highly painful symptoms. It is the case of caseous lymphadenitis (CLA), a common, highly contagious bacterial disease caused by *Corynebacterium pseudotuberculosis* (formerly *C. ovis*) which affects camels as well as sheep and goats. The disease has been reported in dromedary camels in Sudan, Egypt, Somalia, Kenya, Ethiopia, Saudi Arabia, the UAE, Jordan, Iran, India, Kazakhstan and Australia (Wernery and Kinne 2016; Borham et al. 2017).

6.3 Diseases and Disorders with Functional Impact

6.3.1 Respiratory Diseases

Respiratory disease in camels has typically multifactorial aetiology with different clinical manifestations. Provoking breath difficulties, they have a direct impact on the discomfort of the animals. The main causing pathogens can be virus infections as influenza A (Chu et al. 2020) or MERS-CoV (Khalafalla et al. 2015) which produce severe respiratory symptoms including mucous nasal and ocular discharge, bronchitis, dry cough, pneumonia and fever accompanied by mortality (Khalafalla and Hussein 2021). In many studies, however, other respiratory viruses have been detected without clinical respiratory signs (Intisar et al. 2009; Saeed et al. 2015;

Benaissa et al. 2020c). Secondary bacterial infection is also common, especially with bacteria that are prevalent in the environment such as *Streptococcus equi*, *Staphylococcus aureus*, *Pasteurella multocida*, *Streptococcus pyogenes*, *Pseudomonas* spp., *Corynebacterium* spp., *Escherichia coli*, *Haemophilus* spp., *Proteus* spp., *Klebsiella pneumoniae*, *Arcanobacterium pyogenes*, *Pneumococcus* spp. and *Enterobacter* spp. (Khalafalla and Hussein 2021).

6.3.2 Reproductive Failures

The reproductive failures have a direct impact on one of the most important functions in animal breeding: reproduction. They are not necessarily provoking pain or discomfort, but they can directly cause low fertility and consequently have economic impact and lead to early culling (Al-Qarawi 2005). Malnutrition, infections, congenital defects, management errors and ovulatory or hormonal imbalances can all result in low fertility (Benaissa et al. 2015). Analysis of reproductive abnormalities from more than 7300 barren female dromedaries demonstrated that clinical endometritis, ovarian hydrobursitis and vaginal adhesions were the most common clinical findings (Ali et al. 2015). For male camels, *impotentia generandi* has instead been reported as the most common cause of male infertility (Ali et al. 2014). Infectious causes of reproductive failure include clostridial, streptococcal and parasitic myositis (Waheed et al. 2014).

Different specific diseases can also provoke genital lesions in both male and female or abortion (brucellosis, chlamydiosis, Rift Valley fever, salmonellosis, Q-fever). Those diseases can be frequent. For example, serological prevalence of *Chlamydia abortus* in dromedary camels was estimated at 19.6% in the UAE (Zaher et al. 2017); 4.25% in Algeria (Benaissa et al. 2020a); 19.4% and 10.05 in Saudi Arabia (Hussein et al. 2008; Khalifa et al. 2018), respectively; and 7.6% in Tunisia (Burgmeister et al. 1975). Among the most painful reproductive disorders in female camels, uterine prolapse can provoke important pain and, in case of rupture of the uterine mucosa, can lead to the death of the animal (Gutierrez et al. 2001).

6.3.3 Udder Troubles

Mastitis, like in other dairy animals, is one of the most important production diseases in dairy camels. It is generally admitted that one of the main symptoms of mastitis is a painful udder.

Mastitis can be divided into groups depending on clinical signs; subclinical mastitis, acute mastitis and chronic mastitis. Subclinical mastitis usually causes lower loss of productivity and less severe symptoms. Clinical signs associated with acute mastitis are severe pain on palpation and swollen mammary tissue with varying consistency ranging from firm to moderately soft. Mastitis was present in 45.66% of animals in one camel population (Aqib et al. 2022). Somalia is the country with the lowest prevalence (16%; Mohamud et al. 2020), whereas the highest prevalence (90.5%) was reported in Pakistan (Qamar et al. 2011). The average percentage of she-camels affected with mastitis in other countries ranged from a maximum prevalence of 57.5% (in Pakistan) to a minimum of 42.8% (in Ethiopia).

Camel udder can be also affected by different lesions or inflammations (teat tear, oedema, crushing) having usually very painful consequences during milk ejection.

6.3.4 Digestive Disorders

Most of the abdominal pains are linked to disorders of the digestive tract. Diarrhoea, colic, stopping of rumination, intestinal obstructions, ruminal acidosis, simple indigestion, enteritis, bloat, constipation and foreign bodies, linked to gastrointestinal infectious diseases or not, can be observed in camels (Megersa 2010; Kumar et al. 2014). Parasitic, metabolic or infectious diseases can also affect essential abdominal organs as liver.

6.3.5 Skin Diseases

Camels are especially sensitive to skin disorders (ringworm, mange, dermatosis, abscess, etc.) provoking uncomfortable itching (Kamili et al. 2019). Infectious diseases affecting skin in the face (camel pox, papillomatosis, contagious ecthyma) cause epidermal lesions around the eye, nose and mouth. These diseases cause physical insult leading to difficulties in browsing/foraging, mastication, chewing and swallowing forage and dry foods. If those problems are severe enough, anorexia will occur.

Moreover, the flank of camels is without skin muscle and the tail is short contrary to horses or cattle. So, in case of abundant attacks by biting flies, it is more difficult for the animal to keep the aggressors away.

6.3.6 Nutritional and Metabolic Diseases

It is, generally, complex diseases, due to lack or excess of one or several components of the diet. Mineral deficiencies in camels are known to lead to infertility, non-infectious abortion, anaemia and other metabolic diseases including in camel (Abdelrahman et al. 2022).

Among mineral deficiencies in camels, the most common, especially in Middle-East, is selenium deficiency, responsible for the 'white muscle disease' (WMD), i.e. nutritional muscular dystrophy or dystrophic myodegeneration. This disease leads to a peri-acute to subacute degenerative myocarditis and discolouration of the skeletal muscle which can lead to death (Faye and Seboussi 2009). Clinical manifestations occur mainly in young animals but can also occur in older camels. Clinical disorders attributed to a possible deficit of selenium vary from mild stiffness to obvious pain upon walking to an inability to stand (Faye and Bengoumi 2018). Vitamin D and calcium deficiency could provoke many bone disorders in camelids, which leads to skeletal disorders (El Khasmi and Faye 2011); in severe cases it can result in neuromuscular, cardiovascular and kidney failures (Faye and Bengoumi 2018).

In male camels, urolithiasis is a relatively common metabolic disease typically caused by an imbalance of dietary calcium and phosphorus levels. Among toxic diseases, copper intoxication is the most important, but excess intake of selenium is also problematic (Seboussi et al. 2012). Dromedary camels are extremely sensitive to the toxic effects of Narasin (Abu Damir et al. 2013) and salinomycin (Wernery et al. 1998).

Bhat et al. (2013) noticed that metabolic disorders had an increased frequency when large groups of camels were fed fibre-deficient rations or when camels were kept in temperate climates with low levels of solar radiation or during periods of drought (Uhl 2018). In harsh environmental conditions, camels that suffered from hunger and thirst developed nutritional deficiencies and stress, which are predisposing risk factors causing metabolic diseases such as hypocalcaemia and pregnancy toxaemia, as well as infectious diseases.

6.3.7 Musculoskeletal Disorders, Injuries and Miscellaneous Diseases

Lameness, fractures and leg injuries affecting camels may lead to discomfort and pain at the individual level, which in turn results in a high economic loss to the camel industry. Lameness is caused mainly by foot disorders (59.05%) and fetlock and metacarpus (MC)/metatarsal (MT) disorders (40.94%) (Mostafa 2020). In racing camels, lameness related to leg weakness is frequent (9.39% incidence; Gahlot 2007). The principal causes of lameness are cut padfoot (43.1%) and laminitis (Gahlot 2007; Faye et al. 2022). Lameness is associated with signs of pain, reduced appetite, restlessness, altered standing and moving patterns, including limping, stiffness and reduced flexion of the affected leg (Al-Juboori 2013).

Most studies of camel injuries have related their incidence to the type of transport and restraint. These and other inadequacies of handling can result in injuries, foot lameness, tail-tip necrosis and various diseases (Gregory and Grandin 2007). In the mating season, fighting between males causes injuries; such behaviours must be reduced. Hump and back injuries are also frequent with misaligned saddles or packs.

6.4 Surgery Intervention, Medication and Welfare

Generally, castration is a painful procedure in camel pastoral communities. Usually, two main technics are applied: (1) bloody castration with a skin incision to extract the testicle and (2) non-bloody castration by crushing the testicular cord. Many animals' welfare over the world requires that castration be done with the appropriate use of anaesthesia and analgesia so that pain and stress are minimized (Stafford and Mellor 2005). Castration techniques that have been developed in other food animal production species may not be transferable because camels have quite different anatomy of the external reproductive system compared with other mammalian species and camels also exhibit a slow and variable rate of maturation of their reproductive system (Tibary and Vaughan 2006). Chemical castration is not yet developed in this species.

Branding is the most common method of identification and establishment of ownership of camels; it is used also as therapy methods (Dioli 2022). Concern is rapidly growing worldwide by various associations regarding the impact of branding on animal welfare, and branding is increasingly being discouraged and/or recommended to be used together with pain relief therapy for the animal (Jepson 2012). The practice of branding/firing/thermocautery as a treatment for pathological conditions of animals is used in many countries (Dioli 2022), but all medical practices/procedures that are painful and cause very visible injury should be considered detrimental/unethical practice and should be discontinued. Alternative identification of animals by ear tags, bolus or subcutaneous implants (electronic chips) can be suggested although all the techniques require coercive manipulation of the animal (Caja et al. 2016).

The medications commonly used in veterinary clinics are not licensed for camelids. Data on the safety and efficacy of these medications are sparse, and for that reason, their use should be discussed. Notably, the pharmacokinetic of drugs are different in camels than in bovine and other ruminants (Ali et al. 1996). So, medical treatments should be the result of an understanding of the physiological and pharmacological differences between camels and other domestic species.

6.5 Pain Recognition and Prevention

The procuration of a good state of welfare in animals passes through the early recognition and alleviation of pain, distress and suffering. The experience and dimensions (intensity, frequency, duration and quality) of such conditions can be assessed by coupling the observation of objective and subjective behavioural expressions with diverse measures of physiological function (Rutherford 2002). Facial expressions, vocalizations, bodily movements and posture, nonspecific behaviours and general signs of organic disorders are integrated in a variable proportion into multidimensional evaluation scales. Additionally, the level of genomic damage associated with physiological stress conditions can be assessed by the buccal micronucleus assay, a relatively low-cost and non-invasive technique (Santovito et al. 2022). These assessment tools lead to the characterization of the physical and psychological discomfort that an individual animal may be experiencing and the programming of alleviation techniques (Ashley et al. 2005). Further, this conglomerate of information allows the critical appraisal of the acceptability of some practices on account of their measurable impact on animal health and welfare.

For the particular case of camels, no sensitive and specific pain-scoring systems are available. Hence, the welfare of camels might be compromised given the potential perpetuation of neglected practices and the disregard of animal health indicators, which are sometimes challenging to identify since camels do not overtly express their affective state. Similarly to donkeys, camels have evolved as stoic animals that express signs of fear, pain and distress in subtle ways. In fact, a sensible identification of such conditions tends to occur when the animal is at an advanced degree of disease/pain, which renders the palliative efforts impractical (Orth et al. 2020). The larger content of metenk neurons throughout the superior colliculus of the camel brain in comparison with other mammal species has been proposed to be associated with a pain-inhibiting opioid pathway in these animals. Such an anatomical particularity may be an additional life-history adaptive trait in response to extreme temperatures and discomfort in harsh desert environments (Mensah-Brown and Garey 2006). Contemporaneous studies have further documented the anatomy of pain in camels with the identification of nociceptive markers in the lumbar dorsal root ganglion and spinal cord (Javed et al. 2021) and the examination of neuroepithelium at the nasal septal island (Abo-Ahmed et al. 2021) in dromedaries.

In this scenario, based on the evidence that most mammalian species display similar behavioural and physiological responses to pain/discomfort (Hay et al. 2003; Ashley et al. 2005; Sneddon et al. 2014; de Oliveira and Keeling 2018; McLennan et al. 2019), we selected a list of behaviours and physiopathological parameters that could serve as reliable indicators to identify and manage pain/discomfort statuses in camels. Specifically, some of the body language signals and facial expressions compiled as potentially related to pain/discomfort statuses in dromedary camels are illustrated in Fig. 6.1 to provide the evaluators with a clear visual differentiation between them. Moreover, a composite pain scale (Table 6.2) has been conceived based on the literature available on other species.

A composite pain scale is composed of several items, which are scored in two- or three-point scales. The ordinal scale is individually designed for each potential indicator so that it can best capture the variability at intensity, frequency, duration and quality of pain/discomfort. Pain scales have succeeded in the identification of behavioural and physiopathological indicators of pain/discomfort in laboratory animals (Langford et al. 2010; Sotocina et al. 2011; Keating et al. 2012; Häger et al. 2017), companion animals (Morton et al. 2005; Thompson 2013; Dunbar et al. 2016; Evangelista et al. 2019), livestock species (Stafford 2013; Gleerup et al. 2015a, b; Di Giminiani et al. 2016; Guesgen et al. 2016; Orth et al. 2020) and wildlife species (Machin 2013; Posner and Chinnadurai 2013; Reijgwart et al. 2017; MacRae et al. 2018). The proposed composite pain scale for camels needs scientific studies to be applied, validated and refined. At a practical level, when the scale is going to be applied, the registration of some of the physiopathological indicators currently proposed could result not be feasible given the fact that close contact with some animals could not be possible or even dangerous for the operators. So they may be excluded in the future refined scale. Finally, remote and non-invasive vital monitoring using different sensor devices could be suggested to assess better the

Fig. 6.1 Ethogram depicting body postures and facial expressions to be assessed during evaluation of pain/discomfort status in dromedary camels

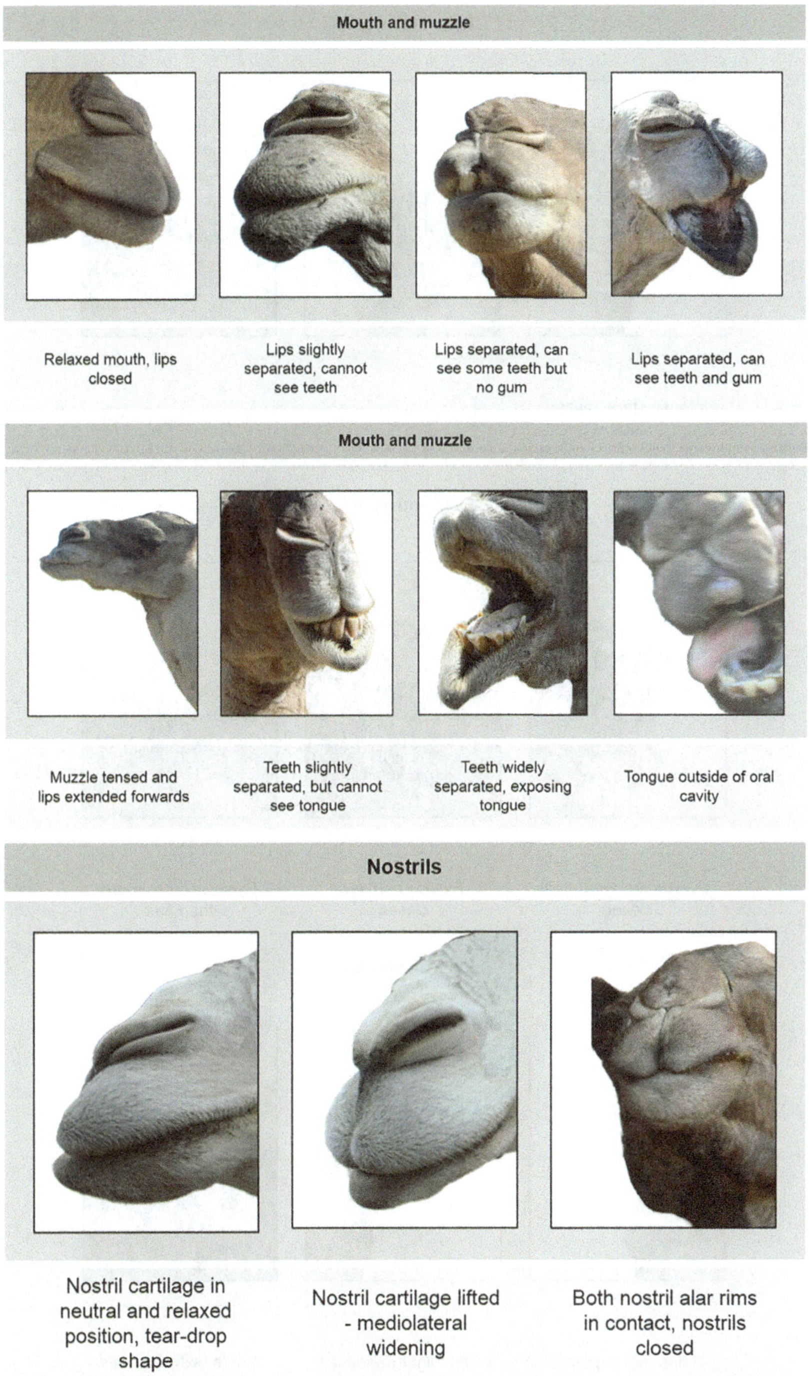

Fig. 6.1 (continued)

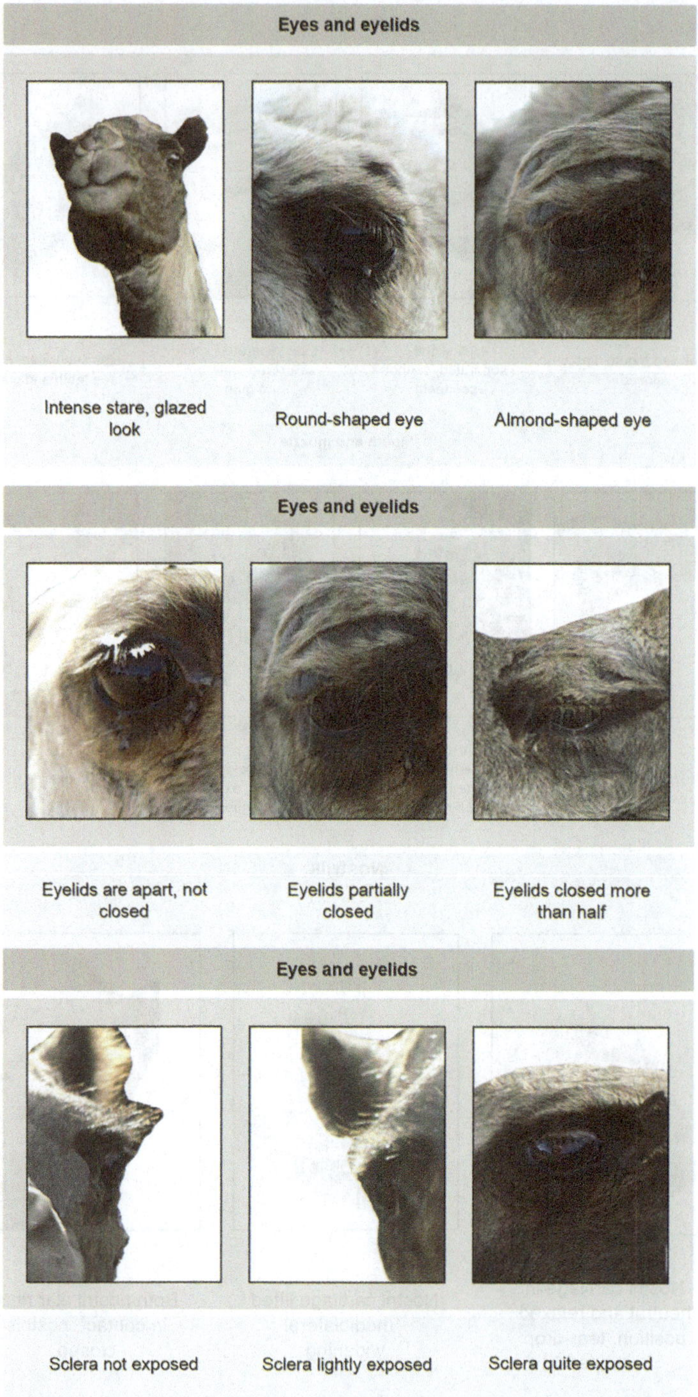

Fig. 6.1 (continued)

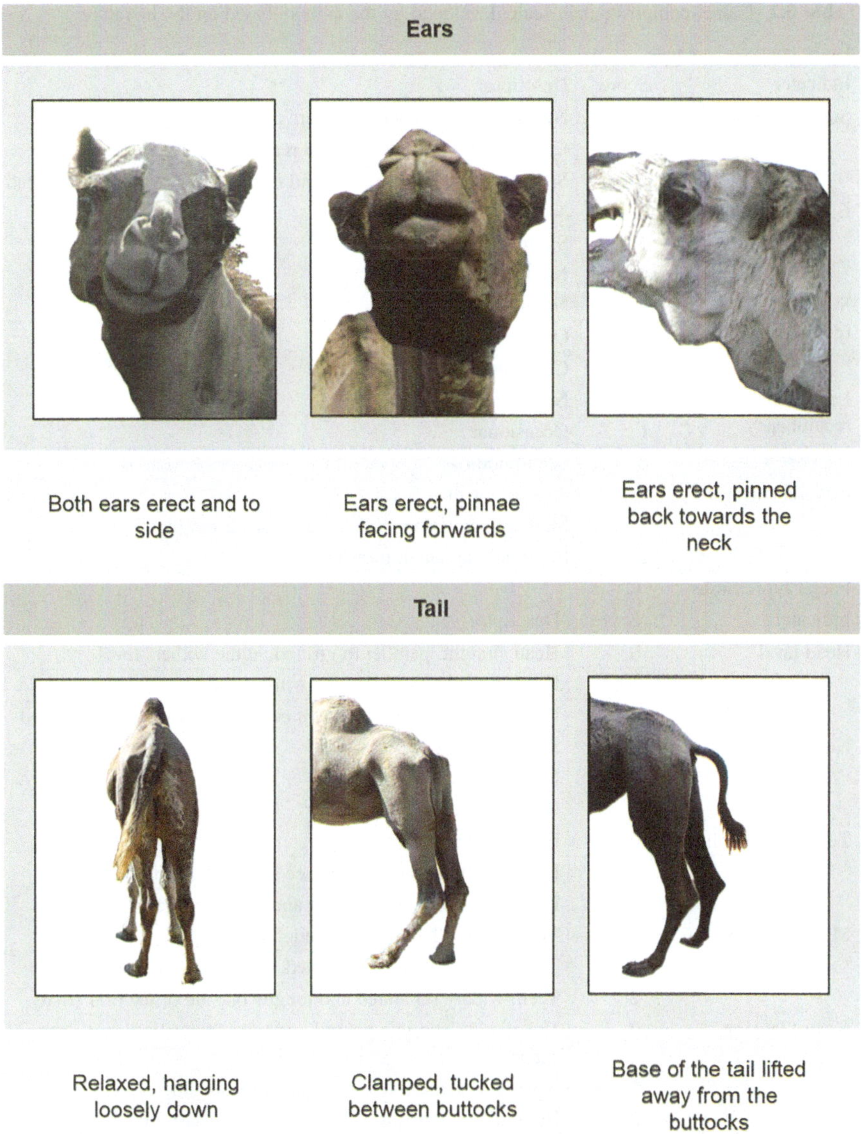

Fig. 6.1 (continued)

Table 6.2 Camel composite pain scale developed by the authors based on the literature

Body language signals		
Indicator	Score	Descriptor
Standing posture	0	Normal, neutral stance: legs are approximately vertical
	1	Canted-in stance: the fore and rear hooves are closer together
	2	Splayed-out stance: the fore and rear hooves are relatively spread
Back position	0	Normal
	1	Slightly arched back
	2	Notably arched back
Sternal recumbency	0	None
	1	Occasional
	2	Continuous
Lateral recumbency	0	None
	1	Occasional
	2	Continuous
General comfort	0	Asleep or calm
	1	Moderate agitation: restless and uncomfortable
	2	Extremely agitated: thrashing
Facial expressions		
Indicator	Score	Descriptor
Head level	0	Head straight, parallel to ground, at the withers level
	1	Head line over or below the withers level
	2	Neck and head are bent down completely towards the ground
Neck position	0	Semi-flexed
	1	Neck extension
	2	Neck retraction
Tension of the lips	0	Relaxed mouth, lips closed
	1	Lips slightly separated, can see some teeth but no gum
	2	Lips separated, can see teeth and gum
Muzzle tension	0	Muzzle tensed and lips extended forwards
	1	The teeth are slightly separated, but the tongue is not visible
	2	Teeth widely separated and tongue outside of the oral cavity
Nostril tension	0	Nostril cartilage in a neutral and relaxed position; tear-drop shape
	1	Nostril cartilage lifted; mediolateral widening
	2	Both nostril alar rims in contact, nostrils closed
Eye shape	0	Intense stare, glazed look
	1	Round- or almond-shaped eye
	2	Eye narrow shape
Orbital tightening	0	Orbital tightening is not present; eyelids are apart, not closed
	1	Orbital tightening is moderately present; eyelids are partially closed
	2	Orbital tightening is obviously present; eyelids are closed more than half
Sclera exposition	0	Sclera not exposed

(continued)

Table 6.2 (continued)

Facial expressions		
Indicator	Score	Descriptor
	1	Sclera moderately exposed
	2	Sclera obviously exposed
Ears position	0	Ears erect, pinnae facing forwards
	1	Ears erect pinned back towards the neck
	2	Both ears are erect and to the sides
Tail position/ movement	0	Tail relaxed, hanging loosely down
	1	Tail clamped, tucked between buttocks
	2	The base of the tail lifted away from the buttocks

Non-specific behaviours		
Indicator	Score	Descriptor
Spontaneous movement/locomotion	0	Moves freely, normal gait
	1	Staggering/laboured gait
	2	Reluctance to move
Attention towards the surroundings	0	Active, attentive, curious
	1	Quiet/depressed
	2	Avoiding eye contact may move away from the observer
Response to approach	0	Looks at observer, head and ears movement, occupied with any other activity
	1	Looks at the observer, head and ears hardly move, and leave slowly when approached
	2	May/may not look at the observer, head an ears hardly move, leaves when slightly approached
Social isolation from congeners	0	No clear signs of social isolation
	1	Avoids active behaviours of congeners
	2	Avoids any contact with congeners; complete social isolation
Resistance to handling	0	Looks at/moves to handler, no offers opposition when touched
	1	Offers opposition when handled, may vocalize
	2	Violent reaction when handled
Aggressiveness	0	None
	1	Displays aggressive behaviours only when approached
	2	Displays aggressive behaviours continuously
Piloerection	0	None
	1	Occasional
	2	Continuous
Attention towards the painful area	0	Does not pay attention to the painful area
	1	Brief attention to the painful area
	2	Biting, nudging or looking constantly at the painful area
Pawing	0	None
	1	Occasional
	2	Continuous
	0	None

(continued)

Table 6.2 (continued)

Non-specific behaviours		
Indicator	Score	Descriptor
Nibbling inanimate elements	1	Occasional
	2	Continuous
Vacuum chewing	0	None
	1	Occasional
	2	Continuous
Self-mutilation	0	None
	1	Occasional
	2	Continuous
Body rubbing/ scratching	0	None
	1	Occasional
	2	Continuous
Body licking	0	None
	1	Occasional
	2	Continuous
Head pressing	0	None
	1	Occasional
	2	Continuous
Tooth grinding	0	None
	1	Occasional
	2	Continuous
Vocalizations	0	Quiet
	1	Vocalizes only when approached closely
	2	Intermittent vocalization

Behavioural indicators of specific pain		
Indicator	Score	Descriptor
Abdominal pain		
Rolling	0	None
	1	Occasional
	2	Continuous
Stretching out	0	None
	1	Occasional
	2	Continuous
Kicking at abdomen	0	None
	1	Occasional
	2	Continuous
Flank watching	0	None
	1	Occasional
	2	Continuous
Tenesmus	0	None
	1	Occasional
	2	Continuous
Limb and foot pain		

(continued)

Table 6.2 (continued)

Behavioural indicators of specific pain		
Indicator	Score	Descriptor
Weight shifting between limbs	0	Standing up with foot in intermittent contact with the ground
	1	Standing up with the same leg always hitched
	2	Frequent shifting of weight on the four limbs
Limb guarding	0	None
	1	Occasional
	2	Continuous
Abnormal weight distribution	0	None
	1	Occasional
	2	Continuous
Pointing, hanging and rotating limbs	0	None
	1	Occasional
	2	Continuous
Abnormal movement	0	None
	1	Occasional
	2	Continuous
Attempts to lie down	0	None
	1	Occasional
	2	Continuous
Head and dental pain		
Headshaking	0	None
	1	Occasional
	2	Continuous
Altered eating behaviour	0	None
	1	Occasional
	2	Continuous
Physiopathology		
Indicator	Score	Descriptor
Heart rate Physiological range = 35–50 beats/min	0	Normal compared to basal level (increased/decreased by < 10%)
	1	11–50% greater/lesser than physiological standard
	2	>50% greater/lesser than physiological standard
Respiratory rate Physiological range = 9–16 breaths/min	0	Normal compared to basal level (increased/decreased by < 10%)
	1	11–50% greater/lesser than physiological standard
	2	>50% greater/lesser than physiological standard
Visual inspection of respiratory rate	0	Normal
	1	Mild abdominal assistance
	2	Marked abdominal assistance
Blood pressure Physiological range = 76–115 mmHg	0	0–10% greater/lesser than physiological standard
	1	11–50% greater/lesser than physiological standard
	2	>50% greater/lesser than physiological standard

(continued)

Table 6.2 (continued)

Physiopathology		
Indicator	Score	Descriptor
Rectal temperature Physiological range = 34–40 °C	0	Normal compared to basal level (variation <0.5 °C)
	1	Variation between 1 and 2 °C
	2	Variation > 2 °C
Colour of local mucosal area(s)	0	Normal colour (rosy red)
	1	Slightly red/pallid
	2	Notably red/pallid
Mucosal ulceration	0	None
	1	Ulceration without bleeding
	2	Ulceration with bleeding
Capillary refill time Physiological range = <2 s	0	Normal
	1	Increased time (2–3 s)
	2	Increased time (>4 s)
Straining to urinate	0	Normal urination
	1	Urinary moderate hesitation; the animal has difficulty starting or maintaining a urine stream
	2	Adopts position but pee barely comes out
Dyssynergic defaecation	0	Normal defecation
	1	Moderate hesitation to defaecate
	2	Adopts position but faeces barely comes out
Faeces consistency	0	Round-formed, firm, normal
	1	Semi-solid
	2	Liquid to watery
Muscle twitching	0	No spasms
	1	Spasms induced only by stimulation
	2	Occasional spontaneous spasms and easily induced spasms
Drooling	0	Never drools
	1	Moderate drooling; only lips wet
	2	Profuse drooling; drool drips off the mouth
Ocular discharge	0	No presence of ocular discharge
	1	Moderate ocular discharge; only eye and eyelids wet
	2	Profuse ocular discharge; drips off the eye and eyelids
Nasal discharge	0	No presence of nasal discharge
	1	Moderate nasal discharge; only nostrils wet
	2	Profuse nasal discharge; drips off the nostrils
Pupil diameter	0	Normal pupils
	1	Slightly constricted/dilated pupils
	2	Severe constriction/dilation of pupils
Enophthalmos or 'sunken eyes'	0	None
	1	Moderately present
	2	Obviously present
Exophthalmos	0	None
	1	Moderately present

(continued)

Table 6.2 (continued)

Physiopathology		
Indicator	Score	Descriptor
	2	Obviously present
Wound infection/ suppuration	0	Normal healing
	1	Erythema and other signs of inflammation
	2	Pus/purulent discharge

level of pain minimizing the human-camel interactions in the future, as currently happening in other species (Wang et al. 2020).

From a laboratorial point of view, the concentrations of hormones (in blood, urine or saliva; ACTH, glucocorticoids, adrenaline and noradrenaline), metabolites (in blood; glucose, lactate and free fatty acids) and inflammatory markers (in blood; haptoglobin and fibrinogen) constitute additional valuable parameters for the measurement of pain/discomfort. However, composite pain scales and welfare assessment tools do not include any laboratory tests, and this is the reason why they have not been included in the proposed one.

6.5.1 Applications of Thermography for Pain Detection and Monitoring

It is well known that the detection of cutaneous temperature changes due to functional variations in local microvasculature helps with the monitoring and follow-up of surgical and therapeutic processes in which analgesia is provided.

Apart from that, infrared thermography can constitute a valuable tool for the investigation of animal behaviour in response to pain and negative stimuli (Casas-Alvarado et al. 2020). Concretely, it would permit the non-invasive, preliminary evaluation of some behaviours and postural expressions that could be misinterpreted if appraised in a blinded manner (Fig. 6.2). Although thermography detects changes in surface temperature, it does not mean clinicians are only able to diagnose dysfunction in superficial organic structures with this technique. Indeed, the pathology of visceral organs is reflexed through surface temperature changes in the area of skin with which shares the autonomic innervation ('referred pain'). In this scenario, it is important to take into consideration the exact segmental relationships between visceral and cutaneous sensory distribution when mapping out referred pain (Goodman and Marshall 2015). However, while there are still no studies on the use of thermography in camels, this methodology should be considered promising to assess health and welfare in this species.

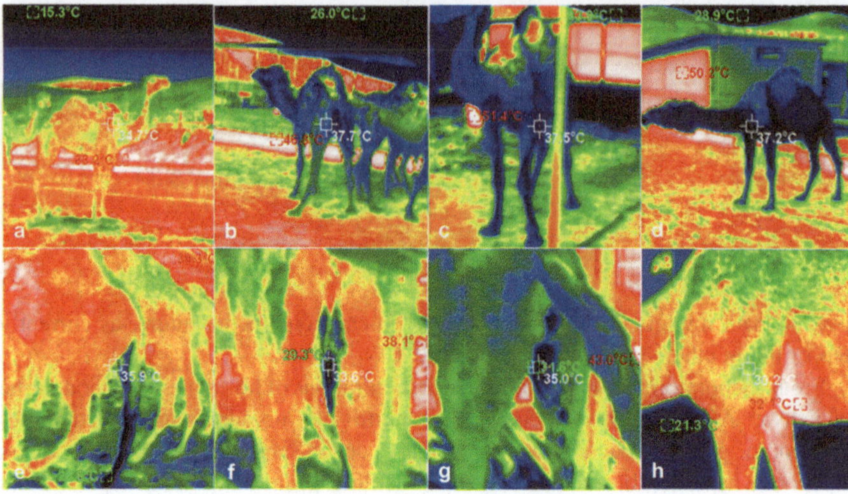

Fig. 6.2 Infrared thermographic images of dromedary camels with normal and altered stance/aplombs and surface temperature changes. (**a**) Dromedary camel with balanced normal square stance and homogeneous distribution of surface temperature. (**b–d**) Dromedaries with base wide stance and aplomb defects but not apparent local changes in surface temperature. (**e**) Dromedary camel with altered stance (left rear limb angled forward) and temperature changes on the distal part of the right rear limb. (**f, g**) Dromedaries with surface temperature changes in the dorso-medial part of the thigh. (**h**) Dromedary camel with surface temperature changes along the cardiac zone and the medial part of forelimbs

6.5.2 Potential Risk Factors for the Poor Physical and Psychological Health of Dromedary Camels

The concept of animal welfare recognizes that an animal's positive or acceptable welfare status draws on the continuous care of its physical health and emotional/mental state (Ohl and Van der Staay 2012). The promotion of both aspects in dromedary camels can be targeted by the early and effective identification and alleviation of the following risk factors/practices (Le Neindre et al. 2009; World Organisation for Animal Health 2022):

1. **Potential sources of camel pain/discomfort associated with farming practices:**
   (a) Veterinary care/training:
      (I) Inappropriate veterinary attention leads to unsuccessful diagnosis, treatment and prevention of organic dysfunction/pathology, as well as increased risk of disease transmission.
      (II) Improper veterinary skills could lead to increased pain and suffering in animals subjected to surgery practices that are routinely performed for management practices (i.e. castration).

(III) Lack of biosecurity plans impedes the effective control of the major sources and pathways for spread of pathogenic agents.

(b) Physical environment:

(I) Inadequate maintenance of fences (i.e. sharp edges and protrusions) or the abuse of electric fencing can compromise physical wellbeing of animals.

(II) The absence of shaded and shelter areas compromise the thermal comfort of camels.

(III) Not well-drained and relatively concrete substrates (walking and resting surfaces) will increase the risk of injury and transmission of diseases and difficult adequate resting behaviour.

(IV) Low space allowance or high stocking density does not allow animals to easily stand up, lie down, turn around, adopt normal resting postures and display natural behaviours.

(V) Indoor environments increase the risk of airborne disease transmission. In this regards, it is crucial to control air quality, temperature and humidity conditions to avoid potential alterations of the camel respiratory health and thus increased susceptibility to respiratory disease.

(VI) Confined camels that have limited access to natural light need to be provided with supplementary lighting to maintain circadian rhythmicity.

(VII) Relative discomfort can appear when camels have not the chance for grazing at free ranging areas. Foraging behaviour and moderate physical exercise should be promoted.

(VIII) The continuous exposure to sudden or loud noises should be minimized as much as possible to prevent stress and fear.

(c) Handling/restraining protocols:

(I) Owners and handlers without sufficient skill and knowledge of camel husbandry, behaviour, biosecurity and general signs of disease/pain/discomfort could be neglecting animal health and wellbeing.

(II) Tattooing or ear-tagging for identifying animals are painful stimuli. Alternative identification systems such as microchip implants are preferable.

(III) Inserting nose rings to avoid undesirable behaviours can cause severe stress in animals.

(IV) Hydraulic, pneumatic and manual restraining devices should be adjusted to the size of camel to be handled to prevent injuries (i.e. excessive pressure).

(V) Mechanical and electrical devices should not be overused for management practices on farm.

2. **Potential sources of camel pain/discomfort associated with genetic selection schemes:**

The historical selection of animals based solely on high levels of productivity and quality of products has brought about negative consequences on the metabolism, reproduction, behaviour and/or the general health of the animals. The underlying reason is the existence of pleiotropic genes that influence two or more apparently unrelated phenotypic traits. In fact, certain disorders that have a multifactorial aetiology and are partly heritable appear with higher incidence in genotypes selected for productive traits (Le Neindre et al. 2009).

Under this paradigm, in addition to productivity, welfare and health parameters (i.e. pathogen resistance, tolerance to heat/cold stress, nutritional maintenance requirements, etc.), as well as behavioural outputs (i.e. maternal behaviour, temperament, aggressiveness, etc.), need to be carefully considered when selecting breeding animals for a particular location and/or production regime.

In the specific case of reproduction management techniques, to prevent dystocia and other adverse pregnancy outcomes, the genetic selection of sires should be done on the account of the maturity and size of the female, given highly heritable effect that sire has on final calf size and thus on ease of calving.

6.5.3 Additional Factors Potentially Modulating Camel Reactivity to Pain/Discomfort

Factors such as age, sex (Guesgen et al. 2011; Sorge et al. 2014; Winston et al. 2014; Prusator and Meerveld 2016), personality (Ijichi et al. 2014), social support (Guesgen et al. 2014), productive purpose, stress levels, familiarity with the environment (Tracey and Mantyh 2007; Hernández-Avalos et al. 2021) and pre-natal (Rutherford et al. 2009; Sandercock et al. 2011) and early life experience of pain (Benatti et al. 2009; Lim et al. 2009; Beggs et al. 2012; Clark et al. 2014) can significantly impact the animal response to a noxious stimulus in mammal species. Hence, these factors could add extra complexity when assessing and managing camel pain.

The reported literature suggests that males have a reduced pain sensitivity when compared to their female counterparts as they age. Personality, social and productive context and frequent exposure to a stressor have a notable influence on the verbal report and facial expression in response to pain. Furthermore, previous experience of pain may lead to the sensitization of peripheral and central neurons or nociceptors, thus significantly reducing pain thresholds and increasing the expression of pain-related behaviours.

6.6 Limitations of Traditional Farming and Medicine for Good Health Promotion in Camels

Although it is patent a contemporaneous increase on the social and economic interests on camel rearing and production at intensive and semi-intensive systems, these livestock species continue to be markedly present at traditional low-input systems. Concretely, camels can be found along the African and Asian continents in extensive pastoralist nomadic environments, in which the husbandry practices are far different from those used in modern farming systems. In particular, many of the husbandry practices at pastoral systems do not fit the general recommendations or guidelines on animal welfare provided by the World Organization of Animal Health (WHOA) (Dioli 2022). Firstly, from a purely functional perspective, considering that camels reared in pastoralist communities serve as packing and riding animals, the welfare procuration at these farming regimes is notably compromised since no empirical data does exist to determine maximum admissible loading capacity.

In regard to the general principles of freedom from hunger and discomfort, the animals reared under this extensive livestock keeping regimes are more prone to suffer from periods of malnutrition or severe starvation since they are dependent to a great extent on the availability and quality of local foraging resources and water, which in turn are controlled by both the periodical and punctual variations at climatic and orographic conditions. Furthermore, this dependence on the external environment is reflected in the affectation of the thermal comfort of camels, a condition of paramount importance for the general wellbeing of these animals at desert habitats, as well as at their increased exposition to potential predators (Schwartz and Dioli 1992).

Concerning the methods used for physical restraining of animals, a wide variety of practices are documented at free-grazing camel keeping systems. The most commonly practised method to immobilize camels is the restriction of anterior third (neck and forelimbs) to freely move. Moreover, a simple rope tied to the lower jaw or a more sophisticated halter is used when a precise control of the animal movements is needed. During rutting season, given the fact that male camels can display severe aggressive behaviour towards congeners and handlers, a permanent tourniquet around the throat area of the animal is frequently applied by pastoralists. A more precise control of animal movement is exerted with invasive techniques such as the implantation of nose rings or the piercing of nasal septum. Similarly, relatively painful procedures are practised for the induction of milk-down reflex but also for the privation of calves to suckle their mothers (Dioli 2022). Overall, the implications of such restraining/control practices on the violation of camel welfare, with special focus on the effects of duration of application, the pressure exerted on the restrained areas by the tension of the rope/halter, the abrasive character of the materials used and the stress and pain pre- and post-implantation of mechanical elements at nasal region still remain unexplored.

As a further limitation, pastoralist communities hardly have access to clinical veterinary services for the medical management and prevention of diseases and injuries. In addition, the scientific documentation of traditional medical practices implemented in these systems is notably scarce (Lamuka et al. 2017), both in clinical terms (pharmacology, dosing and complications) and collateral effects (consequences on environmental and public health). Thus, the prevalence and incidence of infectious diseases at these livestock communities are significantly high, as well as the therapeutic management of pain and distress practically ineffective.

6.7 Recommendations to Enhance Physical and Psychological Health in Camels

The minimization of poor health and pain in camels should be carefully procured through the enhancement of the following practices and strategies:

- Adapted definition of preventive medicine programmes (vaccination and deworming), feeding regime and biosecurity plans.
- Accurate design of housing facilities to promote species-specific natural behaviours and avoid physical injuries.
- Rapid diagnosis of injury or disease based on epidemiology, physical examination and medical testing.
- Solid understanding of the principles of camel behaviour to comprehend the motivation(s) of a punctual emotional state in an animal so that any sign of pain or discomfort can be early detected. Moreover, experienced and competent camel handlers should be the responsible agents for animal moving and restraining.
- Administration of specific and effective treatment (medical, surgical or non-drug therapies) when needed and therapeutic follow-up of animals under specific care until they are recovered completely.
- Ethical definition of selection strategies to avoid negative impacts of breeding criteria on animal health and performance (pleiotropic genes).
- Governmental support to implement education policy on animal care, strengthen participatory research and facilitate access to resources at camel-based pastoral communities.

6.8 Conclusions

There have been many significant developments in camel medicine to prevent and treat dromedary disease. The prevention and control of epizootic disease, in all species, makes a major contribution to animal welfare and veterinary services. Veterinarians are fundamental pieces of actively improving animal welfare. This is not the case in all countries, however. In many poor countries veterinary care of camels remains inadequate, and dromedary camels still suffer from major infectious diseases and parasites that are rare or well-controlled in other species. Camel shelters, keepers and veterinarians may be able to recognize animals that are diseased and ensure that they are given veterinary attention and treatment. The practices of veterinary preventive measures and minimizing the risk of the source of infection also keep animals healthy and free of disease and injury. Proper training of veterinary medical personnel can reduce the stress of each medical intervention. More studies in the quantification of the impact of infections on camel welfare are needed; animal welfare should be considered in each infectious disease eradication programme.

Given the fact that dromedary camels are progressively gaining socioeconomic recognition in the panorama of livestock production for their environmental sustainability, camel farming is becoming more prevalent into semi-intensive and intensive systems. Hence, the probability of these animals being exposed to environmental conditions that limit the fulfilment of their basic natural biological functioning and species-specific needs may be substantially increasing. Consequently, the development and validation of pain/discomfort scoring systems that are specific for camels is an essential prerequisite for the procuration and improvement of welfare in these animals. Both behavioural and physiopathological responses are feasible to be used for pain/discomfort recognition and management in camels. Through the objective and early diagnosis and treatment of pain/discomfort, linked with the avoidance or minimization of potential risk factors, the personnel responsible for animal handling and rearing will be qualified to manage camels in such a way that maintains and promotes their wellbeing and thus their general fitness.

References

Abdalla MI et al (2017) Microscopic and molecular detection of camel anaplasmosis and piroplasmosis in Banadir region, Somalia. Open Access J Vet Sci Res 2(1)

Abdelrahman MM et al (2022) Blood trace element status in camels: A review. Animals 12(16): 2116

Abo-Ahmed AI et al (2021) Unique nasal septal island in dromedary camels may play a role in pain perception: microscopic studies. Saudi J Biol Sci 28(7):3806–3815

Abu Damir H et al (2013) Narasin poisoning in the dromedary camel (Camelus dromedarius). Comp Clin Pathol 22(3):305–311

Adugna M et al (2013) Quality and safety of camel milk along the value chain in Eastern Ethiopia. Int J Food Stud 2(2)

Ahmed MS et al (2020) Molecular characterization and diagnostic investigations of rabies encephalitis in camels (Camelus dromedaries) in Oman: A retrospective study. Trop Anim Health Prod 52(4):2163–2168

Al-Ani F, Roberson J (2005) Fungal infection of camelids. In: Faye B, Esenov P (eds) Desertification combat and food safety—the added value of camel producers. IOS Press, Amsterdam, pp 70–84

Alebie A et al (2021) Prevalence, isolation, identification, and risk factors of major bacterial cause of camel subclinical mastitis. BioMed Res Int. https://doi.org/10.1155/2021/5522331

Ali BH, Oukessou M, Bashir AK (1996) Pharmacokinetic considerations in the camel (Camelus dromedarius): a review. Comp Biochem Physiol C Pharmacol Toxicol Endocrinol 115(1):1–9

Ali A et al (2014) Impotentia generandi in male dromedary camels: Clinical findings, semen characteristics, and testicular histopathology. Theriogenology 82:890–896

Ali A, Derar R, Al-Sobayil F, Al-Hawas A, Hassanein K (2015) A retrospective study on clinical findings of 7300 cases (2007–2014) of barren female dromedaries. Theriogenology 84(3): 452–456

Al-Juboori A (2013) Prevalence and etiology of lameness in racing camels (*Camelus dromedarius*) in Abu Dhabi Emirate. J Camelid Sci 6:116–121

Almuzaini AM et al (2016) An outbreak of dermatophytosis in camels (*Camelus dromedaríus*) at Qassim Region, Central of Saudi Arabia. J Appl Anim Res 44(1):126–129

Al-Qarawi AA (2005) Infertility in the dromedary bull: a review of causes, relations and implications. Anim Reprod Sci 87(1–2):73–92

Alshahrani MY et al (2020) Molecular detection of Candidatus Anaplasma camelii in camels (Camelus dromedarius) from Asir Province, Saudi Arabia. Trop Biomed 37(3):587–598

Antoine-Moussiaux N, Faye B, Vias GF (2007) Tuareg ethnoveterinary treatments of camel diseases in Agadez area (Niger). Trop Anim Health Prod 39:83–89

Aqib AI et al (2022) Pathological insights into camel mastitis. Acta Trop 231:106415

Ashley F et al (2005) Behavioural assessment of pain in horses and donkeys: application to clinical practice and future studies. Equine Vet J 37(6):565–575

Azwai SM et al (1995) Immune responses of the camel (Camelus dromedarius) to contagious ecthyma (Orf) virus infection. Vet Microbiol 47(1–2):119–131

Bahrami S et al (2017) Prevalence and phylogenetic analysis of Theileria equi in Iranian dromedaries. Arch Razi Inst 71(3):169–175

Bahrami S et al (2018) First molecular detection of Anaplasma phagocytophilum in dromedaries (Camelus dromedarius). J Zoo Wildl Med 49(4):844–848

Balkhy HH, Memish ZA (2003) Rift Valley fever: an uninvited zoonosis in the Arabian peninsula. Int J Antimicrob Agents 21(2):153–157

Beggs S et al (2012) Priming of adult pain responses by neonatal pain experience: maintenance by central neuroimmune activity. Brain 135(2):404–417

Bellabidi M et al (2020) Coxiella burnetii in camels (Camelus dromedarius) from Algeria: Seroprevalence, molecular characterization, and ticks (Acari: Ixodidae) vectors. Acta Trop 206: 105443

Benaissa MH et al (2015) Reproductive abnormalities in female camel (Camelus dromedarius) in Algeria: Relationship with age, season, breed and body condition score. J Camel Pract Res 22(1):67–73

Benaissa MH et al (2017) Seroprevalence and risk factors for Coxiella burnetii, the causative agent of Q fever in the dromedary camel (Camelus dromedarius) population in Algeria. Onderstepoort J Vet Res 84(1)

Benaissa MH et al (2020a) First report of Chlamydophila abortus infection in the dromedary camel (Camelus dromedarius) population in eastern Algeria. Comp Immunol Microbiol Infect Dis 73: 101557

Benaissa MH et al (2020b) Seroprevalence and risk factors for Trypanosoma evansi, the causative agent of surra, in the dromedary camel (Camelus dromedarius) population in Southeastern Algeria. Onderstepoort J Vet Res 87(1):1–9

Benaissa MH et al (2020c) First serological evidence of BHV-1 virus in Algerian dromedary camels: Seroprevalence and associated risk factors. Comp Immunol Microbiol Infect Dis 76: 101638

Benatti C et al (2009) Early neonatal inflammation affects adult pain re-activity and anxiety related traits in mice: genetic background counts. Int J Dev Neurosci 27(7):661–668

Bhat TK et al (2013) Value addition of feed and fodder by alleviating the antinutritional effects of tannins. Agric Res 2(3):189–206

Borham M et al (2017) Caseous Lymphadenitis in Sudanese and Somalian Camels Imported for Meat Consumption in Egypt. Alexandria J Vet Sci 55(2)

Boushaki D, Adel A, Dia ML, Büscher P, Madani H, Brihoum BA, Sadaoui H, Bouayed N, Issad NK (2019) Epidemiological investigations on Trypanosoma evansi infection in dromedary camels in the South of Algeria. Heliyon 5(7)

Burgmeister R et al (1975) Untersuchungen über Vorkommen von parasitosis, bacteriellen und viralen Infektions-krankheiten bei Dromedaren in Südtunesien. Dtsch Med Wochenschr 82: 352–354

Caja G, Díaz-Medina E, Salama AAK, Salama OAE, El-Shafie MH, El-Metwaly HA et al (2016) Comparison of visual and electronic devices for individual identification of dromedary camels under different farming conditions. J Anim Sci 94(8):3561–3571

Casas-Alvarado A et al (2020) Advances in infrared thermography: Surgical aspects, vascular changes, and pain monitoring in veterinary medicine. J Therm Biol 92:102664

Chu DK et al (2020) Influenza A virus infections in dromedary camels, Nigeria and Ethiopia, 2015–2017. Emerg Infect Dis 26(1):173

Clark C et al (2014) Long-term and trans-generational effects of neonatal experience on sheep behaviour. Biol Lett 10(7):20140273

Coetzee JF (2011) A review of pain assessment techniques and pharmacological approaches to pain relief after bovine castration: Practical implications for cattle production within the united states. Appl Anim Behav Sci 135:192–213

de Oliveira D, Keeling LJ (2018) Routine activities and emotion in the life of dairy cows: Integrating body language into an affective state framework. PLoS One 13(5):e0195674

Desquesnes M et al (2008) First outbreak of Trypanosoma evansi in camels in metropolitan France. Vet Rec 162(23):750–752

Devaux CA et al (2020) Coxiella burnetii in dromedary camels (Camelus dromedarius): A possible threat for humans and livestock in North Africa and the Near and Middle East? Front Vet Sci 7: 558481

Di Giminiani P et al (2016) The assessment of facial expressions in piglets undergoing tail docking and castration: toward the development of the piglet grimace scale. Front Vet Sci 3:100

Diall O, Desquesnes M, Faye B, Lamine Dia M, Jacquiet P, Sazmand A, Otranto D, Touratier L (2022) Development of a control strategy towards elimination of *Trypanosoma evansi* infection (surra) in camels in Africa. Acta Trop 234:106583

Dioli M (2022) Observation on dromedary (Camelus dromedarius) welfare and husbandry practices among nomadic pastoralists. Pastoralism 12(1):1–23

Doosti A et al (2012) PCR detection of leptospirosis in Iranian camels. Bulg J Vet Med 15(3): 178–183

Dunbar ML et al (2016) Validation of a behavioral ethogram for assessing postoperative pain in guinea pigs (Cavia porcellus). J Am Assoc Lab Anim Sci 55:29–34

Eckstein S et al (2022) Viral and bacterial zoonotic agents in dromedary camels from Southern Tunisia: a seroprevalence study. Microorganisms 10(4):727

El Khasmi M, Faye B (2011) Parathyroid hormone-related peptide and vitamin D in phosphocalcic metabolism for dromedary camel. Iran J Appl Anim Sci 1:203–211

El Wathig M, Faye B (2013) Surveillance of camel trypanosomosis in Al-Jouf región, Saudi Arabia. Camel 1(1):65

El Wathig M, Faye B (2016) Camel calf diarrhoea in Riyadh region, Saudi Arabia. J Camel Pract Res 23(2):283–285

Evangelista MC et al (2019) Facial expressions of pain in cats: the development and validation of a Feline Grimace Scale. Sci Rep 9(1):1–11

Faye B (2019) TADs in the dromedary (chapter 6). In: Kardjadj M, Diallo A, Lancelot R (eds) Transboundary animal diseases in Sahelian Africa and connected regions. Springer, Cham, pp 91–110

Faye B, Bengoumi M (2018) Camel clinical biochemistry and hematology. Springer, Cham

Faye B, Seboussi R (2009) Selenium in camel—a review. Nutrients 1:30–49

Faye B, Ratovonanahary M, Chacornac JP, Soubre P (1995) Metabolic profiles and risk of diseases in camels in temperate conditions. Comp Biochem Physiol 112A:67–73

Faye B, Waltner-Toews D, Mc Dermott J (1999) From "ecopathology" to "agroecosystem health". Prev Vet Med 39:111–128

Faye B, Konuspayeva G, Magnan C (2022) L'élevage des grands camélidés. Ed. QUAE, Versailles, coll. Guide pratique, 204p, ISBN 978-2-7592-3499-8

Gahlot TK (2007) Lameness in camels. In: Proceedings of the international camel conference "recent trends in camelids research and future strategies for saving camels", Rajasthan, India, 16–17 February 2007, pp 158–165

Ghaemi M et al (2019) 'Candidatus Bartonella dromedarii' in the dromedary camels of Iran: Molecular investigation, phylogenetic analysis, hematological findings, and acute-phase proteins quantitation. Vet Microbiol 237:108404

Gleerup KB et al (2015a) Pain evaluation in dairy cattle. Appl Anim Behav Sci 171:25–32

Gleerup KB et al (2015b) An equine pain face. Vet Anaesth Analg 42(1):103–114
Goodman CC, Marshall C (2015) Recognizing and reporting red flags for the physical therapist assistant. Elsevier Health Sciences, St. Louis, MO
Gregory NG, Grandin T (2007) Animal welfare and meat production. CABI, Wallingford, UK
Guesgen MJ et al (2011) The effects of age and sex on pain sensitivity in young lambs. Appl Anim Behav Sci 135(1–2):51–56
Guesgen M et al (2014) Social context and other factors influence the behavioural expression of pain by lambs. Appl Anim Behav Sci 159:41–49
Guesgen M et al (2016) Coding and quantification of a facial expression for pain in lambs. Behav Process 132:49–56
Gutierrez C, Corbera JA, Morales I, Morales M, Navarro R (2001) Uterine prolapse in 2 dromedary camels. Can Vet J 42(10):803
Gutierrez C et al (2010) Trypanosoma evansi: recent outbreaks in Europe. Vet Parasitol 174(1–2): 26–29
Häger C et al (2017) The sheep grimace scale as an indicator of post-operative distress and pain in laboratory sheep. PLoS One 12:e0175839
Hay M et al (2003) Assessment of pain induced by castration in piglets: behavioral and physiological responses over the subsequent 5 days. Appl Anim Behav Sci 82(3):201–218
Hernández-Avalos I et al (2021) Nociceptive pain and anxiety in equines: Physiological and behavioral alterations. Vet World 14(11):2984
Higgins AJ (1986) The camel in health and disease. Baillière Tindall, London. 168 p
Hussein MF (2021) Infectious diseases of dromedary camels. Springer, Cham
Hussein MF et al (2008) Serological prevalence of Q fever and chlamydiosis in camels in Saud Arabia. J Anim Vet Adv 7(6):685–688
Iglesias Pastrana C et al (2020) Effect of research impact on emerging camel husbandry, welfare and social-related awareness. Animals 10(5):780
Ijichi C et al (2014) Pain expression is linked to personality in horses. Appl Anim Behav Sci 152: 38–43
Infantes-Lorenzo JA et al (2020) Tuberculosis caused by Mycobacterium caprae in a camel (Camelus dromedarius). BMC Vet Res 1(16):435
Intisar KS et al (2009) Respiratory infection of camels associated with parainfluenza virus 3 in Sudan. J Virol Methods 163(1):82–86
Javed H et al (2021) Co-localization of nociceptive markers in the lumbar dorsal root ganglion and spinal cord of dromedary camel. J Comp Neurol 529(17):3710–3725
Jepson P (2012) Firing and thermocautery. In: Proceedings of the 51st British Equine Veterinary Association BEVA Congress, September 12–15, 2012 Birmingham, UK
Jones BA et al (2013) Zoonosis emergence linked to agricultural intensification and environmental change. Proc Natl Acad Sci USA 110(21):8399–8404
Kamili A, Faye B, Bengoumi M, Tligui NS (2019) Invited review: Camel skin diseases survey in Morocco. J Camelid Sci 12:1–16
Keating SC et al (2012) Evaluation of EMLA cream for preventing pain during tattooing of rabbits: changes in physiological, behavioural and facial expression responses. PLoS One 7(9):e44437
Khalafalla AI, Abdelazim F (2017) Human and dromedary camel infection with Camelpox virus in eastern Sudan. Vector Borne Zoonotic Dis 17:281–284
Khalafalla, A. I., and M.F. Hussein (2021). "Infectious Diseases of Dromedary Camels". Springer International Publishing.
Khalafalla AI et al (2015) MERS-CoV in upper respiratory tract and lungs of dromedary camels, Saudi Arabia, 2013–2014. Emerg Infect Dis 21(7):1153
Khalifa IA et al (2018) Serological prevalence of abortifacient agents in female Mijaheem camels (Camelus dromedarius) in Saudi Arabia. J Anim Res 8(3):335–343
Koenig R (2007) Camel scientists ask: what's sinking the ships of the desert? Science 318:1372–1373

Kumar S et al (2014) Occurrence and histopathology of intestinal obstruction in camel (Camelus dromedarius). Vet Pract 15(2):303–304

Lamuka PO et al (2017) Camel health management and pastoralists' knowledge and information on zoonoses and food safety risks in Isiolo County, Kenya. Pastoralism 7(1):1–10

Langford DJ et al (2010) Coding of facial expressions of pain in the laboratory mouse. Nat Methods 7(6):447–449

Le Neindre P et al (2009) Animal pain: identifying, understanding and minimising pain in farm animals. Multidisciplinary scientific assessment, Summary of the expert report. INRA, Paris

Lim EJ et al (2009) Long-lasting neonatal inflammation enhances pain responses to subsequent inflammation, but not peripheral nerve injury in adult rats. Int J Dev Neurosci 27(3):215–222

Lorusso V et al (2016) Canine and ovine tick-borne pathogens in camels, Nigeria. Vet Parasitol 228: 90–92

Machin KL (2013) Recognition and treatment of pain in birds. In: Egger CM, Love L, Doherty T (eds) Pain management in veterinary practice. Wiley-Blackwell, Hoboken, NJ, pp 407–415

MacRae AM et al (2018) Initial evaluation of facial expressions and behaviours of harbour seal pups (Phoca vitulina) in response to tagging and microchipping. Appl Anim Behav Sci 205: 167–174

McLennan KM et al (2019) Conceptual and methodological issues relating to pain assessment in mammals: The development and utilisation of pain facial expression scales. Appl Anim Behav Sci 217:1–15

Megersa B (2010) An epidemiological study of major camel diseases in the Borana lowland, Southern Ethiopia. Drylands Coordination Group 58, Oslo

Mensah-Brown E, Garey L (2006) The superior colliculus of the camel: a neuronal-specific nuclear protein (NeuN) and neuropeptide study. J Anat 208(2):239–250

Mirzaei M et al (2018) Morphological and molecular characterization of Onchocerca fasciata (Nematoda, Onchocercidae) from dromedary camels (Camelus dromedarius) in Iran. Parasite 25:50

Mohamud AI et al (2020) Prevalence and major pathogens associated with clinical and subclinical mastitis in dairy camel (Camelus dromedarius) in Benadir Region of Somalia. Vet Sci Res Rev 6(2):132–137

Moltumo S et al (2020) Assessment of welfare problems on working donkeys in Hosaena district, Hadiya zone, southern Ethiopia. Vet Med Open J 5(1):14–20

Morton CM et al (2005) Application of a scaling model to establish and validate an interval level pain scale for assessment of acute pain in dogs. Am J Vet Res 66(12):2154–2166

Mostafa MB (2020) Distal limb lameness in draught dromedary camels. J Camel Pract Res 27(3): 359–366

Musa MT et al (2008) Brucellosis in camels (Camelus dromedarius) in Darfur, western Sudan. J Comp Pathol 138(2–3):151–155

Nagy PP et al (2022) Intensification of camel farming and milk production with special emphasis on animal health, welfare, and the biotechnology of reproduction. Anim Front 12(4):35–45

Narnaware SD et al (2015) Pathology and diagnosis of Mycobacterium bovis in naturally infected dromedary camels (Camelus dromedarius) in India. Trop Anim Health Prod 47:1633–1636

Ogden NH, Lindsay LR (2016) Effects of climate and climate change on vectors and vector-borne diseases: ticks are different. Trends Parasitol 32(8):646–656

Ohl F, Van der Staay FJ (2012) Animal welfare: At the interface between science and society. Vet J 192(1):13–19

Orth EK et al (2020) Development of a donkey grimace scale to recognize pain in donkeys (Equus asinus) post castration. Animals 10(8):1411

Ouchene-Khelifi NA et al (2018) Fasciolosis due to Fasciola hepatica in ruminants in abattoirs and its economic impact in two regions in Algeria. Trop Biomed 35(1):181–187

Padalino B, Menchetti L (2021) The first protocol for assessing welfare of camels. Front Vet Sci 7: 1230

Posner LP, Chinnadurai SK (2013) Recognition and treatment of pain in reptiles, amphibians, and fish. In: Egger CM, Love L, Doherty T (eds) Pain management in veterinary practice. Wiley-Blackwell, Hoboken, NJ, pp 417–423

Prusator DK, Meerveld GV (2016) Sex-related differences in pain behaviors following three early life stress paradigms. Biol Sex Differ 7(1):1–12

Qamar MF et al (2011) Prevalence of sub-clinical mastitis in one humped camel of lesser cholistan, Pakistan. In: International workshop on Dairy Science Park, Peshawar, Pakistan

Rahimi E et al (2014a) Prevalence of Clostridium difficile in raw beef, cow, sheep, goat, camel and buffalo meat in Iran. BMC Public Health 14(1):1–4

Rahimi E et al (2014b) Incidence of Listeria species in bovine, ovine, caprine, camel and water buffalo milk using cultural method and the PCR assay. Asian Pac J Trop Dis 4(1):50–53

Reijgwart ML et al (2017) The composition and initial evaluation of a grimace scale in ferrets after surgical implantation of a telemetry probe. PLoS One 12(11):e0187986

Rutherford KM (2002) Assessing pain in animals. Anim Welf 11(1):31–53

Rutherford KM, Robson SK, Donald RD, Jarvis S, Sandercock DA, Scott EM, Nolan AM, Lawrence AB (2009) Pre-natal stress amplifies the immediate behavioural responses to acute pain in piglets. Biol Lett 5(4):452–454

Saeed IK et al (2015) Mixed infection of peste des petits ruminants virus (PPRV) and other respiratory viruses in dromedary camels in Sudan, an abattoir study. Trop Anim Health Prod 47(5):995–998

Said MB et al (2016) First molecular evidence of Borrelia burgdorferi sensu lato in goats, sheep, cattle and camels in Tunisia. Ann Agric Environ Med 23(3):442–447

Sazmand A, Joachim A, Otranto D (2019) Zoonotic parasites of dromedary camels: so important, so ignored. Parasit Vectors 12:1–10

Sandercock DA et al (2011) The impact of prenatal stress on basal nociception and evoked responses to tail-docking and inflammatory challenge in juvenile pigs. Physiol Behav 104(5): 728–737

Santovito A et al (2022) Buccal micronucleus assay as a useful tool to evaluate the stress-associated genomic damage in shelter dogs and cats: new perspectives in animal welfare. J Vet Behav 47: 22–28

Schwartz HJ, Dioli M (1992) The one-humped camels (Camelus dromedarius) in Eastern Africa: A pictorial guide to diseases, health care and management. Verlag Josef, Scientific Books, Weikersheim, Germany

Seboussi R et al (2012) Selenium toxicity in the dromedary camels clinical symptoms and lesions. In: 3rd ISOCARD international conference ISOCARD

Selmi R et al (2022) Epidemiology and genetic characteristics of tick-borne bacteria in dromedary camels of the world. Acta Trop 234:106599

Sharifiyazdi H et al (2017) Molecular investigation of Anaplasma and Ehrlichia natural infections in the dromedary camel (Camelus dromedarius) in Iran. Comp Clin Pathol 26(1):99–103

Sneddon LU et al (2014) Defining and assessing animal pain. Anim Behav 97:201–212

Sorge RE et al (2014) Olfactory exposure to males, including men, causes stress and related analgesia in rodents. Nat Methods 11(6):629–632

Sotocina SG et al (2011) The Rat Grimace Scale: a partially automated method for quantifying pain in the laboratory rat via facial expressions. Mol Pain 7:1744–1755

Stafford KJ, Mellor DJ (2005) The welfare significance of the castration of cattle: a review. N Z Vet J 53(5):271–278

Stafford KJ (2013) Recognition and assessment of pain in ruminants. Pain Manag Vet Pract:349–357

Thompson L (2013) Recognition and assessment of pain in small exotic mammals. In: Egger CM, Love L, Doherty T (eds) Pain management in veterinary practice. Wiley-Blackwell, Hoboken, NJ, pp 391–397

Tibary A, Vaughan J (2006) Reproductive physiology and infertility in male South American camelids: a review and clinical observations. Small Rumin Res 61(2–3):283–298

Tibary A et al (2006) Infectious causes of reproductive loss in camelids. Theriogenology 66(3): 633–647

Tonouhewa ABN et al (2017) Toxoplasma gondii infection in meat animals from Africa: Systematic review and meta-analysis of sero-epidemiological studies. Vet World 10(2):194

Tracey I, Mantyh PW (2007) The cerebral signature for pain perception and its modulation. Neuron 55(3):377–391

Uhl EW (2018) The pathology of vitamin D deficiency in domesticated animals: an evolutionary and comparative overview. Int J Paleopathol 23:100–109

Waheed MM, Ghoneim IM, Hassieb MM, Alsumait AA (2014) Evaluation of the breeding soundness of male camels (Camelus dromedarius) via clinical examination, semen analysis, ultrasonography and testicular biopsy: a summary of 80 clinical cases. Reprod Domest Anim 49(5):790–796

Wang P et al (2020) Non-contact vital signs monitoring of dog and cat using a UWB radar. Animals 10(2):205

Wang L et al (2021) Molecular identification and biological characterization of Cryptosporidium muris from camels (Camelus bactrianus) in China. Parasit Vectors 14(1):1–10

Wernery U, Kinne J (2016) Caseous lymphadenitis (Pseudotuberculosis) in camelids: a review. Austin J Vet Sci Anim Husbandry 3(1):1–6

Wernery U et al (1998) Salinomycin poisoning in a dromedary breeding herd in the United Arab Emirates. J Camel Pract Res 5:275–279

Wernery U, Kinne J, Schuster RK (2014) Camelid infectious disorders. OIE, Paris. 500 p

Winston JH et al (2014) Chronic prenatal stress epigenetically modifies spinal cord BDNF expression to induce sex-specific visceral hypersensitivity in offspring. Neurogastroenterol Motil 26(5):715–730

Woo PC et al (2014) New hepatitis E virus genotype in camels, the Middle East. Emerg Infect Dis 20(6):1044

World Organization for Animal Health (WHO) (2022) Terrestrial animal health code, 28th edn. World Organisation for Animal Health, Paris, France

Younan M et al (2005) Identification and molecular characterization of Streptococcus equi subsp. zooepidemicus isolated from camels (Camelus dromedarius) and camel milk in Kenya and Somalia. J Vet Med Ser B 52(3):142–146

Younan M et al (2021) Ehrlichia spp. close to Ehrlichia ruminantium, Ehrlichia canis, and "Candidatus Ehrlichia regneryi" linked to heartwater-like disease in Kenyan camels (Camelus dromedarius). Trop Anim Health Prod 53(1):1–5

Zaher HA, Swelum AAA, Alsharifi SA, Alkablawy AH, Ismael AB (2017) Seroprevalence of chlamydiosis in Abu Dhabi dromedary camel (Camelus dromedarius) and its association with hematobiochemical responses towards the infection. J Adv Vet Anim Res 4(2):175–180

Zakian A et al (2016) The first report of peste des petits ruminants (PPR) in camels (Camelus dromedarius) in Iran. Trop Anim Health Prod 48(6):1215–1219

Zakian A et al (2017) An acute outbreak of natural Trypanosoma evansi infection in camel (Camelus dromedarius) herds in the southwestern Iran. Comp Clin Pathol 26(1):51–59

Zhu S et al (2019) A review of zoonotic pathogens of dromedary camels. EcoHealth 16(2):356–377

Behaviour: Behavioural Repertoire and Behavioural Needs of Camels

7

Meriem Fatnassi and Barbara Padalino

Abstract

Each species has a specific behavioural homeostasis. Camels are diurnal animals, and in nature, they spend a large proportion of the day in browsing behaviour, while they show more resting and rumination behaviour at night. Behaviour depends not only on genetics but also on the environment, so variations in behavioural patterns may be due to different factors, such as season, feed and water availability, and housing conditions. This chapter aims to critically review the literature to describe the behaviour of dromedary camels and the factors which may affect it. During the last few years, research has used behavioural responses to understand the effects of different housing systems and husbandry practices on the welfare of camels. It is worth noting that there is no standardized ethogram for camels yet, so whether behaviour can be used in an objective way to assess welfare in camels is still a matter of debate. Deep knowledge of the behavioural repertoire and behavioural needs of dromedary camels is crucial to improve management system conditions and safeguarding welfare. A large space allowance, appropriate feeding and watering practices, sufficient shaded area, more opportunity for social contact and a positive human-camel relationship have been proposed as useful and applicable practices to improve the camel management system and welfare. However, further studies on behaviour are needed for a better understanding of the behavioural needs of camels to provide evidence to the policymakers to issue regulations aiming at safeguarding the welfare of camels.

M. Fatnassi
Wildlife and Livestock Laboratory, Arid Lands Institute, Medenine, Tunisia

B. Padalino (✉)
Department of Agricultural and Food Sciences, University of Bologna, Bologna, Italy
e-mail: Barbara.padalino@unibo.it

B. Padalino, B. Faye (eds.), *Dromedary Camel Behavior and Welfare*, Animal Welfare 24, https://doi.org/10.1007/978-3-031-48600-5_7

Keywords

Behaviour · Dromedary camels · Time budget · Behavioural needs

7.1 Behaviour

Behaviour is commonly defined as an action carried out by an individual (Lidfors et al. 2005). It has also been described as what animals do for interacting with other animals or objects (Power 2000; Lindshield 2017). Behaviour has also been defined as the relationship of adjustment between animals and their environment to maintain homeostasis (Mench 1998; Boussely 2003). Thus, careful observations of behaviour can provide us with much information about animals' needs, likes and dislikes, as well as their internal states, providing, hence, an idea to caretakers about the welfare of the animal (Mench and Mason 1997). Based on duration and frequency, Martin and Bateson (1993) identified two categories of behaviours, namely 'behavioural states' and 'behavioural events'. States are behaviours of relatively long duration, such as a measure of prolonged activity (body posture, eating, etc.), and they are expressed by their duration or mean duration, while events are behaviours of instantaneous duration (e.g., vocalizations, urinations) and they are often expressed by their frequencies of occurrence in a specific time window (Altmann 1974). The description of behaviour is made using an ethogram (McGreevy 2002), which is simply a list of behaviours that are of interest. The ethogram, in an observational study, can be used for data collection and analysis (Lindshield 2017). It is worth noting that a standard ethogram exists for some species, such as horses (McDonnell 2003), but there is no standardized ethogram for camels yet.

7.2 Behavioural Repertoire

The behavioural repertoire is a complete compilation of behaviours that are exhibited by an individual, group, population, or species (Lindshield 2017). However, the behavioural repertoires of animals are never fully constant because the behaviour of a species depends on their genetics for 50% and the environments where they live for the other 50%. This means that the behavioural repertoire of animals living in a free-ranging environment is different from the behavioural repertoire of animals living in captivity.

According to ethologists, the behavioural repertoire of animals consists of numerous categories: feeding, rumination, resting, standing and locomotor behaviour, etc. (Kilgour et al. 2012; Hoyer 2013). However, in a captive environment, the behavioural repertoire includes not only normal or natural behaviours but also other abnormal or stereotypical ones (Mason 1993).

7.3 The Behavioural Repertoire of Camels

Behaviour can be categorized into active and inactive behaviour. Inactive behaviour is defined as resting or sleeping, while all the other behaviours are classified as active behaviours. Time budget means the amount of time that camels spend in their various activities daily (i.e., 24 h). Daily activities are distributed accordingly to time factors, in German 'zeitgeber', such as the presence or absence of sunlight. The light/dark cycle is the most important zeitgeber, and two main phases of circadian rhythms have been described based on it: photophase, or daylight; and scotophase, or dark period.

The daily rhythm of an animal corresponds to a regular oscillation of behavioural and physiological variables for 24 h (Refinetti et al. 2007) and this rhythmicity is linked to its nutritional condition, its social status and its stress condition (Kaczensky et al. 2006). Thus, the study of time budgets and activity patterns of re-introduced animals (i.e. animals which go back to their natural environment) can provide important information about their well-being and adaptation status (Boyd and Bandi 2002; Aubè et al. 2017). When an animal shows most of the active behaviours during the photo-phase, or daylight, it is considered a diurnal animal. Camels are diurnal animals, but, in camels, there have been very few studies on their circadian rhythms (El Allali et al. 2005; Aubè et al. 2017).

7.4 Inactive Behaviour

7.4.1 Resting or Lying

Resting behaviour is one of the most common behaviours in captive and free-ranging animals (Dahlgren 2010; Schwan 2011; Hoyer 2013). It was described as a relaxed state characterized by a general lack of attention, including both relaxation and sleep, usually, in a standing or recumbent position (Ransom and Cade 2009). This behaviour is then a form of physical and energy conservation (Berger and Phillips 1995). It also plays a role in thermoregulation (Berger and Phillips 1995) and memory consolidation (Smith 1995). Given the importance of this behaviour, the reduction in the time spent resting is associated with an increase in plasma cortisol concentration (Gonzalez et al. 2003), reflecting a situation of discomfort that subsequently may lead to poor animal welfare (Fregonesi et al. 2007; Uzal Seyfi 2013).

Nevertheless, the expression of resting behaviour is affected by several factors; mainly housing conditions which include the nature of habitat open area vs individual box (Munksgaard et al. 2005), group size or stocking density (Fregonesi et al. 2007) and the size of the enclosure (Raabymagle and Ladewig 2006). Moreover, resting behaviour is influenced by the working hours of animal caretakers (Dahlgren 2010; Hoyer 2013).

In camels, resting behaviour was described as sitting up position on the brisket with the legs tucked under the body (i.e., the natural sitting position in camels) with the head up or the head on the floor (Fatnassi et al. 2014a, 2014b; Aubè et al. 2017). Camels can sleep also in lateral decubitus, with all body, neck and head touching the floor. This position has been described for camels kept in shaded areas with an adequate space allowance and comfortable bedding/flooring (Zappaterra et al. 2021).

Resting or lying down behaviour in dromedary camels occurs predominantly at night and covers about 50% of the camel time budget (Sambraus 1994). In contrast, Khan et al. (1998) stated camels spent only 4.37% of 24 h in sleeping. This difference could be because in the first case, the authors did not make the difference between resting in a sitting position and sleeping, while the second author considered only sleeping. Dromedary camels spend indeed the majority of their time, when are not grazing, in sternal decubitus, with the head up, resting just looking around, but more often ruminating. More studies on the quality (slow-wave sleep (SWS) and rapid eye movement (REM) sleep) and quantity of sleep in camels would be useful, also to understand whether intensive farming may lead to sleeping disorders as reported in other animals (Houpt et al. 2019).

7.4.2 Standing

In the literature, standing behaviour was described as an inactive position in an upright posture on four or three feet (Fatnassi et al. 2014b, 2021; Padalino et al. 2014). In animals, the duration of standing is affected by the housing conditions and a large proportion of standing within the time budget could indicate boredom, a general discomfort status and poor animal welfare (Fureix et al. 2012; Uzal Seyfi 2013). When camels were housed for 24 h in a single box, they spent 20% of the observation period standing, but this time decreased significantly in camels exposed to females (Fatnassi et al. 2014b). Similarly, over 2 days of observations, Aubè et al. (2017) stated that standing behaviour occupied 13.29% of the time of individually housed camels. However, very less time was reported by Iqbal (1999), who indicated that the time spent by camels standing varied from 2.40 to 3.10% when camels were kept in extensive systems in Pakistan.

However, it is important to highlight that standing without performing any other behaviour must be considered differently from a standing posture associated with other behaviours, such as feeding (Waring 2003). In this context, El Shoukary et al. (2021) reported that camels spent 25% of their time standing when they were fed twice per day, and this time reduced when they were fed once daily favouring the standing doing nothing, which can be considered as a negative welfare animal-based measure (ABM).

7.5 Active Behaviours

7.5.1 Feeding

Feeding behaviour, through both diet selection and food intake, is the predominant way that an animal attempts to fulfil its metabolic requirements and achieve homeostasis (Ginane et al. 2015). Browsing/grazing behaviour is a collection of activities linked to the ingestion of feed (searching, choosing, swallowing and absorption) (Pagot 1992). Camels are predominantly browsers than grazers, taking a bite from one plant and moving to another, covering vast areas each day in search of feed with a split upper lip well suited for this purpose (Yagil 1990; Iqbal and Khan 2001) (see Chap. 4).

Camels usually take a variety of vegetation that presumably provides optimal nutrition; they spent 37% of their time over 24 h in grazing (Iqbal and Khan 2001). They show selection behaviour, irrespective of the availability of feed and quantity (Moaeenuddin et al. 2004). Yagil (1990) declared that camels are selective feeders and eat the freshest vegetation available; they prefer to browse only from bushes and trees, especially during hottest hours of the day (Faye and Tisserand 1989). More often camels tend to be non-selective in their diet during the wet season when forage is plentiful, but they become indiscriminate in their forages' choice during the dry season due to forage scarcity (Shaheen 2009).

In their natural habitats, camels graze a broad spectrum of fodder plants, including thorny species, halophytes and aromatic species generally avoided by other domestic herbivores (Iqbal and Khan 2001; Khaskheli 2020). So, lack of diversity in their diet may affect their welfare (see Chap. 4 for details). Like other livestock, pasture preference in camels depends on the plant species present in the range, the amount of forage available and the nutritional quality of the plant (Dereje and Uden 2005). Additionally, the physical environment, plant environment and animal behaviour all interact to influence the selection process during grazing (Moaeenuddin et al. 2004). In natural conditions, the actual feeding time varied from 60% to 68% of the time budget (Iqbal 1999). In semi-intensive systems, Khorchani et al. (1992) observed the feeding behaviour of four lactating female camels in the arid ranges of Tunisia. These animals were allowed to spend 600 min/day out of their paddock in the desert, and the authors reported that the she-camels spent 464 min of the total time grazing, with a percentage of 77%.

Feeding time seems to be different among seasons; significantly less time was spent on feeding and walking during the green season compared to the dry season (Dereje and Uden 2005). This may be because feed is more available, so they need to browse less to reach their feeding requirement. During very hot weather camels instead tend to avoid feeding and they spend lots of time in a standing position reducing heat gain and thus conserving energy (Dereje and Uden 2005). When it is very hot, camels also change their feeding patterns, becoming selective and eating a little only from trees, while they tend to browse and eat more and without selecting feed, early in the morning and late afternoon, which are the coolest times of day (Faye and Tisserand 1989; Iqbal and Khan 2001; Dereje and Uden 2005). Feeding behaviour can be affected also by diet and animal-related characteristics. Dereje and Uden (2005) indicated that feeding time was negatively correlated with the dietary crude protein (CP) and dry matter digestibility (DMD) levels but positively correlated with total crude fibre (NDF), cellulose and lignin (ADF) and lignin (ADL) (Dereje and Uden 2005). Feeding time was also different according to the age of camels; it has been reported that young camels spent more time browsing compared to adults. While adult camels spent more time resting and doing other activities (Dereje and Uden 2005). Finally, feeding behaviour is affected also by the watering availability and frequency. When camels were watered daily, they spent about 60% of their time budget on feeding-related behaviour, but less time was recorded when dehydration became more severe (Bekele et al. 2011). Overall, it seems clear that husbandry systems may affect a lot of the feeding patterns in camels, as in other species. It becomes crucial, therefore, to safeguard camel welfare, that diets do not only cover nutritional requirements but also the feeding behavioural needs of this species (see Chap. 4).

7.5.2 Rumination

Rumination behaviour is the time spent in the process of regurgitation, chewing, salivation and swallowing of ingested food to reduce food particle size and improve fibre digestion (Beauchemin 1991). This is a specific characteristic of all ruminants (Schirmann et al. 2012), and camels show 2/3 of rumination in sternal decubitus and the rest in a standing position (Beauchemin 1991; Schwan 2011). In poor housing conditions, such as overcrowding conditions, it is possible to observe many camels ruminating in standing, so this could be used as an ABM indicating poor welfare.

Rumination is typically monitored by direct visual observation (Couderc et al. 2006) or from a jaw movement vibrograph (Khorchani et al. 1992) or video recordings (Lindström et al. 2001). In a study, Kaske et al. (1989) reported that rumination in camels started after midnight and lasted, with breaks of 30–60 min, until the next feeding time at 8:00 h. Each rumination cycle consisted of 57 ± 11 jaw movements (observation of 3 animals, each animal 50 cycles). After swallowing the bolus, the next cycle started 12 ± 5 s later and eructation was observed during the whole day and occurred after a contraction of the caudal part.

In a 24-h observation, Khorchani et al. (1992) stated that rumination lasted 573 min and occurred only when the female camels were in their enclosure (during the night), taking 68.1% of the total nighttime, while the remaining time was spent resting and sleeping. In India and Kenya, camels spent about 25% of 24-h time in rumination, with the peak time for rumination activity between 4 and 7 am (Sambraus 1994). Eight hours was suggested as the normal rumination time (Iqbal and Khan 2001). Mengli et al. (2006) instead found that camels spent up to 39% of their day ruminating with 44% of the ruminating activity occurring during the day time and the rest at night. Rumination patterns are affected not only by the light but also by the feed availability. When camels were fed ad libitum, they spent 8 h/day eating, 11 h/day ruminating and 5 h/day resting. Rumination activity occurred mainly during the night with maximum values between 1:00 and 6:00 h. When feed was rationed, rumination was observed mainly in the early morning, rarely during the day (Kaske et al. 1989). Iqbal and Khan (2001) observed the ruminating behaviour of Bactrian Camel during a 24-h period with visual observation under free grazing in the grassland of Inner Mongolia and their results showed that ruminating is affected mainly by grassland conditions, management and weather conditions. Rumination, as feeding, is indeed affected by season; the total rumination time is longer during autumn than in spring and summer.

Commonly, rumination time is associated with increased salivary production and improved rumen health (Beauchemin 1991; Schirmann et al. 2009). Therefore, decreased rumination time is interpreted as an indicator of stress (Schirmann et al. 2009), anxiety (Bristow and Holmes 2007) or illness (DeVries et al. 2009). It is therefore considered an important indicator of animal health and welfare (Schwan 2011) and recently ruminating while sitting was suggested as a positive welfare ABM (Padalino and Menchetti 2021).

7.5.3 Locomotion

Locomotion or locomotor behaviour is defined as the voluntary movement of the whole body, observable from the first day of birth (McDonnell 2003). It is carried out as a result of an innate motivation, where animals move mainly to seek food, water and shelter (Albright and Arave 1997). Walking behaviour was described for camels kept in a single box as 'the camel does more than 2 complete steps' (Fatnassi et al. 2014b; Aubè et al. 2017). Walking is strongly affected by husbandry conditions.

In natural conditions, camels walk about 25–27% of their time budget, but this percentage is higher during the wet season compared to the dry season (Dereje and Uden 2005). Camels usually walk a lot during the day, grazing 8–12 h daily and walking at an average speed of 2 km/h, but, if necessary, they can walk 150 km/day in the desert (Gauthier-Pilters and Dagg 1981). However, when camels are housed in a single box, they spent only 2% of their time walking (Aubè et al. 2017). El Shoukary et al. (2020) reported that camels spent 75.83 ± 1.17 s/20 min walking when they were housed in groups compared to those housed individually (54.89 ± 1.17 s/20 min). This is not surprising considering that camels are social animals. Housing conditions should meet the locomotory and the social behavioural needs to protect camel welfare (see Chap. 5).

7.6 Social and Play Behaviour

In natural conditions, camels are social animals living in groups/herds, consisting of a dominant adult male, females and their offspring. Like horses, other young males which have been chased out of their natal group form bachelor herds (Schulte and Klingel 1991, McDonnell, 2000).

In the herd, camels communicate with each other in different ways; for example, rutting camels interact with females in oestrus by different types of vocalizations such as blathering or grinding teeth (Fatnassi et al. 2014b, 2021). During parturition, the mother turns towards her neonate soon after giving birth interacting closely with it by sniffing, nursing and vocalizing (Hammadi et al. 2021). However, the management system of camels has been changing to more intensive management for semen collection (Monaco et al. 2013) or artificial insemination (Skidmore et al. 2013) and mechanical milking programme (Hammadi et al. 2010; Nagy et al. 2022). Under this system, male camels, for example, are kept isolated in a single box (15 m^2) far from the females' herd and this management affects their locomotion and social activities and limits the expression of normal and rutting behaviours (Fatnassi et al. 2014a, 2021) causing, therefore, the expression of locomotor (head-shaking and pacing in a circle) and oral (self-biting and bar-mouthing) stereotypies (Padalino et al. 2014: Fatnassi et al. 2016). Given the importance of social behaviour, Padalino and Menchetti et al. (2021) reported that the expression of social behaviour must surely be considered in welfare assessment as a positive welfare ABM.

In 1993, Al-Hazmi and Brain described the social behaviour in dromedary camel as a list of behaviours comprising agonistic gestures, approaching, following, genital sniffing, grooming, mutual grooming, naso-nasal investigation, sexual interaction and partner sniffing. The later authors showed that male and female camels housed in the different husbandry systems (partly guarded and unguarded conditions) showed no significant differences in terms of time allocated to social behaviours.

Play behaviour is also categorized as social, object or locomotor play (Burghardt 1999). Social play involves two or more animals (Spinka et al. 2001) and the play partners are usually of similar age and size (Thompson 1996; Spinka et al. 2001). Play behaviour has been identified in many species like horses (McDonnell 2003). It has been defined as a behaviour with no immediate function (Bekoff and Byers 1998; Burghardt 2005) and the proposed roles varied according to animal species (Vieira and Sartorio 2002), intraspecific characteristics, age and gender (Gomendio 1988). In young animals, most authors suggested that play behaviour promotes the ability to handle unexpected situations and allows them to develop social skills and facilitates their integration into groups (Spinka et al. 2001). Most studies reported that play is more present in young, healthy, well-fed and securely attached animals (Hausberger et al. 2007). Hausberger et al. (2012) stated that, in horses, play was more frequent in groups kept outdoors than indoors, suggesting the possible role of play behaviour as an indicator of positive emotions reflecting good welfare (Held and Spinka 2011). In camels, play behaviour has not been defined and more studies are needed to understand the role of playing concerning their welfare.

7.7 Stereotypy or Abnormal Behaviour

Stereotypic behaviour belongs to the category of abnormal behaviour, and it has been defined as 'repetitive, invariant behaviour patterns without apparent goal or function' (Mason 1991; Mason and Latham 2004). Stereotypies are often expressed in animals reared in unsuitable living conditions (McGreevy et al. 1995; Christie et al. 2006) or faced with any sources of chronic stress or poor health conditions (Sarrafchi and Blokhuis 2013). Stereotypical behaviour could be used as an alert since behavioural changes are often the earliest signs that can be found to indicate suboptimal living conditions (Christie et al. 2006; Keeling and Jensen 2009). Stereotypies are sometimes, but not always, linked to physiological changes indicative of stress; and it seems that the performance of stereotypies sometimes appears to be rewarding because it causes the release of endorphins (Cronin et al. 1985; Crockett et al. 2007). Overall, stereotypies are usually considered an animal-based measure of poor welfare.

Stereotypies vary in form and duration depending on the species and the provoking factors. However, even in the same species, they may differ between individuals (Padalino et al. 2014). They exist in various forms, the most common forms being oral and locomotor stereotypies (Mason and Rushen 2006). Many oral stereotypies, for instance, are associated with the lack of opportunity to perform components of feeding behaviour, including foraging (Bayne et al. 1992; Redbo and Nordblad 1997). Instead, locomotory stereotypies seem to be more associated with the inability to express locomotory behaviours, as was demonstrated in horses kept in limited space allowance (McGreevy et al. 1995; Nicol 1999). Thus, the main causes of performing stereotypies are generally attributed to the following factors: inappropriate feeding practices, limited social contact and lack of locomotion because of a restrictive stable environment (Cooper and Albentosa 2005; Wickens and Heleski 2010).

For dromedary camels, the first identification and description of stereotypical behaviours were conducted by Padalino et al. (2014), who identified two general forms of stereotypes: locomotor and oral stereotypies also in this species (Table 7.1). The latter authors found that housing male camels in single boxes led to the development of stereotypical behaviour, but the incidence of this behaviour was reduced by allowing 30 min of social contact with females and 1 h of freedom in a paddock. However, there is still a gap in knowledge regarding the effects of other management systems on camel welfare and the manifestation of stereotypy.

Since stereotypical behaviours often appear to arise from frustration and boredom, studies have established that stereotypies can be reduced when animals are placed in enriched environments. For example, in dromedary camels used for semen collection, most stereotypies decreased when camels were exposed to females, for only 30 min/daily (Fatnassi et al. 2014b) or all day long (Fatnassi et al. 2016, 2021), and when they were given the possibility to exercise for 1 h in a large paddock (Fatnassi et al. 2014b; Padalino et al. 2014). However, there is a lack of knowledge on what can be used as enrichment materials for dromedary camels.

Table 7.1 List of stereotypical behaviour described in dromedary camels housed in individual boxes

Categories	Behaviour	Description
Locomotor	Head-shaking	Camel raises his head to the vertical with a very fast movement (this behaviour included a movement of the head by up to 90°). This stereotypy was considered a behavioural event because it lasts about 1 s
	Pacing in a circle	Camel walks to the other side of his box (stops and tries to look through a small window in the wall) and walks back to his initial position (in doing so, the camel always follows the same path, namely walks in a circle). The camel repeats this movement several times without any clear motivation 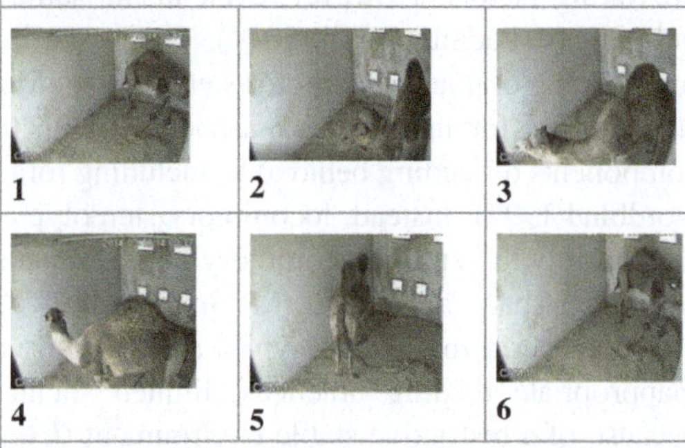 1 2 3 4 5 6
	Tripodal standing	Camel stands on three legs lifting repetitively a foreleg
	Weaving	Camel remains stationary but shifts its weight from one foreleg to the other and swings its head from side to side
Oral stereotypy	Self-mutilation	Camel bites different parts of his forelegs (right or left) from the shoulders to the feet. This stereotypical behaviour was considered a behavioural state, since the affected camel could bite his legs for a variable length of time, ranging from a few seconds to several minutes
	Bar-mouthing	Camel licks, bites or plays with the lips on the bars of the box's gate 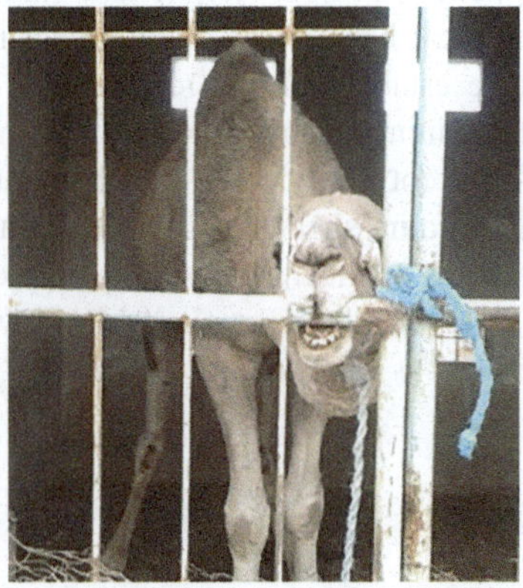
	Wall licking	

(continued)

Table 7.1 (continued)

Categories	Behaviour	Description
		Camel licks the walls with the lips or tongue 

Adapted from Padalino et al. (2014); Fatnassi et al. (2016, 2021)

7.8 Sexual and Mating Behaviour

Male camel is described as a seasonal breeder with a marked peak in sexual activity (the rut) during the breeding season. During this period, camels exhibit morphological, behavioural and endocrinological peculiarities (Fatnassi et al. 2014a).

The beginning of the rutting season was generally marked by a notably profuse secretion of the poll glands of the neck, which contains androgen concentration, like blood, and pheromones which serve to attract females and to mark their territory (Ebada et al. 2012). During the rut, camels increase pacing and anxiety becoming very aggressive towards other males and humans showing all typical sexual behaviour; soft palate 'dulaa', blathering, tail beating or flapping, grinding and urination (Bhakat et al. 2005; Fatnassi et al. 2014a) (Table 7.2). The expressing of rutting behaviour of camels reared for semen collection was evaluated through a female parade; this method consists of bringing a female in estrus, near the box, for a limited time window (e.g., 12 min) (Padalino et al. 2013; Fatnassi et al. 2014a; Fig. 7.1). During female presence, the observers must note down the occurrence of following behaviours: sniffing, flehmen, whistling, urination, *dulaa* extrusion, tail flapping. Furthermore, the intensity of salivary production, nervousness and poll gland secretion should be scored (absent, low, high, very high). At the end of the female passage, the observers may score the camel's sexual behaviour according to the score proposed by Padalino et al. (2013) and Fatnassi et al. (2014a).

Thus, it seems that the assessment of male camel behaviour during female parades using an ethogram represents a useful tool to monitor camel male sexual behaviour and stimulate libido. Therefore, it could be adopted by the camel industry in other countries (Fatnassi et al. 2014a).

Table 7.2 Description of typical sexual behaviour in male dromedary camel

Sexual behaviour	Description
Grinding of teeth/ whistling	The male moves the lower jaw on the left and right side, with a closed mouth, grinding the teeth and producing a typical squeaking/whistling sound
Urination	The dromedary male assumes the urinating position, spreads his hind leg and emits small quantities of urine
Tail flapping/ beating	The tail is held under the prepuce and then it is beaten up and down four or five times, spreading urine over the croup and surrounding areas
Blathering	Emission of typical metal and gurgling sounds
Dulaa extrusion	Exteriorization of the soft palate, usually named *Dulaa*
Poll gland secretion	The occipital poll glands become thick and large and produce a tarry and dark secretion that colours the occipital area and the first part of the neck
Scratching occipital glands	The male rubs or scratches the neck, particularly the occipital area, on the wall or the windows of the box as well as on the gate bars
Froth on the mouth	The froth is generally attributed to increased secretion of the salivary glands by continuous grinding of teeth and by the frequent exteriorization of the soft palate
Nervousness	Increased pacing, anxiety and vocalizations

Adapted from Fatnassi et al. (2014a)

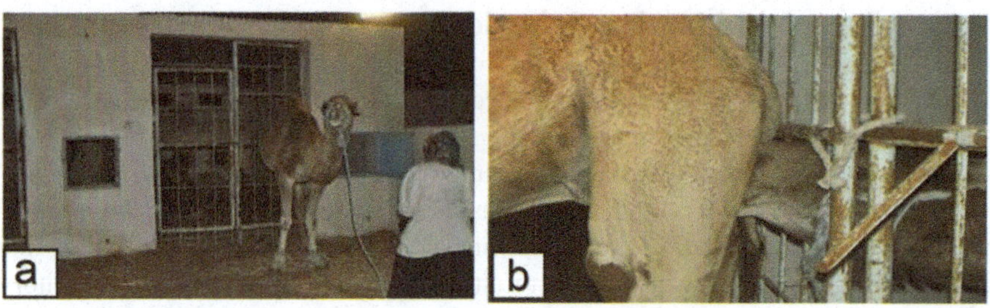

Fig. 7.1 Evaluation of dromedary sexual behaviour through the female parade. (**a**) Position of the female camel behind the box door; (**b**) sniffing the perineal region of the female

Mating behaviour is trigged by hormonal state, which may, in turn, depend on seasonal and other environmental factors as well as the accessibility and readiness of appropriate sexual partners. Like all animal species, mating behaviour is divided into two phases: a pursuit phase and a mating phase (Fernandez-Baca 1993; Vaughan et al. 2003; El-Bahrawi 2005).

During the first phase, the male chases the female, sniffs her genital region, urine or faeces, then performs the typical flehmen gesture of artiodactyls by lifting his head and curling back his upper lip (Gauthier-Pilters and Dagg 1981; Hanzen et al. 2014) (Fig. 7.2). The first phase lasts only a few seconds when the female is in oestrus but can last a few minutes if the female is non-receptive (Fernandez-Baca 1993). The second phase, namely the copulatory phase, begins when the male forces the female to adopt a sternal recumbency position (Taha-Ismail 1988; Hanzen et al. 2014). In this phase, the male mounts the female by extending his forelimbs to each side with

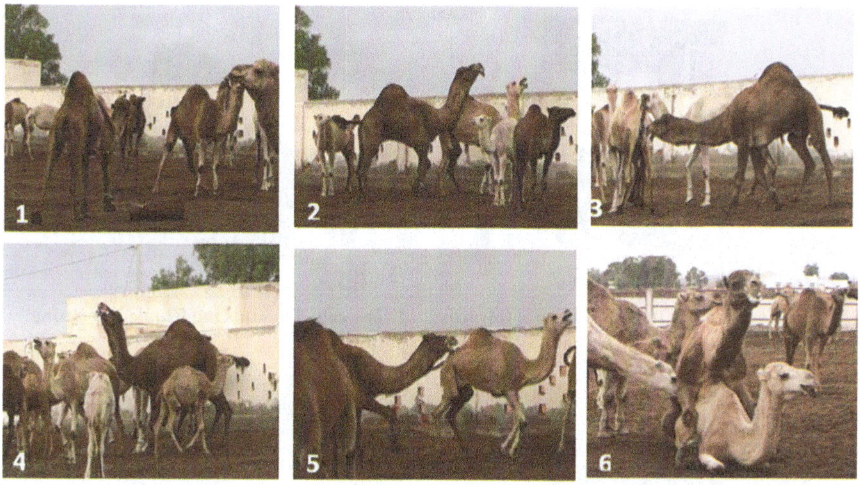

Fig. 7.2 Mating behaviour sequences in bulls reared with females (Fatnassi 2017). (**1**) The male searches the female, (**2**) approaches her, (**3**) sniffs her vulvar region, (**4**) shows flehmen behaviour, (**5**) pursues the female and forces her to adopt a recumbency position, and (**6**) mounts and mates the female

the head and neck kept straight. The hind legs of the male are fully fixed, with the heel-to-hock area resting on the ground (Abdel-Rahim and El-Nazier 1992; Hanzen et al. 2014). During copulation, the dromedary shows multiple pelvic thrusts and muscle contractions followed by a discharge of sperm (Fig. 7.2).

Ejaculation begins within minutes of penile intromission and occurs several times throughout the mating period. Copulation is accompanied by frequent grinding of teeth and abundant salivary secretion. At the end of copulation, the dromedary dismounts the female. The duration of this phase varies from male to male depending on their libido and sexual arousal (El-Bahrawi 2005; Tibary et al. 2014).

In nature, the males show a pursuit behaviour to mate with the females. However, in intensive management such as the case of the semen collection centre, this behaviour is absent; the male is brought to a female who usually is restrained in a couched position with ropes. In this situation, the male successfully achieves intromission and ejaculation regardless of the physiological state of the female (oestrus or not) (Fig. 7.3) (Fatnassi 2017). However, for the welfare point of view of the she-camels, it is strongly recommended that the she-camel is in heat, restrained properly and that she is not overused for the semen collection of multiple bulls (see Chap. 9).

In camels reared for semen collection, a procedure to collect semen respecting the natural mating of the camels was proposed (Padalino et al. 2015). The latter is more welfare friendly than other methods, in particular of electroejaculation (for details see Chap. 9 on reproduction). Using this method, mating behaviour and libido can be directly monitored and scored using a standard method with fixed times (Padalino et al. 2015; Fatnassi et al. 2021). In this case, the mating time was divided into three

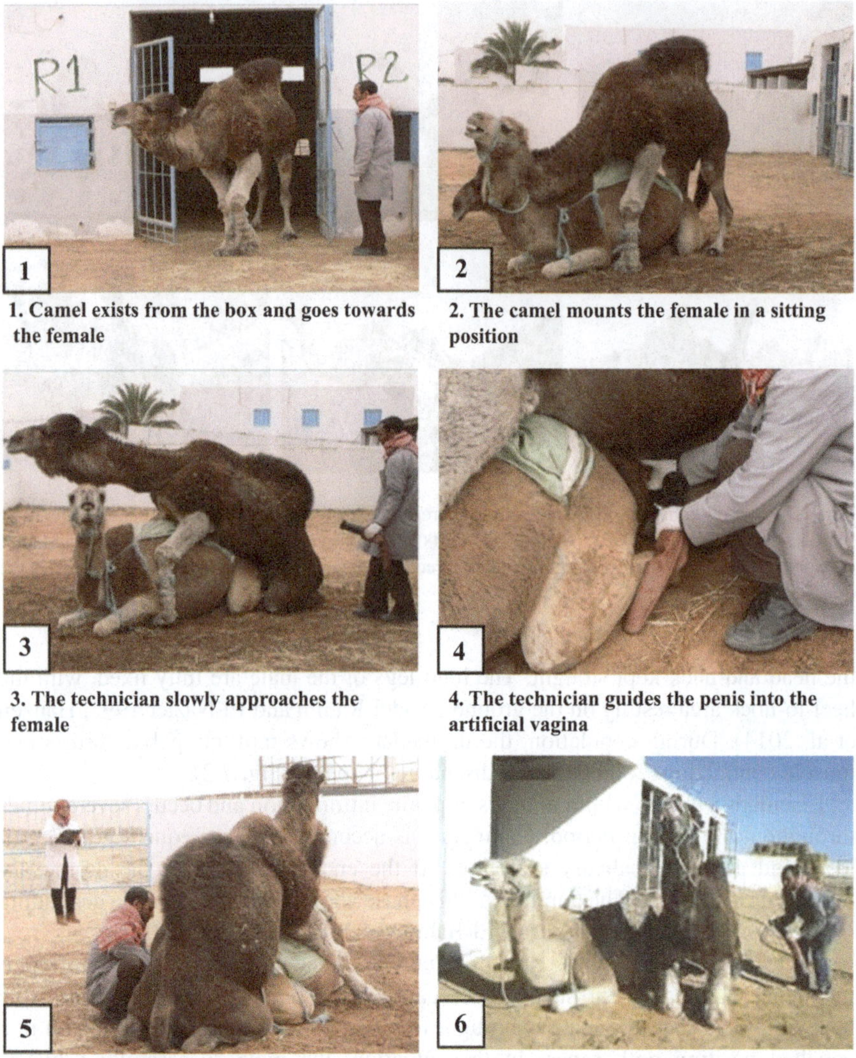

Fig. 7.3 Semen collection steps in camels reared for artificial insemination: from the exit from the box until the dismount and the end of ejaculation (Fatnassi 2017). (**1**) Camel exists from the box and goes towards the female. (**2**) The camel mounts the female in a sitting position. (**3**) The technician slowly approaches the female. (**4**) The technician guides the penis into the artificial vagina. (**5**) Intromission and ejaculation. (**6**) The camel dismounts the female

parts according to the different states: copulation/service time, standing over/near the female time and walking around time. The maximal timing used during the semen collection procedure of dromedary camel bulls is shown in Table 7.3 (Fatnassi et al.

Table 7.3 Maximal timing used during semen collection procedure of dromedary camel bulls

Parameters	Description	Maximal time (min)
Latency time	Time from the exit of camel from the box until its mount for the first time	15
Time between two services	Time from withdrawal of penis from artificial vagina until its intromission again	20
Standing on/over the Female time	Time spent when the male camel is near female expressing their sexual behaviours (sniffing, flehmen, dulaa)	30
Walking around time	Time when the camel is walking in the collection area, being not interested in, and far from, the female	4 × 3
Mating time	The time from first sitting on female to the return in the box = service/ejaculation time + standing over the female + walking around	45

Adapted from Padalino et al. (2015), Fatnassi et al. (2021)

2021). If one of these maximal times is reached, the mating session should be stopped (Padalino et al. 2015).

During this more welfare-friendly way of semen collection session, to properly score the libido of a bull, it is also important to note down the frequency of the following events: mounting attempts, number of mounts, blathering, defecation, *dulaa* extrusion, flehmen, jumping, neck-touching, sniffing, sound emission, tail flapping and teeth-grinding. At the end of each session, the libido score of the male camel can be consequently scored from 0 (absent: camel shows no sexual interest with latency time more than 15 min) to 5 (excellent: camels copulate more times, walk rarely with complete mating time) (Padalino et al. 2015).

7.9 Heat Behaviour in She-Camels

Female camels are seasonal polyestrous animals. Oestrus occurs at regular intervals in the year, with a period of 4–6 days. In this context, several studies have been established to detect the signs of oestrus behaviour in camels. The she-camel becomes restless and shows a slight vulval discharge. Seeking the male was described as a good behavioural indicator of camels in estrus (Atigui et al. 2013). Indeed, when a she-camel is in oestrus, she frequently approaches the bull and tries to solicit the attention of the male by adopting a dog-sitting posture (Padalino et al. 2016). Restlessness and urination have been suggested as oestrous camel behavioural patterns (Homeida et al. 1988; Atigui et al. 2013). Ghoneim et al. (2015) evaluated the behaviour of the female in the presence of a vasectomized bull in a large space, and sexual receptivity was marked only by an acceptance score, ranging from 'abstinence' (the female did not stand close to the male and escaped from him) to 'completely receptive' (stood close to the male and got into the sitting position). More recently, Padalino et al. (2016) used, for the first time, a standardized behavioural approach to detect signs of oestrus in she-camels through 15-min visual observation of females in the presence of a restrained bull. These authors reported that a female camel behaves differently when she has at least one mature follicle in her ovaries. During the ovulatory phase, female camels showed a marked interest in the bulls, interacted with the male more frequently, lay down and spent longer periods in this position in front of the bulls.

Consequently, visual observation of the behavioural signs of a female camel in the presence of a restrained bull might be useful for dromedary camel breeders, farmers and scientists to detect the ovulatory ovarian phase in camels and improve pregnancy rates in breeding camels (Padalino et al. 2016). Considering that in some intensive farms, behavioural observations have been considered time-consuming and moderately sensitive (Musa and Abusineina 1978, see Chap. 9 for details), more studies on the possible use of non-invasive smart technologies for the automatic detection of heat in dairy camel industry should be implemented.

7.10 How Management May Affect the Daily Rhythm of Camels

Male camels kept 24 h in a single box spent 42% of their time budget lying down and only 2% walking. Stereotypy occupied 15% of the time budget, with the dominance of swaying and wall-licking behaviours (Table 7.4).

During the scotophase period, camels spent their time lying down (63%) and rumination (21%). However, the time budget of housed camels during the photophase appeared to be different from the night. Indeed, lying down and rumination occupied only 14.26% and 2.08% of the time, whereas feeding, stereotypy and standing were observed, respectively, 30.33, 24.25 and 25.16% of the time (Aubè et al. 2017).

As mentioned before, the distribution of the different behaviours within the time budget changes with housing systems: at pasture, for example, camels were more active during the day and spent 24% of the time walking and 61% feeding with a total of 30 km covered per day (Chaibou 2005). Allowing camels to graze for 8 h/day, they spent 37.41% grazing, 31.7% rumination, 26.52% idling and 4.32% resting during a period of 24 h (Khan et al. 1998). Moreover, the feeding practices could affect the time spent in feeding and rumination by camels. When camels were

Table 7.4 Time budget and behaviours distribution during 24 h in dromedary camels housed in single box

Behavioural states	Time budget (%)	Time budget during scotophase (%)	Time budget during photophase (%)
Feeding	14.27	2.80	30.33
Rumination	12.92	20.66	2.08
Walking	2.03	0.68	3.92
Stereotypy	15.02	8.43	24.25
Standing	13.29	4.80	25.16
Lying down	42.47	62.62	14.26

Adapted from Aubè et al. (2017)

fed twice a day, they spent 29, 23, 22 and 25% in feeding, ruminating, lying and standing, respectively (El Shoukary et al. 2021).

Camels housed in an individual box showed less time feeding, walking and ruminating over the whole day compared to those kept in barns, fed with hay-based diets (with hay ad libitum), feeding took up 23.3% of the day and rumination 34.6% (Von Engelhardt et al. 2006). Likewise, Bactrian camels fed with hay ad libitum ruminated 22% of their time (Cahill and McBride 1995) and camels housed in open barns with straw ad libitum ruminated 40 and fed 30% (Hedi and Khemais 1990).

Therefore, housing system and feeding practices could explain the difference in the time budget of dromedary camels. Many studies conducted in dromedary camels, indicated that housing camels in a single box for 24 h should not be recommended since this system does not allow them to express their natural behavioural patterns (Fatnassi et al. 2014b; Padalino et al. 2014; Aubè et al. 2017).

7.11 Behavioural Needs of Camels

Behavioural needs are generally conceptualized as those behaviours that the animal must perform regardless of environmental circumstances. They are indeed primarily internally motivated behaviours that may occur even in the absence of appropriate external stimulation, although sometimes in an aberrant form (Mench 1998).

To ensure a good welfare status, a management system should meet the behavioural needs of the animals. Based on previous studies, to meet the feeding behavioural needs, providing more hay or a diet with a high fibre content, which are longer to ingest and digest than concentrate, could help to increase feeding and rumination times in camels. This may help in decreasing feeding frustration and the expression of oral stereotypies (Padalino et al. 2014; Aubè et al. 2017). In camels housed in boxes, providing straw bedding should be also recommended, to fulfil their behavioural needs to forage and feed. Given the negative effects of housing in a single box on behavioural repertoire (Fatnassi et al. 2014b, 2021; Padalino et al. 2014), giving camels access to a paddock (or even better put them on pasture) during daylight (more particularly around midday), where they can graze and walk seems necessary to allow them to express their natural behavioural patterns. According to previous studies, housing male camels for 24 h a day should not be recommended because of the high incidence of abnormal behaviours (Padalino et al. 2014; Fatnassi et al. 2016). Consequently, giving access to a paddock for at least 1 h and the possibility to interact with females for at least 30 min a day or to interact with other conspecific animals might be recommended to meet the physiological and behavioural needs of male camels and to enhance their reproductive performance, health and welfare (Fatnassi et al. 2014b).

7.12 Human-Camel Relationship

Human-animal relationships can be defined as 'the degree of relation or distance that exists between an animal and a human being, perceived, developed and expressed through their mutual behaviour' (Ellingsen et al. 2014). Human-animal relationship has been defined as an important criterion of the welfare principle of 'appropriate behaviour' and could be considered an interesting indicator of animals' internal states (Grandgeorge and Hausberger 2011). Camels have been domesticated many years ago and they have a long history of relationship with humans (see Chaps. 1 and 2). Indeed, camels had 'social competencies and needs' that could be used for establishing human-animal attachment. Thus, the assessment of the quality of human-camel attachment is an important means of improving animal welfare. In general, animals may perceive interaction with humans as: (a) negative, when they are afraid of people, avoiding contact with them; (b) neutral, when the fear level is low, but animals still avoid contact; and (c) positive, when fear is absent, and animals allow physical contacts (des Roches et al. 2016). Human-animal relationships are mainly affected by handling procedures. In general, and according to Ellingsen et al. (2014), handling that includes abrupt movements, pushing and the use of prods, shouts and kicks is considered negative, while handling characterized by slow movements, whispers and petting has positive effects on animals. Therefore, it is important to evaluate animals' reactions during handling procedures (Napolitano et al. 2013) because this factor may be used to select animals less afraid of humans and easier to manage (Windschnurer et al. 2009). In camels reared in intensive or semi-intensive systems, Padalino and Menchetti et al. (2021) have proposed an approaching test to measure the camel-human relationship. However, it is well known that for reaching a 'Good human-animal relationship' education of animal caretakers on animal learning theory is crucial (see Chap. 8). Consequently, it has been suggested that it is possible to assess the quality of the human-animal relationship directly by looking at caretakers' attitudes and handling practices (Spoolder 2007). However, there are other caretaker characteristics, affecting the relationship with the animals, that might be considered during a welfare assessment like the caretaker's knowledge, training and familiarity with the animals. According to the literature, these factors seem to improve empathy, attitudes and, ultimately, the quality of the handling of the camel caretakers and consequently they have a positive effect on the welfare and productivity of camels (Menchetti et al. 2021).

7.13 Recommendations for Ensuring the Principle of Appropriate Behaviour

Based on the literature, the following recommendations may be drawn to meet the behavioural needs of the camels:

- Camels should express feeding behaviour for at least 60% of their time budget/day.
- Camels should have access to a clean water source to cover their drinking needs.
- Camels should be kept in a group to express social behaviours.
- Camels should have access to pasture or be free to walk for about 30 km/day.
- Camels should have enough space to move around and rest comfortably (19–40 m^2/camel) and a sufficient shaded area (2.5–7 m^2/camel).

References

Abdel-Rahim SE, El-Nazier AE (1992) Studies on the sexual behaviour of the dromedary camel. In: 1st international camel conference, Dubai, UAE, pp 115–118

Albright JL, Arave CW (1997) The behaviour of cattle. CABI, Walingford, p 306

Al-Hazmi MA, Brain PF (1993) A preliminary study on the behaviour of the Dromedary (Camelus dromedarius) in the mid-western region of Saudi Arabia. J King Abdul Aziz Univ Sci 5:71–76. https://doi.org/10.4197/Sci.5-1.7

Altmann J (1974) Observational study of behavior: sampling methods. Behaviour 49:227–266. https://doi.org/10.1163/156853974X00534

Atigui M, Hammadi M, Khorchani T (2013) Effects of oestrus on milk yield and composition in Tunisian Maghrebi camels (Camelus dromedarius). Emir J Food Agric 25:291–295. https://doi.org/10.9755/ejfa.v25i4.15497

Aubè L et al (2017) Daily rhythms of behavioral and hormonal patterns in male dromedary camels housed in boxes. PeerJ 5:e3074. https://doi.org/10.7717/peerj.3074

Bayne K, Dexter S, Suomi S (1992) A preliminary survey of the incidence of abnormal behavior in rhesus monkeys (Macaca mulatta) relative to housing condition. Lab Anim 121:38–46

Beauchemin KA (1991) Ingestion and mastication of feed by dairy cattle. Vet Clin N Am Food Anim Pract 7:439–462. https://doi.org/10.1016/s0749-0720(15)30794-5

Bekele T, Lundeheim N, Dahlborn K (2011) Milk production and feeding behavior in the camel (Camelus dromedarius) during 4 watering regimens. J Dairy Sci 94:1310–1317. https://doi.org/10.3168/jds.2010-3654

Bekoff M, Byers JA (1998) Play: evolutionary, comparative and ecological perspectives. Cambridge University Press, Cambridge

Berger RJ, Phillips NH (1995) Energy conservation and sleep. Behav Brain Res 69:65–73. https://doi.org/10.1016/0166-4328(95)00002-b

Bhakat C, Raghavendra S, Sahani MS (2005) Effect of different management conditions on rutting behavior of Indian dromedary camel. Emir J Food Agric 17:1–13

Boussely L (2003) Étude bibliographique du bien-être chez le cheval. Thèse de doctorat: Ecole nationale de Médicine Véterinaire, Alfort, France, 139 p

Boyd L, Bandi N (2002) Reintroduction of takhi, Equus ferus przewalskii, to Hustai National Park, Mongolia: Time budget and synchrony of activity pre- and post-release. Appl Anim Behav Sci 78:87–102

Bristow DJ, Holmes DS (2007) Cortisol levels and anxiety-related behaviors in cattle. Physiol Behav 90:626–628. https://doi.org/10.1016/j.physbeh.2006.11.015

Burghardt GM (1999) Play. In: Greenberg A, Haraway M (eds) Comparative psychology: a handbook. Garland, New York, pp 725–735

Burghardt GM (2005) The genesis of animal play: testing the limits. Bradford Books, MIT Press, Cambridge, MA

Cahill L, McBride B (1995) Effect of level of intake on digestion, rate of passage and chewing dynamics in hay-fed Bactrian camels. Proc Nutr Advisory Group 1:3–35

Chaibou M (2005) Productivité zootechnique du désert: Le cas du bassin laitier d'Agadez au Niger. Montpellier 2

Christie JL et al (2006) Management factors affecting stereotypies and body condition score in nonracing horses in Prince Edward Island. Can Vet J 47:136–143. PMID: 16579039; PMCID: PMC1345728

Cooper JJ, Albentosa MJ (2005) Behavioral adaptation in the domestic horse: potential role of apparently responses including stereotypic behavior. Livest Prod Sci 92:177–182. https://doi.org/10.1016/j.livprodsci.2004.11.017

Couderc JJ et al (2006) Silage chop length and hay supplementation on milk yield, chewing activity, and ruminal digestion by dairy cows. J Dairy Sci 89:3599–3608

Crockett CM et al (2007) Beta-endorphin levels in longtailed and pigtailed macaques vary by abnormal behaviour rating and sex. Peptides 28:1987–1997. https://doi.org/10.1016/j.peptides.2007.07.014

Cronin GM, Wiepkema PR, Van Ree JM (1985) Endogenous opioids are involved in abnormal stereotyped behaviours of tethered sows. Neuropeptides 6:527–530. https://doi.org/10.1016/0143-4179(85)90114-3

Dahlgren I (2010) A comparison of group housing and individual for dairy bulls kept for breeding. Fakulteten för veterinärmedicinochhusdjursvetenskap. Institutionen för husdjurensmiljöochhälsaAvdelningen för etologiochdjurskydd, p 27

Dereje M, Uden P (2005) The browsing dromedary camel: I. Behaviour, plant preference and quality of forage selected. Anim Feed Sci Technol 121(3–4):297–308

des Roches ADB et al (2016) A prospective exploration of farm, farmer, and animal characteristics in human-animal relationships: An epidemiological survey. J Dairy Sci 99:5573–5585. https://doi.org/10.3168/jds.2015-10633

DeVries TJ et al (2009) Repeated ruminal acidosis challenges in lactating dairy cows at high and low risk 56 for developing acidosis: Feeding, ruminating, and lying behavior. J Dairy Sci 92:5067–5078. https://doi.org/10.3168/jds.2009-2102

Ebada S, Helal A, Alkafafy M (2012) Immunohistochemical studies on the poll gland of the dromedary camel during the rutting season. Acta Histochem 114:363–369. https://doi.org/10.1016/j.acthis.2011.07.005

El Allali K et al (2005) Seasonal variations in the nycthemeral rhythm of plasma melatonin in the camel (Camelus dromedarius). J Pineal Res 39:121–128. https://doi.org/10.1111/j.1600-079X.2005.00224.x

El-Bahrawi KA (2005) Reproductive studies on desert animals: sexual behaviour and semen characteristics and freezability of male dromedary camels. Ph. D. thesis, Fac. Agric. Alex. Univ., Egypt

Ellingsen K et al (2014) Using qualitative behavior assessment to explore the link between stockperson behavior and dairy calf behavior. Appl Anim Behav Sci 153:10–17. https://doi.org/10.1016/j.applanim.2014.01.011

El Shoukary RD, Osman AS, Mohamed AA (2020) Effects of stocking density on some behavioral and some blood biochemical parameters in camel during the rut period. Egypt J Vet Sci 51:253–262. https://doi.org/10.21608/ejvs.2020.24526.1153

El Shoukary RD et al (2021) Time-budget and blood parameters of the dromedary camels (Camelus dromedarius) under different feeding management strategies. Adv Anim Vet Sci 9(6):895–903. https://doi.org/10.17582/journal.aavs/2021/9.6.895.903

Fatnassi M (2017) Behavioral repertoire, sperm quality and identification of ovulation inducing factor and its role in ovulation in camel (Camelus dromedarius). Unpublished PhD thesis, University of Sousse, Sousse, Tunisia

Fatnassi M et al (2014a) Evaluation of sexual behavior of housed male camels (Camelus dromedarius) through female parades: correlation with climatic parameters. Trop Anim Health Prod 46:313–321. https://doi.org/10.1007/s11250-013-0489-x

Fatnassi M et al (2014b) Effect of different management systems on rutting behavior and behavioral repertoire of housed Maghrebi male camels (Camelus dromedarius). Trop Anim Health Prod 46: 861–867. https://doi.org/10.1007/s11250-014-0577-6

Fatnassi M et al (2016) Effect of continuous female exposure on behavioral repertoire and stereotypical behaviors in restrained male dromedary camels during the onset of the breeding season. Trop Anim Health Prod 48:897–903. https://doi.org/10.1007/s11250-016-1028-3

Fatnassi M et al (2021) Effects of two different management systems on hormonal, behavioral, and semen quality in male dromedary camels. Trop Anim Health Prod 53(2):275. https://doi.org/10.1007/s11250-021-02702-6

Faye B, Tisserand JL (1989) Problèmes de la détermination de la valeur alimentaire des fourrages prélevés par le dromadaire. Séminaire sur la nutrition et l'alimentation du dromadaire, Ouargla, Algérie. Options méditerranéennes 2:61–65

Fernandez-Baca S (1993) Manipulation of reproductive functions in male and female New Word camelids. Anim Reprod Sci 33:307–323. https://doi.org/10.1016/0378-4320(93)90121-7

Fregonesi JA, Tucker CB, Weary DM (2007) Overstocking reduces lying time in dairy cows. J Dairy Sci 90:3349–3354. https://doi.org/10.3168/jds.2006-794

Fureix C et al (2012) Towards an ethological animal model of depression? A study on horses. PLoS One 7:e39280. https://doi.org/10.1371/journal.pone.0039280

Gauthier-Pilters H, Dagg AI (1981) The camel. University of Chicago Press, Chicago, IL

Ghoneim IM et al (2015) Relationship between the size of the dominant follicle, vaginal electrical resistance, serum concentrations of oestradiol and progesterone and sexual receptivity during the follicular phase of the dromedary camel (Camelus dromedarius). Anim Reprod Sci 154:63–67. https://doi.org/10.1016/j.anireprosci.2015.01.003

Ginane C et al (2015) Feeding behaviour in ruminants: A consequence of interactions between a reward system and the regulation of metabolic homeostasis. Anim Prod Sci 55:247–260. https://doi.org/10.1071/AN14481

Gomendio M (1988) The development of different types of play in gazelles-implications for the nature and functions of play. Anim Behav 36:825–837

Gonzalez M, Yabuta AK, Galindo F (2003) Behaviour and adrenal activity of first parturition and multiparous cows under a competitive situation. Appl Anim Behav Sci 83:259–266. https://doi.org/10.1016/S0168-1591(03)00037-6

Grandgeorge M, Hausberger M (2011) Human-animal relationships: from daily life to animal-assisted therapies. Ann Ist Super Sanità 47(4):397–408. https://doi.org/10.4415/ANN_11_04_12

Hammadi M et al (2010) Training period and short time effects of machine milking on milk yield and milk composition in Tunisian maghrebi camels (Camelus dromedarius). J Camel Pract Res 17(1):1–7

Hammadi I et al (2021) Dam parity and calf sex affect maternal and neonatal behaviors during the first week postpartum in stabled Maghrebi dairy camels. Animal 1–8. https://doi.org/10.1016/j.animal.2020.100149

Hanzen C et al (2014) Anatomie et physiologie de la reproduction du male. In: Hanzen C (ed) Principes de reproduction des petits camélidés sud-américains. International Veterinary Information Service, Ithaca, NY

Hausberger M, Gautier E, Müller C, Jego P (2007) Lower learning abilities in stereotypic horses. Appl Anim Behav Sci 107:299–306. https://doi.org/10.1016/j.applanim.2006.10.003

Hausberger M et al (2012) On the significance of adult play: What does social play tell us about adult horse welfare? Naturwissenschaften 99:291–302. https://doi.org/10.1007/s00114-012-0902-8

Hedi A, Khemais K (1990) Intake, digestion and feeding behaviour of the one-humped camel stall-fed straw-based diets. Livest Res Rural Dev 2

Held SDE, Spinka M (2011) Animal play and animal welfare. Anim Behav 81:891–899. https://doi.org/10.1016/j.anbehav.2011.01.007

Homeida AM, Khalil MGR, Taha AA (1988) Plasma concentrations of progesterone, oestrogens, testosterone and LH-like activity during oestrous cycle of the camel (Camelus dromedarius). J Reprod Fertil 83:593–598. https://doi.org/10.1530/jrf.0.0830593

Houpt KA, Erb HN, Coria-Avila GA (2019) The sleep of shelter dogs was not disrupted by overnight light rather than darkness in a crossover trial. Animals 9(10):794. https://doi.org/10.3390/ani9100794

Hoyer BH (2013) 'Environmental enrichment': strategies to improve the housing conditions of breeding bulls. Impact on time budget, physical activity, rumination, sexual behavior and semen quality. Unpublished PhD thesis, University of Veterinary Medicine Hannover

Iqbal A (1999) Studies on some of the productive, reproductive and behavioural aspects of camel in Pakistan. Ph.D. Thesis, Univ. Agri., Faisalabad

Iqbal A, Khan BB (2001) Feeding behaviour of camel—review. Pak J Agric Sci 38:58–63

Kaczensky, P et al (2006) Activity patterns of brown bears (Ursus arctos) in Slovenia and Croatia. Journal of Zoology 269(4):474-485.

Kaske M et al (1989) Circadian changes of forestomach motility and of rumination in camels. Asian Australas J Anim Sci 2:301–302. https://doi.org/10.5713/ajas.1989.301

Keeling L, Jensen P (2009) Abnormal behaviour, stress and welfare. In: Jensen P (ed) The ethology of domestic animals, 2008, 2nd edn. CABI, Wallingford, pp 85–101

Khan B et al (1998) A study on some of the activity patterns of Camelus dromedarius maintained in Thal area of the Punjab Pakistan. Pak J Agric Sci 33:67–72

Khaskheli AS (2020) A review on several important aspects of the camels. Aceh J Anim Sci 5(2): 129–135. https://doi.org/10.13170/ajas.5.2.17580

Khorchani T et al (1992) Nutrition of the one-humped camel. II. Intake and feeding behaviour on arid ranges in southern Tunisia. Anim Feed Sci Technol 39:303–311. https://doi.org/10.1016/0377-8401(92)90049-C

Kilgour RJ et al (2012) The behavior of beef cattle at pasture. Appl Anim Behav Sci 138:12–17. https://doi.org/10.1016/j.applanim.2011.12.001

Lidfors L, Berg C, Algers B (2005) Integration of natural behavior in housing systems. Ambio 34: 325–330

Lindshield SM (2017) Behavioral flexibility. In: Fuentes A (ed) The inter-national encyclopedia of primatology. Wiley, Hoboken, NJ

Lindström T, Redbo I, Uvnäs-Moberg K (2001) Plasma oxytocin and cortisol concentrations in dairy cows in relation to feeding duration and rumen fill. Physiol Behav 72:73–81. https://doi.org/10.1016/s0031-9384(00)00381-4

Martin P, Bateson P (1993) Measuring behaviour: An introductory guide, 2nd edn. Cambridge University Press, Cambridge, UK

Mason GJ (1991) Stereotypies: a critical review. Anim Behav 41:1015–1037. https://doi.org/10.1016/S0003-3472(05)80640-2

Mason GJ (1993) Age and context affect the stereotypies of caged mink. Behaviour 127:191–229

Mason G, Latham N (2004) Can't stop, won't stop: is stereotypy a reliable animal welfare indicator? Anim Welf 13:57–69

Mason G, Rushen J (2006) Stereotypic animal behaviour: fundamentals and applications to welfare, 2nd edn. CABI, Wallingford

McDonnell SM (2000) Reproductive behavior of stallions and mares: comparison of free-running and domestic in-hand breeding. Anim Reprod Sci 60–61:211–219. https://doi.org/10.1016/s0378-4320(00)00136-6

McDonnell S (2003) A practical field guide for horse behavior: the equid ethogram. Blood-Horse Inc., Lexington, KY

McGreevy PD (2002) Development and resolution of behavioural problems in the ridden horse. In: Havemeyer workshop on horse behaviour and welfare, Holar, Iceland, June 13–15, 2002, pp 110–115

McGreevy PD et al (1995) Management factors associated with stereotypic and redirected behaviour in the Thoroughbred horse. Equine Vet J 27:86–91. https://doi.org/10.1111/j.2042-3306.1995.tb03041.x

Mench J (1998) Why it is important to understand animal behavior. ILAR J 39(1):20–26. https://doi.org/10.1093/ilar.39.1.20
Mench JA, Mason GJ (1997) Behaviour. In: Appleby MC, Hughes BO (eds) Animal welfare. CABI, Wallingford, pp 127–142
Menchetti L et al (2021) Application of a protocol to assess camel welfare: scoring system of collected measures, aggregated assessment indices, and criteria to classify a pen. Animals 11: 494. https://doi.org/10.3390/ani11020494
Mengli Z et al (2006) Bactrian camel foraging behaviour in a Haloxylon ammodendron (C.A. Mey) desert of Inner Mongolia. Appl Anim Behav Sci 99(3–4):330–343
Moaeenuddin M et al (2004) Feeding behavior of camel under stall feeding. J Anim Plant Sci 14(3–4):74–76
Monaco D et al (2013) The experimental semen collection centres for dromedary camels in Egypt and Tunisia: Current situation and future developments. In: Proceeding of the XI congress of Italian Society of Animal Reproduction (SIRA), Ustica, 2013, pp 132–136
Munksgaard L et al (2005) Quantifying behavioral priorities: Effects of time constraints on the behavior of dairy cows, Bos Taurus. Appl Anim Behav Sci 92:3–14
Musa BE, Abusineina ME (1978) The oestrous cycle of the camel (Camelus dromedarius). Vet Res 103(25):556 557. https://doi.org/10.1136/vr.103.25.556
Nagy PP, Skidmore JA, Juhasz J (2022) Intensification of camel farming and milk production with special emphasis on animal health, welfare, and the biotechnology of reproduction. Anim Front 12(4):35–45. https://doi.org/10.1093/af/vfac043
Napolitano F et al (2013) Public knowledge, attitudes, and experience regarding the use of antibiotics in Italy. PLoS One 8(12):e84177. https://doi.org/10.1371/journal.pone.0084177. PMID: 24376793; PMCID: PMC3871686
Nicol C (1999) Understanding equine stereotypies. Equine Vet J 28:20–25. https://doi.org/10.1111/j.2042-3306.1999.tb05151.x
Padalino B, Menchetti L (2021) The first tool for assessing welfare of camels. Front Vet Sci 7: 631876. https://doi.org/10.3389/fvets.2020.631876
Padalino B et al (2013) A new ethogram for evaluating housed male dromedary camel behavior and libido. In: Proceeding of the XI congress of Italian Society of Animal Reproduction (SIRA), Ustica, Italy, pp 137–142
Padalino P et al (2014) Could dromedary camels develop stereotypy? The first description of stereotypical behaviour in housed male dromedary camels and how it is affected by different management systems. PLoS One 9:e89093
Padalino B, Monaco D, Lacalandra GM (2015) Male camel behavior and breeding management strategies: how to handle a camel bull during the breeding season? Emir J Food Agric 27:338–349. https://doi.org/10.9755/ejfa.v27i4.19909
Padalino B et al (2016) Behavioral indicators to detect ovarian phase in the dromedary she-camel. Theriogenology 85(9):1644–1651. https://doi.org/10.1016/j.theriogenology.2016.01.027
Pagot J (1992) Animal production in the tropics. Macmillan, London
Power CN (2000) Regional training seminar on guidance and counselling. Module 4. Behaviour modification UNESCO
Raabymagle P, Ladewig J (2006) Lying behavior in horses in relation to box size. J Equine Vet Sci 26:11–17. https://doi.org/10.1016/j.jevs.2005.11.015
Ransom JI, Cade BS (2009) Quantifying equid behavior: A research ethogram for free-roaming feral horses. US Geological Survey Techniques and Methods 2-A9, 23 p
Redbo I, Nordblad A (1997) Stereotypies in heifers are affected by feeding regime. Appl Anim Behav Sci 53:193–202. https://doi.org/10.1016/S0168-1591(96)01145-8
Refinetti R, Cornélissen G, Halberg F (2007) Procedures for numerical analysis of circadian rhythms. Biol Rhythm Res 38:275–325. https://doi.org/10.1080/09291010600903692
Sambraus HH (1994) Lying down behaviour and its diurnal distribution in dromedaries. Anim Breed Genet 111(5–6):563–471

Sarrafchi A, Blokhuis HJ (2013) Equine stereotypic behaviors: Causation, occurrence, and prevention. J Vet Behav 8:386–394. https://doi.org/10.1016/j.jveb.2013.04.068
Schirmann K et al (2009) Technical note: Validation of a system for monitoring rumination in dairy cows. J Dairy Sci 92:6052–6055. https://doi.org/10.3168/jds.2009-2361
Schirmann K et al (2012) Rumination and its relationship to feeding and lying behavior in Holstein dairy cows. J Dairy Sci 95:3212–3217. https://doi.org/10.3168/jds.2011-4741
Schulte N, Klingel H (1991) Herd structure, leadership, dominance and site attachment of the camel, Camelus dromedarius. Behaviour 118(1–2):103–114. https://doi.org/10.1163/156853991X00229
Schwan L (2011) Social behavior and time budget of breeding bulls. Student report, Swedish University of Agricultural Sciences
Shaheen G (2009) Seasonal variation in nutritional and anti-nutritional components of native shrubs and trees grown in Hazargangi Chiltan National Park, Karkhasa and Zarghoon. PhD dissertation, University of the Balochistan, Quetta
Skidmore JA, Morton KM, Billah M (2013) Artificial insemination in dromedary camels. Anim Reprod Sci 136:178–186. https://doi.org/10.1016/j.anireprosci.2012.10.008
Smith G (1995) Dopamine and food reward. In: Fluharty S, Morrison A, Sprague J, Stellar E (eds) Progress in psychobiology and physiological psychology. Academic, New York, pp 83–144
Spinka M, Newberry RC, Bekoff M (2001) Mammalian play: training for the unexpected. Q Rev Biol 76:141–168. https://doi.org/10.1086/393866
Spoolder HAM (2007) Animal welfare in organic farming systems. J Sci Food Agric 87:2741–2746. https://doi.org/10.1002/jsfa.2999
Taha-Ismail ST (1988) Reproduction in the male dromedary (Camelus dromedarius). Theriogenology 29:1407–1418. https://doi.org/10.1016/0093-691X(88)90022-2
Thompson KV (1996) Play-partner preferences and the function of social play in infant sable antelope, Hippotragus niger. Anim Behav 52:1143–1155
Tibary A, Pearson LK, Anouassi A (2014) Applied andrology in camelids. In: Chenoweth PJ, Lorton SP (eds) Animal andrology: theories and applications. CABI, Wallingford, pp 418–449
Uzal Seyfi S (2013) Seasonal variation of the lying and standing behavior indexes of dairy cattle at different daily time periods in free-stall housing. Anim Sci J 84:708–717. https://doi.org/10.1111/asj.12062
Vaughan J, Galloway D, Hopkins D (2003) Artificial insemination in alpaca (Lama pacos). A report for the rural industries research and development corporation, Barton, Kingston, p 90
Vieira ML, Sartorio R (2002) Motivational, causal and functional analysis of play behavior in two rodent species. Estudios de Psicologia 7:189–196
Von Engelhardt W, Haarmeyer P, Lechner-Doll M (2006) Feed intake, forestomach fluid volume, dilution rate and mean retention of fluid in the forestomach during water deprivation and rehydration in camels (Camelus sp.). Comp Biochem Physiol A Mol Integr Physiol 143:504–507. https://doi.org/10.1016/j.cbpa.2006.01.021
Waring GH (2003) Horse behavior, 2nd edn. William Andrew, New York, p 153
Wickens CL, Heleski CR (2010) Crib-biting behavior in horses: A review. Appl Anim Behav Sci 128:1–9. https://doi.org/10.1016/j.applanim.2010.07.002
Windschnurer I, Boivin X, Waiblinger S (2009) Reliability of an avoidance distance test for the assessment of animals' responsiveness to humans and a preliminary investigation of its association with farmers' attitudes on bull fattening farms. Appl Anim Behav Sci 117:117–127. https://doi.org/10.1016/j.applanim.2008.12.013
World Organization for Animal Health (2022) Terrestrial Code. Available at: https://www.woah.org/en/what-we-do/standards/codes-and-manuals/terrestrial-code-online-access/
Yagil R (1990) Camel's milk-one hump or two? The Economist, October 20
Zappaterra M et al (2021) Do camels (Camelus dromedarius) need shaded areas? a case study of the camel market in Doha. Animals 11(2):1–16. https://doi.org/10.3390/ani11020480

Camel Handling and Training

8

Coralie Le Meur, Barbara Padalino, and Bernard Faye

Abstract

Camel training has been performed only in some countries and mainly using ancestral methods. There is a huge gap between traditional belief and science in relation to animal training. There is a belief that camels are very aggressive animals, difficult to handle, and needs to be trained using aversive methods. However, camels are used for riding and carriage and like all other animals, can learn and can be trained with appropriate methods, namely using learning theory. A good trainer should have a good knowledge of camel behaviour, be able to read their behaviour and communicate clearly with the camel, adapting himself to different kinds of situations according to each animal's character. Camels have great learning potential, adapting their behaviours quickly according to the environment. In order to train them, spending a large amount of time studying their behaviour, and observing how they interact together and towards humans, the trainer will be able to pull off the best part of each camel and develop a particular bond with each of them. This chapter presents scientific knowledge in animal learning and psychology. Specifically, it focuses on associative and non-associative learning and gives some practical guidance on how to train camels from the ground based on the experience of the main author.

C. Le Meur
DromaSud, Frontignan, France

B. Padalino (✉)
Department of Agricultural and Food Sciences, University of Bologna, Bologna, Italy
e-mail: Barbara.padalino@unibo.it

B. Faye
CIRAD, Montpellier, France

B. Padalino, B. Faye (eds.), *Dromedary Camel Behavior and Welfare*, Animal Welfare 24, https://doi.org/10.1007/978-3-031-48600-5_8

Keywords

Dromedary camels · Training · Learning · Handling

8.1 Introduction

Camels, like any other animals, can learn and adapt to new situations. Understanding and interpreting the behaviours of the camel is the key to learning from them and later adapting teaching methods. Learning is not necessarily meaning to be trained. Learning is happening every time an action is done. Thorndike (1898a, b) is one of the founders of the learning theory concept, and inventor of the concept of connectionism (models of mental or behavioural phenomena as emergent processes from a network of interconnected single units). Pavlov (1927) is a precursor as well and proposed the concept of classical conditioning. The non-associative and associative learning rules (see Table 8.1 for the most relevant definitions) should be deeply understood by all individuals working with camels, not only to improve the physical and psychological welfare of camels but also to reduce accidents and the number of camels lost to behavioural problems (i.e., aggressivity) caused by an improper relationship with a person. Learning how to handle camels using learning theory will also reduce camel-related human injuries, and the time spent handling and moving camels. This chapter critically reviews the literature to provide a comprehensive, detailed and deeper understanding of how associative and non-associative learning functions when working with camels. This chapter highlights some examples of practical outcomes in training from the ground based on the main author's experience.

8.2 Associative and Non-associative Learning

Training is based on the proper communication between the trainer and the animal; the trainer should send a clear stimulus to avoid misunderstanding and should reward the wanted behaviour (McGreevy and Boakes 2006). Before starting the training programme, the trainer should have a good knowledge of camel behaviour and learning theory, a key factor to establish a good human-camel relationship. Table 8.1 shows the main definition of animal learning.

Animals can learn through associative and non-associative learning (Table 8.1).

In the non-associative, there is only one stimulus, while in the associative learning a relationship between at least two stimuli becomes established. Non-associative learning is divided into two categories: habituation and sensitisation.

Habituation. When a phenomenon happens frequently, animals will react less to this stimulus, becoming habituated to it and reducing or totally eliminating their behavioural response. Example: When a camel living in the desert is moved to a paddock near a road, his reaction when seeing a car or a truck for the first time should be to stop grazing and run away, being afraid of the noise. By the time, if several cars

Table 8.1 Definitions of animal learning

Associative learning	Process that allows the animal to establish the connection between two events in a relationship of reciprocity between them (Vallortigara 2000)
Classical conditioning	Is a type of associative learning whereby behavioural response becomes elicited from a conditioned stimulus (Pavlov 1927). With classical conditioning, animals learn which environmental cues predict future events so that they can behave accordingly (Cooper 1998). In such cases, the animal has no control over events; and the response is not under the control of the animal. Classical conditioning increases the predictability of environmental stimuli (Vallortigara 2000)
Communication	The activity of conveying information through the exchange of thoughts, messages, or feelings, as by vocal and visuals signals, or behaviour. It is the meaningful exchange of information between two or more living creatures (Barnlund 2008)
Continuous reinforcement	Each correct behaviour of the animal is reinforced (Cooper 1998)
Habituation	The animals decrease their response to a single stimulus (McGreevy and Boakes 2006)
Learning	The information obtained from the interaction between an environmental stimulus and the elicited behaviour will form the experience, according to which the animal will change its behaviour in the presence of that stimulus when it will reoccur in the future (Vallortigara 2000). Broadly, animals learn to use the information coming from the environment to change their behaviour in the most advantageous manner to them (Nicol 2005)
Non-associative learning	Refers to a relatively permanent change in the strength of a behavioural response to a single stimulus due to repeated exposure to that stimulus (Vallortigara 2000)
Operant conditioning	Is a type of associative learning in which an individual's voluntary behaviour is modified by its antecedents and consequences (Skinner 1938). It works by giving or taking away rewards or punishments (discomforts) when the horse performs a desired behaviour through the chain: stimulus—response—reinforcement (Cooper 1998). In operant conditioning it is the animal's behaviour that determines the progression of the reinforcement. Therefore, it allows the animal to associate two events over which it has control (Vallortigara 2000)
Primary and secondary reinforcements	Primary reinforcements are any resources that animals have evolved to seek (food, water, sex, play, freedom, companionship), whereas secondary reinforcement are stimuli which are not intrinsically rewarding but that can be associated with primary reinforcement (through classical conditioning) (Mills 1998)
Punishment	Punishment is any action that makes the occurrence of a behaviour less likely to be performed in the future (Mills 1998) • Positive punishment is to add something undesirable or painful • Negative punishment is to remove something desirable by the animals
Reinforcement	Any event that increases the frequency of a certain behaviour and makes it more likely to occur in the future (Vallortigara 2000). The reinforcement needs to be something biologically relevant for the

(continued)

Table 8.1 (continued)

	animal (the removal of discomfort or the appearance of food), so it is highly motivated to obtain it • Negative reinforcement is the subtraction of something aversive (Thorndike 1898a, b) • Positive reinforcement is the addition of something pleasant (Skinner 1938)
Relationship	The emerging bond from a series of interactions that partners have. It is based on past experiences and expectations of the other individual's responses (Hinde 1979)
Sensitisation	Sensitisation is the opposite of habituation. There is an increase in the response after repeated presentations of the stimulus by itself (McGreevy and Boakes 2006)
Stimulus	Any appreciable change in the environment that causes a behavioural response in the animal (Vallortigara 2000)
Training	Training suppresses undesirable behaviour and enhances desirable natural or new behavioural responses by punishing or reinforcing them with the deliberate or accidental application of learning theory (Cooper 1998). The goal of training is to lead the animal to perform a predictable behaviour as a result of the appearance of specific signals (McGreevy and Boakes 2006)

Adapted from Baragli et al. (2015)

Fig. 8.1 Camel wearing quietly different type of halters

and trucks are passing by, the camel will not run away by habituation because it becomes a normal situation, and he will continue to graze near the road. Another example of habituation is when the camel learns to wear equipment, like a halter or a blanket (Fig. 8.1). The first time the halter is placed on the camel's head, the camel will shake the head and will show several stress-related behaviours, but after a while, he will get habituated and will wear the halter quietly, and even if the halter is removed and put on later, the stress-related behavioural reactions will not be present anymore.

Sensitisation can counteract habituation and can be utilised when a camel has stopped completely to react to a stimulus. It can be used when an animal does not react to any stimulus.

Associative learning is the process whereby things that occur close in time can be associated. In associative learning, the animals make an association between a stimulus and a response. Associative learning is divided into two categories: classical (or Pavlovian) conditioning and operant (or instrumental) conditioning (Table 8.1).

Operant conditioning (OC): OC is the process of learning through reinforcement and punishment. It involves an organism that must first act up on the environment in some way. Thorndike's learning theories came from his study on cats in a puzzle box. In the experiment, one hungry cat was put in a box. On the outside of the box was a fish that the cat could see and smell. The box had a door that could be opened by pressing a lever inside the cage. Sensing the fish, the cat would engage in a variety of behaviours in an attempt to open the door and get the fish. Eventually one of these behaviours (pressing the lever) would result in the door opening and the cat getting the fish. Then, the consequences associated to the behaviour of pressing the lever were freedom and the fish (rewards). Learning for the hungry cat was a matter of making the connection between lever-pressing and door-opening/fish-eating. This learning was incremental, not insightful. This means that the cat was not able to gain sudden insight or make a logical connection between lever-pressing and door-opening/fish-eating. Instead, the cat made small incremental gains towards the lever-open door connection. Each time the cat was put in the puzzle box, it took successively fewer trials to express the right behaviour (pressing the lever). Finally, after many times in the puzzle box, the cat eventually would go directly to the lever. This is called trial and error learning or selecting and connecting. A behaviour was selected (lever-pressing) and a connection was eventually made and strengthened with the door-opening consequence (Johnson 2014).

How to teach an animal to press a lever? Merely by defining a wanted behaviour and rewarding the animal at each time when that behaviour is appearing. At the reverse, by defining an unwanted behaviour and punishing it every time it is appearing. Reinforcement always increases the wanted behaviour; in opposition punishment always decreases the occurrence of unwanted behaviour. There are also two types of reinforcement, positive and negative and two types of punishments, positive and negative (Table 8.1). For rendering a reinforcement or a punishment effective, it should occur immediately after the behaviour to be encouraged or discarded. Indeed, the timing is very important to be taken into account. Different studies on animal learning theory show that a short time between the behaviour and the reinforcement or punishment is more effective than a long time (Gibbon 1977).

Example of training using reinforcement: The trainer wants to teach a camel to move forward using halter pressure (negative reinforcement) in order to lead him for a walk. The camel doesn't know how to react at the first time, so he will try many solutions, pull backwards, jump, and move forward. As trainer, we know what is the wanted behaviour (moving forward), so the pressure (negative reinforcement) must be realised when the camel moves forward. We can also combine training with negative and positive reinforcement, and in that case, when the camel moves forward the pressure is released (negative reinforcement) and some food is given (positive reinforcement). By operant conditioning, the animal learns that pulling the halter forward means that he must move forward. The camel will be willing to give again a similar behavioural response (moving forward) the next time the halter is pulled forward; the camel will show the wanted behaviour to avoid the pressure and obtain additional food.

Example of training using punishment: when a camel is dangerously moving, the trainer will yell at him in order to decrease the unwanted behaviour. In this case, a stimulus is added (yelling) in order to decrease the behaviour (moving dangerously). In negative punishment, a positive stimulus is removed. As an example, when the camel misbehaves, the trainer doesn't give him food. In this case, the stimulus (the food) is removed until the unexpected behaviour decreases. Negative punishment works only with animals trained with positive reinforcements.

There are many studies on dogs which have proved that training using positive reinforcement is more welfare friendly because it is associated with positive neurotransmitters and emotion, training by punishments on the contrary leads to negative emotions (Gal 2017).

Classical conditioning (CC): CC is referring to the behavioural and physiological changes after experiencing of a predictive relationship between a neutral stimulus and a consequent biologically significant event. It involves an organism that is passive, simply responding to a stimulus presented to it. Pavlov noticed that the presentation of meat powder (an unconditioned stimuli or UCS) to his dog caused its salivation (an unconditioned response or UCR). The original stimuli and response are unconditioned because both occurred naturally without any conditioning. During the conditioning, the meat powder (UCS) was paired with a neutral stimulus (NS). The neutral stimulus was a bell.

Here, "neutral" means that there is no particular response of the dog when the bell is used. A plate of meat is presented to the dog simultaneously with the bell ringing several times. Those simultaneous stimuli produced exactly a similar response (UCR), i.e., salivation. The link between the bell ringing and the meat is providing an important strength. Consequently, the bell ringing will produce a similar response i.e., salivation, which becomes the conditioned response (CR). Thus, the dog is conditioned and can respond to the bell ringing at each time. The bell ringing is provoking by itself a conditioned stimulus (CS) (Johnson 2014). The more a certain event or environment is paired with a particular consequence, the stronger the association.

Example: When the trainer asks the camel to stand up using a particular sound before applying the pressure with the rope, after many repetitions, the camel will stand up only with the voice anticipating the pressure.

Clicker training (Feng et al. 2016) is a method based on behavioural psychology that relies on associative learning, combining classical and operant conditioning. A clicker is a mechanical device that makes a short and distinct "click" sound which tells the animal exactly when it is doing the right behaviour. This clear form of communication, combined with positive reinforcement, is an effective, safe and humane way to teach any type of animal any behaviour that it is physically and mentally able to do. Before starting the training, the clicker is loaded, which means that the clicker is played, and food is given to the animals. This first phase is based on classical conditioning, the animal learns that the sound of the clicker is associated with food, like the bell and the meat in Pavlov's dog, and the sound of the clicker can be considered a secondary reinforcer (Table 8.1). There are different methods of clicker training, namely free capture, targeting, and combined with negative reinforcement (Feng et al. 2016). In the case of the free capture, in the second phase of the clicker training, the trainer clicks at the moment when the animal shows the wanted behaviour: for instance, when the camel lifts his foot, the trainer clicks simultaneously. When the camel lies down, the trainer clicks. Clicking is like taking a picture of the behaviour that the trainer wants to reinforce. Immediately after "taking the picture," a reward is given to the animal. The reward can be a break time in the exercise, playing for a short moment or a piece of favourite food. Quickly, the animal is associating the behaviour with the click and the reward. The animal becomes more willing to recall this pleasant experience and will repeat the action that made it hear the click and so the reward. In targeting (Fig. 8.2), during the second phase of the clicker training, a target (e.g., a cone or a stick) is used and when the animal touches the target the clicker is played and food is given, then the target is moved and the animal moves and keeps touching the target. This is particularly used for teaching animals to self-load into vehicles (Dai et al. 2019).

An animal who was clicker trained or operantly trained is more willing to learn new behaviour. Even years later, learned behaviours are still remaining because animals were aware of them as they learned them rather than acquired them without awareness. As they have control over the consequences of their actions, they do develop confidence. Because they expect those consequences to be pleasurable, they become more and more enthusiastic about learning sessions. Basically, all behaviour can be reinforced and learned with all animals following these three steps: seeing the behaviour; marking the behaviour; reinforcing the behaviour.

Clicker-trained animals want to perform behaviours for which they have been rewarded in the past.

They will perform any behaviour if they did understand the meaning of the cue and if the desire of the reward is strong. If they do not perform the behaviour, the animal is not necessarily disobeying, so the trainer should think about the following questions:

- Does the animal know the meaning of the cue?
- Does the animal know the meaning of the cue in the environment in which it was first taught, but not in the environment in which it was given?
- Is the reward for doing the behaviour sufficiently desired by the animal?

Fig. 8.2 Example of targeting training

According to the answers to those questions, the trainer should revise the training process and make sure the animal knows the meaning of the cue in different environments, regardless of distractions and that the desire for reward is strong enough for the behaviour. Clicker trainers who learn the underlying principles have at their disposal a powerful set of tools that enable them to analyse behaviours, modify existing methods for individual animals and create new methods when none previously existed. This flexibility allows the tools of clicker training to be re-invented in new forms that work in a range of situations, and for an infinite variety of animals. Jim Wiltens, co-leader for the "Camels Over the Himalayas Expedition" has successfully experienced clicker training on camels with Karen Pryor technique of clicker training (Pryor 2009).

Animal learning theory can be applied to camels, individually or at the herd level contributing to the improvement of their management, and consequently to their welfare. To manage camels properly, their high capacities to adapt to new situations should be considered. This is why knowing how a camel is learning is important (Iglesias et al. 2020). Their main motivation is to get food in the easiest way as possible and feeling safe in their environment. Unfortunately, there are no studies on the effects of training based on learning theory in camels, so these types of studies

Fig. 8.3 Carolie holding a dromedary camel with a very soft hand

are needed, and the rest of the chapter is based only on the experience of the main author (Fig. 8.3), who is an experienced trainer of camel in France.

8.3 Camel Training Methods

Nowadays, the main objectives in camel training are linked to the purpose of moving them, and keeping them quiet during procedures. Often it is needed to teach them to enter a specific place like for a lactating female in a milking lane or into a crush for a clinical inspection, or for any type of camels teach them to load into a vehicle for transportation. Similarly, it may be needed to teach them to stay quiet during the milking process, or while they are tethered somewhere, or it is important to teach them how to respond to stimuli to ride them, and other similar activities.

8.3.1 Training Tools and Aids

One controversial element in animal training is visual contact as it has been observed in horse and human interaction. The effect of human eye contact on animals has been studied in both dogs (Wallis et al. 2015) and sheep (Beausoleil et al. 2006) and it has been shown that eye contact from humans to dogs can be perceived by the dog as a threat. In the study of Beausoleil et al. (2006), it has been reported that human eye contact with sheep did not provoke fear but still induced a certain nervosity.

However, with horses, the effect of human eye contact is not well understood (Worth 2016). According to professional horse trainers, there are different points of view regarding the effect of human eye contact with horses. According to some reports, the trainer should use soft eye contact when handling horses; it means a soft look at the horse is possible but a wide field of view must be kept. A soft look means not looking straight in the horse's eyes. Some trainers indicate hard contact is preferred to establish dominance on the horse. Another category of trainers say all kinds of eye contacts should be avoided as it will scare the horse; it would think the trainer is stalking him. Due to these different recommendations, a study was performed (Verrill and McDonnell 2008) to determine if making direct eye contact or not making it really influenced reactions from new horses when being first-time catch in a pasture. No difference was shown in the study. Many of the horses used in this study were semi-wild ponies and catching them in the pasture was not easier or harder regardless of eye contact. Some of the horses could be caught and others could not be, eye contact had no effect. Thus, eye contact may not be an important factor in human-horse interaction. Probably, similar conclusions could be done for human-camel interactions, which are similar to human-horse interactions.

The material of training tools is essential when working with camels. All tools must be strong enough to resist camel strength. From the head collar to the saddle, it mustn't hurt the camel. It is impossible to train any animal if it is physically painful for him to be around human. A rope halter adjusted to a camel head may have more beneficial effects, once training is started, rather than a nose peg, or nose ring which can often cut nostrils if there is too much tension on it. Lead rope, physical link between the camel and his handler, should be selected to be light but strong. Camel head being horizontally oriented, weight from the lead rope can quickly create discomfort on camel head. Saddle should consider vertebral bones from the camel but also the hump. Shoulders and hips should be free from their actions too. No hot spots should appear on camel skin after the saddle is taken down. A stick, which is the prolongation from the hand, can be used to help the handler to stay safe at a good distance from the camel.

8.3.2 Safety of Handlers/Trainers

As camels are massive animals, it is preferable to stay safe around them and use some security placement while working around them. Camels have very flexible leg attachments on their body, so they can kick in a very large range around their back

Fig. 8.4 Two handlers during a handling demonstration

legs. Front legs are used in many assaults and can cause strong damage to handler's body. Camels have sharp teeth, and all handlers should always consider it while working with them. In a general manner, camels use to push themselves in the herd, the leader making his own way to the best resource (food, water...). Being a natural behaviour but dangerous for a human, nobody should stand close to camel's shoulders or right in front of them. Also, working with closed-wall enclosures can be dangerous, and the feasibility to be smashed against the wall is high. The open-walled enclosure allows escaping for the handler. It is very important to never in any way lie down around a camel. It's also recommended to have at least two handlers when starting camel training (Fig. 8.4).

8.3.3 Round Pen Training

Round pen training is largely used in horses and the role of ethology in this type of training has been reviewed by Henshall and McGreevy (2014) and should be taken into account also for camels. Approaching a free-roaming camel in a large area can be tricky, consequently round pen training can be a good method to first approach a camel and establish a human-camel relationship. Whatever the training goals, the first step with any camel during round pen training is getting his attention. Round pens create a training environment where this becomes easy. Their small diameters limit the camel's ability to flee or evade the trainer, and their shape limits his activity options. They offer great potential for opening good communication lines between

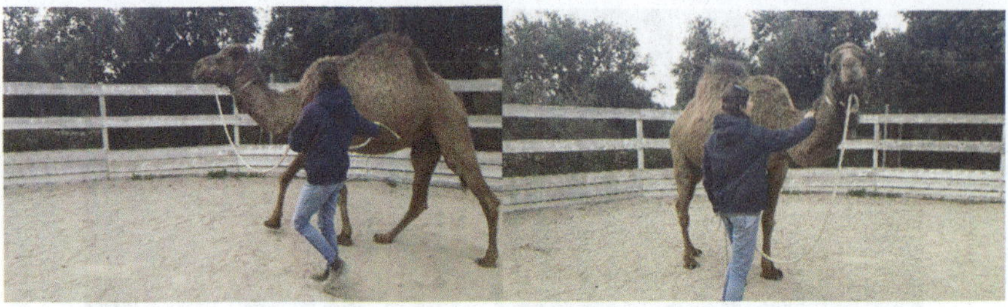

Fig. 8.5 The main author training a camel in the round pen

the trainer and the camel, and they can be used also to help the camel to focus on specific tasks.

Based on the main author's experience (Fig. 8.5), this is how a trainer should proceed during round pen training with a camel:

- Neutral: standing quietly, at the point of camel's shoulders, with shoulders parallel to his body is a "neutral" position.
- Move forward: turning shoulders slightly in order to face the direction in which the trainer wants the camel to move. If the camel does not move off right away, step sideways and in, towards the camel's hip, with shoulders still turned in the desired direction, to encourage him to move forward. Dropping back slightly behind the line of the hip drives the camel forward even more aggressively for an increase in gait.
- Slow down: stepping sideways so being in front of the shoulder line, ask the camel to slow down. If he doesn't respond, taking a step in, towards his head, should make him slow down.
- Stop: stepping towards the camel's head, and a step or two further says "Stop." If the camel instead of stopping, turns away, the trainer should go directly across the circle's diameter and turn him there, and continue until the camel realises he cannot run away left or right. As soon as the camel stops, the pressure should release.
- Turn: turning shoulders parallel to the camel, take a step sideways so the handler is in front of the camel's head, then step in as the handler turns his shoulders in the opposite direction from the camel's direction of travel. As the camel turns towards the wall, stay aware of that kick zone.

In this way, the camel will give attention to the trainer and the trainer will take control of the camel's movements. Getting the head, being able to catch all attention of the camel, will give full control right from his feet. Approaching the camel to catch him can still be tricky as the animal may fear the trainer. A technique from approach/walk away will be useful. This way to work is largely used while desensitising animals. Very good timing in the retraction from the pressure (aversive sensation) will be highly necessary. For a non-trained or mistreated camel, being around humans can be a stressful moment. In the beginning, it will be impossible to touch the camel; the trainer will need to remain patient and learn mimetic, physical signals, camel can show as trainer is entering his comfort zone and adapt his movement forward. A very stressed camel will run away with only a step forward from the trainer. When the camel will stand still and allow trainer to enter his comfort zone for a short instant, trainer should go backwards again to give space to the camel to reward him. If the camel moves, as the trainer is now able to control his feet in case of its movement, the trainer should be able to stop the camel. It is very important for the trainer to not move his feet backwards in those moments. With time and repetitions, the trainer should be able to touch the camel. In the same method of approach/step away, once the camel is not moving anymore, it is time to introduce food to reinforce the training. Camel may not accept food due to the stress at first attempts. High-aroused animals may refuse to eat (McGreevy and Boakes 2006).

The 'comfort zone' of a camel during the initial phase of training may be large. Trainer will have to adapt his distance from the camel; larger the distance is, quieter it should be as there is no immediate danger for him. With a trained camel, one step backwards should be enough to give him space.

A comfort zone is when the camel is fully relaxed, living in the herd, roaming freely in a natural environment. In this comfortable zone, camel's brain is ready to learn as it has lot of free space to process new information. During training, the trainer should try to keep the camel in this mental and physical condition (muscles are relaxed, respiration and pulse rates are low). But as the trainer wants to teach new things to the camel, it will be difficult to stay in this ideal conditions. In the stretch zone, which is when the camel is a little stressed from being outside his habitual pasture, being in a new environment, it will feel more vulnerable. His brain is less free to accept all information as a part is taken by survival mode. Panic zone is when the environment completely takes over the camel. His brain can't focus on any cue; it is fully absorbed by survival reactions. Body is super tense, reactions are not controlled at all. This mental stage should be avoided as much as possible during training process (Palethorpe and Wilson 2011). Each camel has his own nuance between comfort, stretch and panic zone.

When the camel is standing and letting the trainer come close to him, the trainer will be in a position to start the desensitising work (Fig. 8.6). Being able to touch the camel, every place on his body without dangerous reaction is a safety act for the trainer. In a way to avoid dangerous reactions from the camel, a long stick can be used to touch it. Always in the approach/step away method, it's easier to start by touching camel hump as it is not an organ they use to protect first, like the genital area or base from the neck. Moreover, the hump is poorly innerved and consequently, poorly sensitive.

Rewarding the camel with food will highly improve his willingness. Also, it will create jaw movements, which help his relaxation. A camel calmly chewing his cud is a relaxed camel. A "head work" must be done, as the handler will need to put a halter on the camel's head.

A belief says that " if you can touch the head and especially the nose of your camel, you can trust him."

Fig. 8.6 The main author performing desensitisation work

After the camels accept to be touched, the following step could be followed to put the halter. It may be an oppressive sensation for the camel to see the halter coming directly at him. It's better to softly lay halter on his head by coming from sideways and not from the front. Halter should not be buckled up at the first attempt as the camel will probably take it as aggression and will try to escape. Having the halter on his head for the first time may create stress-related behaviour, but as described above he will get used to it quickly.

8.3.4 Use of Halter and Lead Rope

In the beginning, when the handler will put pressure on the lead rope, the camel will probably pull in a reverse movement to free himself, and a sort of fight will start (Fig. 8.7). Many solutions are available to the trainer. A second person can walk behind at a distance from the camel, this can help the camel to move forward and this is when the first handler has to release the pressure on the lead rope. So, the camel will learn to follow the front handler by negative reinforcement. If different trainers are working with the same camel, they should all use the same cue to ask for a behaviour and they should decide what is, or not, allowed as the response from the camel. One command must induce one answer. Many commands for the same answer will disturb the camel. Also, if trainers accept different answers from the camel, it will be confusing.

The camel should always have control of the environment during his training session and the trainer should adapt the time of the training session to each camel. Each animal has a different temperament and learning ability, so the trainers must

Fig. 8.7 Camel rearing as first reaction to the pressure

develop an individual training plan, train calmly and quietly, using minimally invasive pressures, and consider that each training session should not be too long or stressful. A camel with high arousal or when tired will learn less.

The training session should follow the ten principles of training in equitation science (http://www.equitationsciencetraining.net), now adapted to camels:

1. Train accordingly to the camel's ethology and cognition
2. Use learning theory appropriately
3. Train signals that are easy to discriminate
4. Shape response and movements
5. Elicit response one at a time
6. Train only one response per signal
7. Form consistent habits
8. Train persistence of responses
9. Avoid and dissociate flight responses
10. Demonstrate minimal level of arousal sufficient for training

Fig. 8.8 A dromedary camel trained for carriage

Following these principles, everything can be taught to a dromedary camel. One of the first exercises is to teach "parking"; the camel should stay quietly in a spot without any constraints (Fig. 8.8). When a trainer has taught to go forwards, backwards, sideways, stopping and parking, which is the base, the dromedary camel can be trained for more difficult tasks, as riding or driving (Fig. 8.8).

8.4 Are Camels Smart Animals?

When people ride camels they are curious and a little bit suspicious about this animal but when they have to train it, they are astonished how smart the camel can be. Despite their rough appearance, camels are confident animals. They are not so hard to train due to their ability to adapt their behaviour in almost all situations. They also have a very good memory, which allows them to roam in very large areas without trouble, and so gives them possibility to retain situations given by the trainer even years later.

> Camels are the most intelligent creatures I know except for dogs, and I would give them an IQ rating roughly equivalent to eight-year-old children. They are affectionate, cheeky, playful, witty, yes witty, well-possessed, patient, hard-working, and endlessly interesting and charming (Davidson, 2017)

If determining human intelligence can be tricky, it is quite harder to determine animal intelligence (e.g., see Pouydebat 2017). Some scientists have proposed an equation between body weight and brain's size called encephalization quotient (EQ) (Sousa and Wood 2007). On average encephalization quotient in animal is 1. On average the brain weight of the Bactrian camel is 626 g and the encephalization quotient value is 1.3, indicating his high level of intelligence. The rhinencephalon, being a part of the archaic brain having the function of instinctive and emotional behaviour, is mature and well developed, in accordance with his good olfactory sense. The hippocampus, a complex brain structure that has an important docket in learning and memory function, is considerably large concomitant with ability of spatial memory. Adaptive behaviours of the Bactrian camel are corresponding to his anatomical features and are providing morphological evidence for the camel to adapt to his living environment. These anatomical features agree with the corresponding adaptive behaviours of the Bactrian camel and provide morphological evidence of the camel adapting to the arid and semi-arid environment (Chen et al. 2009). Camel hippocampus is nearly similar to humans and elephants (6.3 cm length and 0.9 cm width).

According to camel farmers, those animals tend to remember positive and negative experiences they met in their life. It is very common to hear stories about camels remembering someone who mistreated them, or returning to a land they appreciate after being moved to a foreign place. To our knowledge, there is no data regarding intelligence tests for camels (as maze test, learning ability test or memory test), but one reference involving small camelids (*Lama glama*) concluded, based on a mirror test, that they have an average intelligence for an ungulate species (Tansley 2011). There are some behavioural tests which can be used to study the ability to memorise and the ability to solve problems in animals; from the authors' knowledge these types of studies have not been performed in camels, so they are recommended to give a deeper knowledge on how to train and handle camels.

8.5 Conclusions

Camels are capable to adapt themselves to new environments and situations quickly. There are different methods to train them but only those based on animal learning theory are considered efficient and welfare friendly. Conditioning answers according to different stimuli, and rewarding using correct timing, allows a favourable way of training. A great trainer, with correct knowledge about camel behavioural reactions, should be able to obtain a very good life companion from a camel. In deduction from good training, camels should become easier to manage, as an individual or as a herd animal. If camels are tamed and handled using learning theory, their relationship with humans will be better, and their quality of life from farm to fork and from birth to death will be enhanced. Moreover, appropriate handling of camels will be safer also for camel handlers. Consequently, more studies on the effects of training methods on camels are recommended as well as studies to investigate the camel's cognitive ability.

References

Baragli P, Barbara P, Angelo T (2015) The role of associative and non-associative learning in the training of horses and implications for the welfare (a review). Annali dell'Istituto superiore di sanita 51:40-51.

Barnlund DCA (2008) Transactional model of communication. In: Mortensen CD (ed) Communication theory, 2nd edn. Transaction, New Brunswick, NJ

Beausoleil NJ, Stafford KJ, Mellor DJ (2006) Does direct human eye contact function as a warning cue for domestic sheep (Ovis aries)? J Comp Psychol 120(3):269. https://doi.org/10.1037/0735-7036.120.3.269

Chen J, Bai Z, Gao C et al (2009) Morphology of Rhinencephalon and Hippocampal formation of the Bactrian Camel (*Camelus bactrianus*) with their adaptive features. Vet Res Commun 33:25–32

Cooper JJ (1998) Comparative learning theory and its application in the training of horses. Equine Vet J Suppl 27:39–43

Dai F, Dalla Costa A, Bonfanti L, Caucci C, Di Martino G, Lucarelli R, Padalino B, Minero M (2019) Positive reinforcement-based training for self-loading of meat horses reduces loading time and stress-related behavior. Front Vet Sci 6:350

Davidson R (2017) Tracks. 10/18 publ., Paris

Feng LC, Howell TJ, Bennett PC (2016) How clicker training works: comparing reinforcing, marking, and bridging hypotheses. Appl Anim Behav Sci 181:34–40. https://doi.org/10.1016/j.applanim.2016.05.012

Gal Z (2017) The effects of using aversive training methods in dogs—A review. J Vet Behav 19: 50–60. https://doi.org/10.1016/j.jveb.2017.02.004

Gibbon J (1977) Scalar expectancy theory and Weber's law in animal timing. Psychol Rev 84(3): 279–325

Henshall C, McGreevy PD (2014) The role of ethology in round pen horse training—A review. Appl Anim Behav Sci 155:1–11

Hinde R (1979) Towards understanding relationships. Academic Press, London

Iglesias C, Navas FJ, Ciani E, Arando A, González A, Marín C, Nogales S, Delgado JV (2020) Ethological characterization of the Canarian camel breed. Archivos de zootecnia 69(265):14–21

Johnson AP (2014) Operant conditioning. In: Skinner BF (ed) Education psychology: Theories of learning and human development. National Science Press, Al-Cajon, CA. http://www.nsspress.com/

McGreevy PD, Boakes RA (2006) Carrots and sticks: principles of animal training. Cambridge University Press, Cambridge

Mills DS (1998) Applying learning theory to the management of the horse: the difference between getting it right and getting it wrong. Equine Vet J Suppl 27:44–48

Nicol C (2005) Learning abilities of the horse. In: Mills DS, McDonnell SM (eds) The domestic horse: The origins, development and management of behaviour. Cambridge University Press, Cambridge

Palethorpe R, Wilson J (2011) Learning in the panic zone: Strategies for managing learner anxiety. J Eur Ind Train 35:420–438. https://doi.org/10.1108/0309

Pavlov IP (1927) Conditioned reflexes. Routledge and Kegan Paul, London

Pouydebat E (2017) L'intelligence animale: Cervelle d'oiseaux et mémoire d'éléphants. Odile Jacob, Paris, p 215

Pryor K (2009) Reaching the animal mind. Clicker training and what it teaches us about all animals. Scribner, New York

Skinner BF (1938) The behaviour of organisms. Appleton-Century-Crofts, New York

Sousa DA, Wood B (2007) 4.18—The Hominin fossil record and the emergence of the modern human central nervous system. In: Kaas JH (ed) Evolution of nervous systems. Academic Press, Oxford, pp 291–336. https://doi.org/10.1016/B0-12-370878-8/00018-5

Tansley A (2011) Comparative Intelligence of the Llama (Lama glama): Implications for Welfare. Biosciences Undergraduate Research at Nottingham University of Nottingham, 5 p. https://www.nottingham.ac.uk/biosciences/documents/burn/2011/comparative-intelligence-of-the-llama%2D%2Dabigail-tansley.pdf

Thorndike EL (1898a) An experimental study of the associate processes in animals. Psychol Rev 2(4):1–8

Thorndike EL (1898b) Animal intelligence: An experimental study of the associative processes in animals. Psychol Monogr Gen Appl 2(4):i–109

Vallortigara G (2000) Other minds. Comparative study of animal cognition. Il Mulino, Bologna

Verrill S, McDonnell S (2008) Equal outcomes with and without human-to-horse eye contact when catching horses and ponies in an open pasture. J Equine Vet Sci 28(5):309–312. https://doi.org/10.1016/j.jevs.2008.03.006

Wallis LJ, Range F, Müller CA, Serisier S, Huber L, Virányi Z (2015) Training for eye contact modulates gaze following in dogs. Anim Behav 106:27–35. https://doi.org/10.1016/j.anbehav.2015.04.020

Worth S (2016) Studying visual communication. University of Pennsylvania Press, Philadelphia, PA

9 Welfare Aspects of Reproductive Care and Management of Dromedary Camels

J. A. (Lulu) Skidmore, Judit Juhasz, and Peter Nagy

Abstract

For efficient reproduction, camels have to be healthy, well-housed and well-fed. Under natural conditions, camels tend to roam freely but to improve reproductive efficiency they need to be kept under more intensive systems where breeding practices are closely monitored and controlled. Currently, ultrasound of the reproductive tract is considered the most reliable method to monitor follicular development and decide the correct time to mate. This requires some form of restraint, and padded stocks are the safest method for camel and operator. As genetically superior males are in higher demand, this leads to overuse of the male camels resulting in lower than optimum fertility and females requiring several matings for successful conception. This led to the development of embryo transfer and artificial insemination to reduce the number of matings required and thereby improve the welfare of both male and female camels. Good management of the perinatal and early post-partum period is critical for the success of any breeding programme. This chapter describes reproductive management from a welfare aspect.

Keywords

Dromedary camel · Reproduction · Welfare

J. A. (Lulu) Skidmore (✉)
Camel Reproduction Centre, Dubai, United Arab Emirates

J. Juhasz · P. Nagy
Emirates Industry for Camel Milk and Products, Farm and Veterinary Department, Dubai, United Arab Emirates

B. Padalino, B. Faye (eds.), *Dromedary Camel Behavior and Welfare*, Animal Welfare 24, https://doi.org/10.1007/978-3-031-48600-5_9

9.1 Introduction

There are a lot of publications on reproductive physiology, reproductive care and assisted reproduction in dromedary camels, but the welfare aspects and challenges of reproduction have not been focused on until recently. It is well documented that reproductive efficiency in camels is generally regarded to be low, partly due to the fact they are seasonal breeders with a relatively short breeding season during the cooler, winter months (Wilson 1984), have a long gestation period of approximately 13 months, a prolonged inter-calving interval (2–3 years) and a high rate of embryonic/foetal and neonatal mortality (Nawito et al. 1967; Nagy et al. 2021). From the animal welfare point of view, the purpose of reproductive procedures is to manage breeding activities, pregnancy period, deliveries and the neonatal period effectively and efficiently in the way that best fits the species and ensures the survival of the offspring with minimal discomfort and unnecessary suffering to the animals. Unfortunately, these complex biological processes are frequently handled by untrained staff and animal welfare is not taken into consideration. The aim of this chapter is to describe reproductive care and management of dromedary camels from the welfare point of view and summarize its challenges.

9.2 Reproductive Physiology of Male and Female Dromedary Camels

Female dromedaries are induced ovulators and seasonal breeders (El Wishy 1987). They reach sexual maturity at approximately 2–3 years of age, but breeding maturity is attained later. Therefore, the age at first calving ranges between 60 and 72 months (Beniwal and Chaudhry 1984), although, there have been attempts to advance the calving age by mating young females at an earlier age (Yagil and Etzion 1984; Kamoun and Wilson 1994). However, this is not considered good practice if the female has not reached physical maturity. The breeding season may last from September to May (Nagy and Juhász 2019); however, there are important geographical differences. For example, in Tunisia, Hammadi (2003) reported that it usually lasts from December to March, whereas in most of Arabia it occurs between November and April (Abdel-Rahim and El-Nazier 1990). Outside of the breeding season, mating activity ceases; dromedary bulls have reduced libido and the ovaries are inactive or show a limited number of small follicles with irregular or extended follicular wave patterns (Musa and Abusineina 1978). This is why it has become increasingly important to understand the physiology of reproduction in this species so that efficient management, high welfare standards and the use of assisted reproductive techniques, such as embryo transfer and artificial insemination, can be used to try and improve their reproductive efficiency.

Camels exhibit periodic waves of follicular growth and regression rather than regular oestrous cycles during the breeding season (Tibary and Anouassi 1997a; Skidmore et al. 1995). These follicular waves overlap, so whilst one follicle is regressing another one, or multiple follicles are growing (Manjunatha et al. 2012) which means that unmated females can show almost constant sexual receptivity. However, external signs of heat and typical oestrous behaviour are not consistent or evident in this species (Novoa 1970; Skidmore 2011). Preliminary studies carried out by Padalino et al. (2016) did indicate that female camels show more receptive behaviour towards a restrained male during the mature follicle phase than in the non-mature phase; however, allowing 15 min exposure time to the male per female would not be practical in large scale breeding programmes (Skidmore, personal observation).

Female dromedaries naturally only ovulate when mated, therefore, mating the female at the correct time of the follicular phase is important to achieve pregnancy. Growing, dominant follicles between 1.3 and 1.9 cm in diameter are capable of ovulation, whereas larger or regressing follicles are not, and follicles smaller than 1.3 cm (1.1–1.2 cm) may ovulate but are unlikely to yield fertile oocytes (Skidmore et al. 1995, 1996). Moreover, in comparison with other domestic species, the mated, non-pregnant camel has a very short luteal lifespan of around 8 days (Skidmore et al. 1995). In case of pregnancy, the corpus luteum does not regress and continues its progesterone secretion throughout gestation and therefore is responsible solely for the maintenance of pregnancy in camelid species (El Wishy et al. 1981; El Wishy 1987). Twin pregnancies are relatively frequent (approximately 8%), but most twins undergo natural resorption of both or one of the foetuses before 100 days of gestation. Therefore, twin abortions are rare (approximately 0.7%) and twin births are extremely rare (<0.1%) (Nagy et al. 2021). Total pregnancy wastage in dromedaries is high, as one-quarter of early pregnancies are lost at various stages of gestation. Two-thirds of these losses occur in the first 100 days, whilst one-third of total pregnancy losses are detected later in gestation or at around parturition. More than half of these mid to late pregnancy losses are abortions before 330 days of gestation, and approx. 40% are diagnosed as perinatal mortality (Nagy et al. 2022a). The perinatal and neonatal periods are the most hazardous part in the life of the animals with a lot of animal health and welfare concerns that will be discussed later.

Male dromedaries reach sexual maturity at approximately 3–4 years of age but are used in breeding programmes from approximately 5 to 6 years of age (Wilson 1984). They are also strongly seasonal and manifest seasonal changes in testicular size and consistency, libido, semen quantity and quality (Al-Bulushi et al. 2019). Typical semen parameters of a mature dromedary with good fertility are as follows : 2–8 ml volume, 300–600 × 10^6/ml concentration, and 40–60% total motility (Skidmore and Billah 2006; Al-Bulushi et al. 2019). End-of-season fertility of male dromedaries can reach as high as 90%; however reliable data in the literature are scarce, but in research settings, with proper management of males, pregnancy rates could be >60% after natural mating or artificial insemination (Al-Bulushi et al. 2019).

9.3 Housing and Management of Breeding Animals

For efficient reproduction, the housing and management of breeding animals should take into consideration basic animal welfare principles such as "good housing," "good feeding," "good health" and "appropriate behaviour." Here, we are not going into detail (see Chaps. 4–7), however, we highlight the most important aspects that are relevant to camel breeding. Menchetti et al. (2021) suggested a minimum space allowance of 19 m^2 per camel, whereas, recently Nagy et al. (2022b) suggested a space allowance of about 50 m^2 for lactating dromedaries. The size of the group can also vary considerably from 5 to 50 females kept together depending on the size of the paddock and the number, age and fertility of the males. Irrespective of the size, breeding paddocks should be designed so that camels are comfortable, there is no risk of injury and animals can be herded, selected and handled properly for any procedures (i.e. restraint for examination, hand mating or treatment/blood collection).

The importance of proper housing of dromedary males is often neglected. In general, bulls are considered dangerous and therefore they are frequently kept in small confined areas, behind high walls or in cage-like structures. This kind of housing and lack of appropriate social interaction may lead to the development of stereotypic behaviour (Padalino et al. 2014). Camels are social animals and this is also true for the males (see Chaps. 4 and 7) and despite common beliefs to the contrary, if male dromedaries are trained and handled properly, they do not show aggressive behaviour (Juhasz and Nagy, personal observation). Following sexual maturity, when bulls are then kept separately, they should not be deprived of visual contact with each other and the environment. In hand mating systems, it is common practice to keep males in open, shaded paddocks of approximately 60–100 m^2 that are adjacent and relatively close to one another. On the other hand, in case of group mating, males are usually kept in a small confined area inside the larger paddock of the females that are assigned to this particular male (Juhasz and Nagy, personal observation).

Breeding animals should be fed a balanced diet in order to maintain good body conditions throughout the breeding season. This is especially important for the males that cover a large number of females per season; however, there are conflicting views and practices concerning the feeding of female dromedaries. In traditional management, the feeding of females is restricted in order to decrease body condition during the mating period. However, there are no controlled studies supporting such a practice.

In traditional management systems, female camels are frequently hobbled during the breeding season. The assumption behind this practice is that animals are easier to handle, move less, have better conception and decreased losses during pregnancy. However, none of these claims have been proven scientifically, but according to our experience, hobbles can be useful in cases of mating larger numbers of untrained females because the animals are easier to catch and restrain for mating, so this practice is quicker for the handlers. However, the material used should be soft, durable and its placement and tightness should be carefully checked regularly in order to prevent injuries, pain or unnecessary discomfort to the animal (Juhasz and Nagy, personal observation).

9.4 Restraint for Reproductive Procedures

Reproductive manipulation of dromedaries especially in the case of ultrasonography, assisted reproduction or obstetric intervention requires some level of restraint of the animals. This can be done in various forms. Camels can be examined/manipulated in a sitting (cush) position, in a standing position in stocks or if stocks are not available with tied hind legs. As camels tend to sit spontaneously, examination of animals in the cush position has been widely practised. The advantage of this method is that it does not require installation of any equipment and can be carried out either inside, outside or even on the plateau of a truck. Camels usually do not need any specific extra training if they were already trained to sit down, however, this can be a challenge if the animal had not been trained to sit. Nevertheless, the hind legs, and occasionally the front legs, should be tied with soft, flat ropes to prevent the camel standing up or rolling and injuring the person involved during the procedure. The disadvantage of this method is that it is rather uncomfortable for the examiner, especially if a large number of animals have to be checked routinely.

Camels are best restrained for reproductive examination in specifically designed standing stocks. There are many advantages of this method: firstly, it is safe and easy for the animals and examiner and also camels can easily be trained to enter stocks. Generally they will stand quietly, without sedation, for ultrasound to be performed per rectum (Fig. 9.1) and each examination only takes a maximum of 1–2 min. The use of stocks allows the examination of a large number of female animals within a relatively short period of time, especially if several such stocks are installed side by side. The disadvantage of this method is that the stocks need to be fabricated and camels have to be trained not only to enter the stocks, but also to enter the room where they are installed. Therefore, these stocks are mainly used at reproductive centres, but not at extensive, nomadic farms (Tibary and Anouassi 1997b). Gynaecological and obstetrical procedures require special attention because animals can experience increased discomfort, and therefore, more intensive restraint is necessary in order to prevent injury to the animals and staff. These can also be done either in sitting or in standing position, however, the standing position in stocks is the preferred method of choice. To prevent the animal from jumping out, sitting down or hurting/kicking the examiner, it may have to be sedated, tied down at the level of the neck and/or have a rope suspended under the abdomen to prevent the animal trying to sit. Sedation is most frequently carried out using α2-adrenergic receptor agonists or xylazine and can also be combined with epidural anaesthesia (Lidocaine 2%). It should be emphasized that rectal and vaginal examinations are to be performed slowly and carefully with plentiful lubrication in order to prevent mucosal injury or perforation and to minimize the discomfort to the animal.

The third option to perform a reproductive examination, when stocks are not available, is to perform it in a standing position with both hind legs tied together or

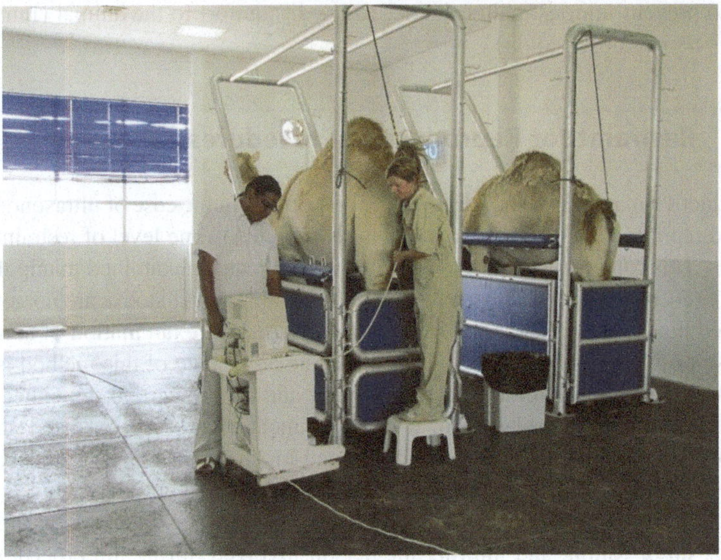

Fig. 9.1 Ultrasound of female camel standing in padded stocks

by lifting and tying up one of the front legs. The decision for this method should be taken based on the behaviour and co-operation of the animal, and also on the experience and skill of the staff and that of the examiner. Camels are frequently restrained not only for reproductive examinations, but also for drug administration (e.g. hormone injections for ovulation induction or superovulation) and blood sample collection. For these procedures, paddocks should have an appropriate facility for selecting and treating the animals (i.e. a treatment corridor, or a railing for tying the camels to).

Overall, the best way for enhancing the camels' welfare and the safety of the operators would be to train the animals (see Chap. 8).

9.5 Monitoring and Controlling Follicular Activity, Mating Methods and Pregnancy Diagnosis

9.5.1 Monitoring and Controlling Follicular Activity

Conventional breeding practices in most domestic species rely on the detection of oestrous behaviour for deciding the correct time to mate the females. However, as mentioned earlier, detection of oestrous behaviour in camels is unreliable. In fact, one of the most important welfare questions in dromedary reproduction is whether the female is mated at the right time of her follicular wave cycle. In traditional breeding systems, females are usually mated randomly, irrespective of their follicular stage. Therefore, there are many unnecessary handlings, matings and increased risk of injury if camels are not mated in the presence of a mature follicle in their ovary. Nevertheless, due to follicular wave characteristics (overlapping waves) pregnancies can be achieved. However, the fertility rate is lower when compared with controlled reproductive management (Nagy and Juhasz 2012). Ultrasonographic examination of the ovaries (per rectum) is the most reliable method to monitor follicular development, so that mating can be performed at the most appropriate time (when a mature follicle of 1.3–1.7 cm is present) and reduce the number of matings required per male to achieve pregnancy. It is not uncommon to mate the female camel twice 12–24 h apart to enhance pregnancy rates. However, as ovulation can be induced with a single injection of gonadotrophin-releasing hormone (GnRH), injection of GnRH 24 h before mating or artificial insemination, has produced equivalent pregnancy rates (Al-Bulushi et al. 2019; Skidmore and Billah 2006) and will therefore improve the welfare of the male camel by reducing the number of matings required. However, the use of ultrasound is not always practical on pastoral or nomadic farms so it is important to develop protocols whereby follicular growth can be controlled and synchronized to produce a dominant follicle capable of ovulating at a known time after treatment. This would make breeding management easier and enable one to schedule fixed-timed matings (Nikjou et al. 2008; Nagy and Juhasz 2012) or inseminations, thereby once again, reducing the number of matings required per pregnancy. One such protocol involves single injections of GnRH at 14-day intervals (Nikjou et al. 2008; Skidmore et al. 2009). After two or three successive GnRH injections this method resulted in the synchronization of follicular waves enabling fixed-time matings to occur on day 28 after the start of treatment without the need of ultrasound, and pregnancy rates of 46% were obtained (Nagy and Juhasz 2012). Manjunatha et al. (2015) carried out a synchronization protocol using a combination of GnRH and Prostaglanin F2α (PGF2α) (GnRH day 0, PGF2α day 7, GnRH day 10, PGF2α day 17) and camels were mated on day 22 after the start of treatment. They obtained pregnancy rates of 60.2 and 53.6% in camels kept at a research facility and under field conditions, respectively.

9.5.2 Mating Practices and Diagnosis of Pregnancy

The breeding of dromedary camels is done predominantly by natural mating. In some reproductive centres, artificial insemination is also practised but not in very high numbers. Natural mating can be done in a controlled manner, by the so-called hand mating method when the female camel is taken to the male individually. It can also be practised in groups, which is called harem mating, when one bull is assigned to a group of female camels (up to approximately 40–50 females at a time) and the animals are kept together (Tibary and Anouassi 1997c, d). In the hand mating system, the whole process is supervised by animal handlers with the advantages being that it is more controlled, usually better recorded, results in less injuries and allows better management of both females and males. It is mainly practised for dromedary males that are in high demand; however, hand mating also has the risk

Fig. 9.2 Pregnancy diagnosis of female camels using "tail cocking method"

that these valuable males are overused, as too many females (as many as 10–15) can be assigned to these animals every day (Tibary and Anouassi 1997d). On the other hand, harem or group mating is usually practised on larger farms or closed herds. This can also be done with some control when the bull is supervised whilst he is amongst the females during the day, but is separated from them during the night, although the male and females can also be kept together without any surveillance at all. The disadvantage of harem mating is that many breedings remain unnoticed and unrecorded and the possibility of injuries is higher. The most frequent injuries are bite wounds on the perianal area or on the hump caused by the male, but occasionally more severe injuries, such as musculus gracilis rupture, can occur. In addition, in harem mating systems, hobbles are frequently applied on the front legs restricting the movement of females, causing skin lesions and potential further injuries.

Pregnancy diagnosis is a crucial part of reproductive management in all species including dromedaries. Traditionally, in camels it is performed by the so-called tail cocking method. This is based on the visual observation of lifting and upward coiling of the tail in the pregnant animal when approached by a male camel (Fig. 9.2; Banerjee et al. 1981; Skidmore 2000). However, the same response has been observed in unmated animals treated with exogenous progesterone and also in younger animals that may be alarmed by the male (Tibary and Anouassi 1997b). On the other hand, older females with calm temperaments may not show any sign of tail cocking even during pregnancy. Therefore, despite common beliefs, this visual method of pregnancy diagnosis is not efficient enough and can result in mating of already pregnant animals which is an important welfare issue in this species. However, according to our observations, such unnecessary matings of pregnant animals, apart from causing unnecessary handling and discomfort to the animal, usually do not have reproductive side effects and abortions generally do not occur. Rather, it is only noticed retrospectively after parturition when breeding records are compared with delivery time and the maturity and health status of the calf (Juhasz and Nagy, personal observation). The behaviour of the female during mating is also controversial as the level of vocalization could be similar, for human ears, irrespective of whether she is in the follicular phase and ready for mating or is pregnant. Nevertheless, despite such vocalization, no increase in serum cortisol concentration was detected in females that were mated during the follicular phase (Nagy et al. 2016). However, it is worth highlighting that vocalizations, their meaning and connection to discomfort have not been studied in camels yet.

The most reliable method of pregnancy diagnosis is transrectal ultrasonography that has been widely used in dromedaries with high efficiency and has been described extensively in the literature (Tibary and Anouassi 1997b). In addition, due to the endocrinology of pregnancy, serum progesterone determination can also be used with high efficiency for pregnancy diagnosis in this species eliminating the need for transrectal examination and its impact on camel behaviour and welfare discussed above, provided accurate mating records are available (Nagy et al. 2021).

9.6 Use of Assisted Reproduction

9.6.1 Assisted Reproductive Techniques

With increasing interest in camel racing, beauty competitions and camel milk, obtaining more offspring from the genetically superior animals in each category has become more desirable. Unfortunately, however, this leads to overuse of the elite males. As there are rarely any stud fees involved everyone wants the genetically superior males for breeding their females so it is not unusual for males to be used to mate up to 10–15 females a day, whereas under natural conditions it is thought that one male camel might cover 70 females in a season (Gauthier-Pilters and Dagg 1981). According to Tibary and Anouassi (1997d,), mating four times daily with one rest day a week had no negative effect on fertility of the males but mating as many as 10–15 certainly would. In another study Al-Bulushi et al. (2018) indicated that semen quality, concentration and activity were greater in camels collected just once a week compared with twice weekly semen collections. However, mating once a week would not be feasible in any breeding programme and more studies are required to determine the optimal use of male dromedaries according to age and fertility of the male, considering not only fertility but also welfare aspects. Currently, the use of assisted reproductive techniques such as artificial insemination (AI) and embryo transfer (ET) come in very useful to reduce the number of matings required by the male camel.

9.6.2 Embryo Transfer

Under natural mating conditions female camels produce one calf after a 13-month gestation period which usually means, at best, they only produce two calves every 3 years. On the other hand, embryo transfer allows the production of multiple offspring from a female camel in one season and would reduce the number of matings required per male to get the same number of females pregnant. However, embryo transfer can be labour intensive as it necessitates stimulating donor females to produce several follicles and synchronizing recipients with the donor. Stimulating the donor is easily achieved with injections of equine chorionic gonadotrophin (eCG) and porcine Follicle Stimulating Hormone (FSH) (McKinnon et al., 1994; Skidmore et al. 2002). Although this method produces good follicular development it does involve handling and injecting the camels on a daily basis for at least 4 days. However, more recently a simplified method of superovulation has been developed by Manjunatha et al. (2020) whereby he used single injection of eCG and found 3000 or 4000 IU administered 2–4 days after ovulation produced comparable follicle development and embryo yield to multiple injections of FSH, thereby reducing the extra handling required and subsequently distress on the female camel.

Embryos are generally recovered 8 days after mating. Embryo recovery is carried out non-surgically with the animals standing in padded stocks, but as the procedure can take up to 15–20 min it is necessary to sedate the camel to prevent undue distress to the camels or injury to personnel. The embryo flushing catheter is placed through the cervix, *per vagina*, and media flows in by gravity so here is no danger of overfilling the uterus (Mickinnon et al. 1994; Skidmore et al. 2002). Similarly, embryos are recovered by gravity flow into sterile beakers as the uterus empties. Because of the stimulatory effect of the hormones on the ovaries donor camels are usually only stimulated twice or maximum three times in a breeding season (Skidmore, personal observations).

The transfer of embryos is also carried out non-surgically, in the standing position in the stocks with the recipient camels sedated as mentioned above. Embryos are loaded into an embryo transfer pipette which is passed, *per vagina*, through the cervix into the uterus where the embryo is deposited (Mckinnon et al. 1994; Skidmore et al. 2002), a procedure taking only 1–2 min. However, there are no studies on the welfare aspects of the procedure, but the high success rate of fresh embryo transfer with good quality embryos and synchronized recipients (70–75%) suggest that the method is efficient and most probably does not cause any serious or long-lasting discomfort (McKinnon et al. 1994; Skidmore et al. 2002). Although there is no data regarding transportation of pregnant camels, it is preferred not to transport the recipients back to their farms until they are confirmed pregnant at 60 days when the pregnancy is well implanted (Skidmore et al. 1996) and, to date, we have not heard of any pregnancy losses occurring due to transportation. Interestingly, in horses, transportation even during early pregnancy (between 3 and 5 weeks) did not affect the incidence of early embryonic death (Baucus et al. 1990).

9.6.3 Artificial Insemination

Artificial insemination is another widely accepted method to improve reproduction efficiency and reduce the number of matings required per male camel. Semen can be collected using one of two methods, namely electro-ejaculation or by artificial vagina (AV).

Electro-ejaculation can be achieved using a standard bovine ejaculator 12 V and 180 mA (Standard Precision Electronics, Denver, CO) inserted into the rectum and applying electrical impulses of increasing intensity until ejaculation occurs (Tingari et al. 1986). This method, however, involves pain and requires general anaesthesia of the animal which can be hazardous, especially with valuable males, and the semen can be contaminated with urine, so this method is not recommended primarily for welfare reasons of the male camel.

Semen collection by artificial vagina generally gives better results. A modified bovine AV (30 cm in length and 5 cm internal diameter, with imitation sponge cervix of about 8 cm in length: Fig. 9.3) can be used. A sexually receptive female teases the male and gets him aroused before he is allowed to mount her. Once he has mounted the female his penis is redirected into the artificial vagina which is held alongside the female (Fig. 9.4). Once ejaculation is finished the male stands up and is led back to his pen. This whole process can take between 5 and 10 min, or sometimes longer, so it can be advantageous to add 1–2 ml of semen extender to the collection vessel to protect the sperm during the collection procedure (see Skidmore et al. 2013).

For the welfare of the female it is generally accepted to use a different female for each collection procedure; however, a phantom camel has been developed to avoid

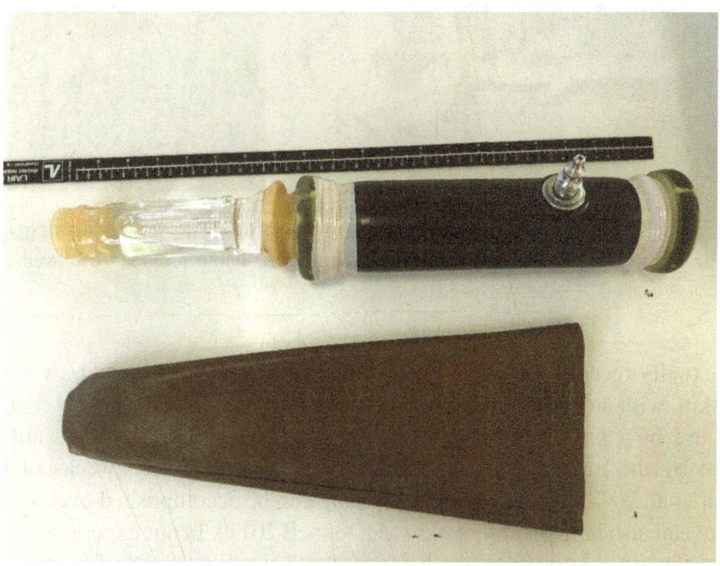

Fig. 9.3 Artificial vagina used for semen collection from male camels

Fig. 9.4 Semen collection using an oestrus female

a b

Fig. 9.5 Inside the phantom camel for semen collection showing the positioning of the artificial vagina (**a**). Semen collection using a phantom female (**b**). Photo provided by Sweihan Camel Hospital

using sexually receptive females altogether. It is a mould of a camel, overlaid with camel skin, with an artificial vagina placed inside (Fig. 9.5a, b: Ziapour et al. 2014).

The use of a phantom does, however, have its drawbacks as it is not always accepted by the male camels and they will need several days/weeks of training (Ziapour et al. 2014). In stallions, this training can be accomplished over a couple of days, but can also take several weeks (McDonnell 2011). In our experience, only one of five males would use the phantom even after several attempts of training and the semen quality was very variable (Skidmore and Nagy, personal opinion). Other methods using a condom are under research and development (Mansour 2022) but further work is required to address the welfare concern of this method and design a better fixation method, thereby maximizing the volume and quality of semen produced.

Good ejaculates collected by artificial vaginas vary between 2 and 8 ml in volume, but the concentration can be highly variable (between 200 and 600 × 10^6/ml). Literature indicates that a total of 150–300 × 10^6 live spermatozoa are required for successful insemination (Skidmore and Billah 2006; Al-Bulushi et al. 2019). The number of doses will vary according to the initial concentration and volume of the semen sample, but on average two to –three insemination doses are obtained per ejaculate. This would reduce the number of mating required per male by 50–66% and thereby improve the welfare of the male camels. Insemination of the female is a simple procedure taking only 1–2 min to perform and is carried out non-surgically, with the female standing in stocks under mild sedation, as discussed previously. The insemination catheter is passed through the cervix and the semen is deposited in the body of the uterus or at the tip of the uterine horn ipsilateral to the mature follicle. Pregnancy rates of between 50 and 70% have been obtained with fresh semen (Skidmore and Billah 2006; Al-Bulushi et al. 2019) and 20 and 60% with cooled semen (Al-Bulushi et al. 2019; Malo et al. 2020); however very few pregnancies have been achieved with frozen semen. Deep freezing of camel spermatozoa still remains a major challenge for the advancement of AI in this species but once achieved it will further improve the welfare of male camels by reducing the need to transport them within and between countries to dissipate their genetics.

9.7 Management of Pregnant Animals, Parturitions and Newborn Calves

The pregnancy, perinatal and early post-partum periods are the most critical part of the life cycle of dromedaries when the chance for discomfort, pain and disease is increased. Therefore, camels need extra attention and care during these periods. Despite this, traditionally, pregnant camels receive limited attention and are usually left roaming around in the desert without any supervision. Therefore, reproductive problems (mainly abortions) during this time remain unnoticed, and reliable data on the extent of these disorders is scarce (Khalafalla et al. 2017). Towards the end of their gestation period camels are usually kept in confined areas which allow for more supervision and assistance if required. However, even during this time untrained people or camel herders generally provide the first line of assistance, and professional veterinary care is rarely available, which can lead to increased losses during the perinatal period (Menchetti et al. 2021). These losses are not only important in pastoral systems, but also in semi-intensive and intensive production systems. Earlier, perinatal calf mortality was regarded as an indicator of management quality, but recently it is also considered a crucial animal welfare indicator and became the most commonly used population-level welfare index in bovine dairy farms (Mee 2013).

The proper feeding and meeting of the energy and protein requirements of pregnant dromedaries are important issues to ensure optimal foetal and mammary gland development (Wardeh 2004). In particular, the latter is crucial in providing good quality colostrum to the newborn after birth. In addition, vitamin and mineral supplementation of pregnant dromedaries is vital for improving not only the health status of the dam but also that of the calf. Therefore, special attention should be paid to selenium (Se) supplementation as most areas where camels are kept are deficient in Se. Without this supplementation in the last 3 months of gestation, the likelihood of certain calf diseases such as white muscle disease and calf mortality are increased (Seboussi et al. 2009) (see Chap. 6). Dromedaries require constant supervision by trained staff during the peri-parturient period to make sure that deliveries are spontaneous and uneventful, or in case of difficult calving obstetrical assistance is fast, hygienic and efficient. Fortunately, the incidence of dystocia is low in this species (2–5%) and only less than 1% of deliveries require major obstetrical intervention (Tibary et al. 2008). Nevertheless, we would like to emphasize the importance of staff training for delivery assistance and for basic obstetrical procedures. Without proper training, obstetrical manipulation can be unhygienic (e.g. entering with dirty hands into the genital tract) and excessive force might be used causing further injuries such as tears (i.e. recto-vagina, evisceration of internal organs) which would also put the life of the dam in danger.

Following the birth of the calf, the dam should be carefully monitored for the expulsion of foetal membranes and also for any injuries or post-natal bleeding. In addition, the calf should be examined and its airways and umbilical cord cleaned. A healthy dromedary calf should attain sternal sitting position within 30 min of delivery and should be able to lift its head and appear alert looking around the environment. The transfer of passive immunity is of vital importance for the survival of the calf therefore, proper quantity and quality of colostrum must be provided to the newborn within a few hours of delivery. Normally, the calf is able to suckle on its own, or with some minor assistance, within 2–3 h of delivery. Alternatively, it is also possible that the colostrum is milked out from the mammary gland and is fed by bottle to the calf to ensure that the neonate consumes a sufficient quantity (approximately 1 l) in the first 24 h of life (Tibary and Anouassi 1997d). First-delivery animals represent a special problem in neonatal management. Many of these young females reject the calf after delivery and the newborn does not have access to colostrum nor to milk later on. Hammadi et al. (2021) found that the two main reasons for calf rejection were dystocia and the presence of an alien calf at the birth site in stabled dromedaries. There are several traditional practices used (restricting air movement by plugging the nostrils, closing the vagina with a piece of wood) to force such females to accept their calves but they raise serious welfare concerns (Dioli 2022). However, using professional management, patience, a calm environment and using mild sedation, most of these nervous first-delivery animals will accept their calf over time. According to Hammadi et al. (2021), 3 days of confinement and forced contact with the calf while bottle feeding with the milk of the dam is sufficient to establish the proper mother-young relationship. In such a situation, the temporary and controlled use of hobbles can be beneficial and is sometimes needed to prevent unnecessary injuries. In traditional management, newborns are frequently restricted from having free access to suckle in order to prevent excessive milk intake and the development of diarrhoea, but there is no scientific basis for this controversial practice (Dioli 2022). Freshly delivered calves should be kept together with their dam in a calm environment, without any restriction, in order to allow the development of a good bond between dam and calf.

9.8 Conclusion

The welfare of male and female camels should be given more consideration in many breeding establishments to maximize the health and well-being of the animals and thereby improve overall pregnancy rates and minimize early pregnancy loss, abortions and calf mortality. These days, with the intensification of camel breeding, all programmes need to start with proper training not only of the camels but also of the staff and handlers so that they are sympathetic to the camel's temperament and requirements. Good management, knowledge and monitoring of the reproductive cycle and carrying out reproductive procedures in a clean, safe and controlled environment is tantamount to success and will reduce the stress caused to the animals and improve results. Judicious use of embryo transfer and artificial insemination will improve reproductive efficiency and genetic traits of the camel herd, when genetically superior donors are used, and also improve the welfare of the male and female camels by reducing the number of matings required to produce the same number of offspring. In the future, the transfer of frozen-thawed embryos and insemination of cryopreserved semen will further improve the welfare of camels by reducing the need to transport the camels, as embryos and semen can be more easily transported than live animals, and at the same time further increase the spread of the superior genetics worldwide.

References

Abdel-Rahim SEA, El-Nazier AT (1990) Factors affecting camel reproductive performance in the tropics. In: Saint-Martin G (ed) Proc. workshop: "Is it possible to improve the reproduction of the camel". CIRAD-EMVT, Paris, pp 131–148

Al-Bulushi S, Manjunatha BM, Bathgate R, Rickard JP, de Graaf SP (2018) Effect of semen collection frequency on the semen characteristics of dromedary camels. Anim Reprod Sci 197: 145–153. https://doi.org/10.1016/j.anireprosci.2018.08.022. Epub 2018 Aug 17. PMID: 30170879

Al-Bulushi S, Manjunatha BM, de Graaf SP, Rickard JP (2019) Reproductive seasonality of male dromedary camels. Anim Reprod Sci 202:10–20. https://doi.org/10.1016/j.anireprosci.2018.12.013. Epub 2019 Jan 2. PMID: 30658862

Banerjee SR, Vyas KK, Pareek PK, Deora KS, Vyas UK (1981) Note on the clinical diagnosis of early pregnancy in camel. Indian J Anim Sci 51:909–910

Baucus KL, Ralston SL, Nockels CF, McKinnon AO, Squires EL (1990) Effects of transportation on early embryonic death in mares. J Anim Sci 68(2):345–351. https://doi.org/10.2527/1990.682345x

Beniwal BK, Chaudhry AL (1984) Age at first calving in Bikaneri camel. Indian J Anim Sci 54: 598–599

Dioli M (2022) Observation on dromedary (*Camelus dromedarius*) welfare and husbandry practices among nomadic pastoralists. Pastoralism 12:7. https://doi.org/10.1186/s13570-021-00221-5

El Wishy AB (1987) Reproduction in the female dromedary (*Camelus dromedarius*): A review. Anim Reprod Sci 82:587–593

El Wishy AB, Hemeida NA, Omar MA, Mobarak AM, El Sayed MAI (1981) Functional changes in the pregnant camel with special reference to fetal growth. Br Vet J 137:527–537

Gauthier-Pilters H, Dagg AI (1981) Behaviour. In: Gauthier-Pilters H, Dagg AI (eds) The camel, its evolution, ecology, behaviour and relationship to man. Chicago Printing Press, Chicago, pp 78–100

Hammadi M (2003) Charactérisation, modulation nutritionnelle et implication du système IGF dans la fonction de reproduction chez la chamelle (*Camelus dromedarius*. Unpublished PhD thesis, University of Gembloux

Hammadi I, Chniter M, Brahmi M, Atigui M, Bouzaida MD, Seddik MM, Nowak R, Maria GA, Hammadi M (2021) Mismothering and remedying the mother-young relationship in stabled dromedary camels. Appl Anim Behav Sci. https://doi.org/10.1016/j.applanim.2021.105424

Kamoun M, Wilson RT (1994) Improving early reproductive characteristics of Tunisian camels by nutritional and management interventions. J Arid Environ 89–94

Khalafalla AI, Al Eknah MM, Abdelaziz M, Ghoneim IM (2017) A study on some reproductive disorders in dromedary camel herds in Saudi Arabia with special references to uterine infections and abortion. Trop Anim Health Prod 49:967–974. https://doi.org/10.1007/s11250-017-1284-x

Malo C, Crichton EG, Skidmore JA (2020) Preservation of the spermatozoa of the dromedary camel (*Camelus dromedarius*) by chilling and freezing: the effects of cooling time, extender composition and catalase supplementation. Theriogenology 153:9–18. https://doi.org/10.1016/j.theriogenology.2020.04.043. Epub 2020 May 5. PMID: 32416545

Manjunatha BM, Pratap N, Al-Bulushi S, Hago BE (2012) Characterization of ovarian follicular dynamics in dromedary camels (*Camelus dromedarius*). Theriogenology 78(5):965–973. https://doi.org/10.1016/j.theriogenology.2012.05.011. Epub 2012 Jul 21. PMID: 22819287

Manjunatha BM, Al-Bulushi S, Pratap N (2015) Synchronisation of the follicular wave with GnRH and PGF2α analogue for a timed breeding programme in dromedary camels (*Camelus dromedarius*). Anim Reprod Sci 160:23–29. https://doi.org/10.1016/j.anireprosci.2015.06.023. Epub 2015 Jul 2. PMID: 26164524

Manjunatha BM, Al-Hosni A, Al-Bulushi S (2020) Superovulation of dromedary camels with eCG: An effective and practical technique. Theriogenology 151:112–118. https://doi.org/10.1016/j.theriogenology.2020.04.017. Epub 2020 Apr 16. PMID: 32325323

Mansour N (2022) Semen collection from dromedary camel bulls, methodology and a new patented approach. In: Juhasz J, Skidmore JA, Nagy P (eds) Proceedings of the ICAR satellite meeting on camelid reproduction, Bologna, Italy, pp 40–44

McDonnell SM (2011) How to train a stallion to use a dummy mount. In: Reproduction 1, AAEP Proceedings 2011, vol 57, pp 37–47

McKinnon AO, Tinson AH, Nation G (1994) Embryo transfer in dromedary camels. Theriogenology 41:145–150

Mee JF (2013) Why do so many calves die on modern dairy farms and what can we do about calf welfare in the future? Animals 3:1036–1057. https://doi.org/10.3390/ani3041036

Menchetti L, Faye B, Padalino B (2021) New animal-based measures to assess welfare in dromedary camels. Trop Anim Health Prod 53(6):533. https://doi.org/10.1007/s11250-021-02978-8. PMID: 34739606

Musa BE, Abusineina ME (1978) The oestrous cycle of the camel (*Camelus dromedarius*). Vet Rec 102:556–557

Nagy P, Juhasz J (2012) Fertility after ovarian follicular wave synchronization and fixed-time natural mating compared to random natural mating in dromedary camels (*Camelus dromedarius*). Anim Reprod Sci 132(3–4):223–230. https://doi.org/10.1016/j.anireprosci.2012.05.010. Epub 2012 May 24. PMID: 22683027

Nagy P, Juhász J (2019) Pregnancy and parturition in dromedary camels I. Factors affecting gestation length, calf birth weight and timing of delivery. Theriogenology 134:24–33. https://doi.org/10.1016/j.theriogenology.2019.05.017. Epub 2019 May 17

Nagy P, Das Gupta A, Barua R, Juhasz J (2016) Comparison of mating vs. GnRH induced LH and Cortisol release in dromedary camels (*Camelus dromedaries*). Abstract. 18th International conference on animal reproduction (ICAR), Tours, France, 26–30 June 2016

Nagy P, Reiczigel J, Gupta AD, Barua R, Juhász J (2021) Pregnancy and parturition in dromedary camels II. Incidence, timing and factors affecting early pregnancy loss (EPL) and the outcome of twin pregnancies. Theriogenology 172:289–299. https://doi.org/10.1016/j.theriogenology.2021.07.004. Epub 2021 Jul 14

Nagy P, Reiczigel J, Gupta AD, Barua R, Juhász J (2022a) Pregnancy loss in dromedary camels. In: Juhasz J, Skidmore JA, Nagy P (eds) Proceedings of the ICAR satellite meeting on camelid reproduction, Bologna, Italy, pp 26–30

Nagy P, Skidmore JA, Juhasz J (2022b) Intensification of camel farming and milk production with special emphasis on animal health, welfare, and the biotechnology of reproduction. Anim Front 12(4):35–45. https://doi.org/10.1093/af/vfac043. eCollection 2022 Aug. PMID: 35974793

Nawito MF, Shalash MR, Hoppe R, Rakha AM (1967) Reproduction in the female camel. Bull Anim Sci Res Inst (Cairo) 2:82

Nikjou D, Niasari-Naslaji A, Skidmore JA, Mogheiseh A, Razavi K, Gerami A (2008) Synchronization of follicular wave emergence prior to superovulation in Bactrian camel (*Camelus bactrianus*). Theriogenology 69:491–500

Novoa C (1970) Reproduction in the Camelidae: A review. J Reprod Fertil 22:3–20

Padalino B, Aubé L, Fatnassi M, Monaco D, Khorchani T, Hammadi M, Lacalandra GM (2014) Could dromedary camels develop stereotypy? The first description of stereotypical behaviour in housed male dromedary camels and how it is affected by different management systems. PLoS One 9(2):e89093. https://doi.org/10.1371/journal.pone.0089093. eCollection 2014. PMID: 2458652

Padalino B, Rateb SA, Ibrahim NB, Monaco D, Lacalandra GM, El-Bahrawy K (2016) Behavioral indicators to detect ovarian phase in the dromedary she-camel. Theriogenology 85:1644–1651

Seboussi R, Faye B, Askar M, Hassan K, Alhadrami G (2009) Effect of selenium supplementation on blood status and milk, urine, and fecal excretion in pregnant and lactating camel. Biol Trace Elem Res 128(1):45–61. https://doi.org/10.1007/s12011-008-8251-3. Epub 2008 Oct 30. PMID: 18972072

Skidmore JA (2000) Pregnancy diagnosis in camels. In: Skidmore JA, Adams GP (ed) Recent advances in camelid reproduction. International Veterinary Information Service. https://www.ivis.org/library/recent-advances-camelid-reproduction/pregnancy-diagnosis-camels

Skidmore JA (2011) Reproductive physiology in female old world camelids. Anim Reprod Sci 124(3–4):148–154. https://doi.org/10.1016/j.anireprosci.2010.08.023

Skidmore JA, Billah M (2006) Comparison of pregnancy rates in dromedary camels (*Camelus dromedarius*) after deep intra-uterine versus cervical insemination. Theriogenology 66:292–296

Skidmore JA, Billah M, Allen WR (1995) The ovarian follicular wave pattern in the mated and non-mated dromedary camel (*Camelus dromedarius*). J Reprod Fertil Suppl 49:545–548

Skidmore JA, Billah M, Allen WR (1996) The ovarian follicular wave pattern and induction of ovulation in the mated and non-mated one -humped camel (*Camelus dromedarius).* J Reprod Fertil 106:185–192

Skidmore JA, Billah M, Allen WR (2002) Investigation of factors affecting pregnancy rate after embryo transfer in the dromedary camel. Reprod Fertil Dev 14:109–116

Skidmore JA, Adams GP, Billah M (2009) Synchronization of ovarian follicular waves in dromedary camels (*Camelus dromedarius*). Anim Reprod Sci 114(1–3):249–255

Skidmore JA, Morton KM, Billah M (2013) Artificial insemination in dromedary camels. Ani Reprod Sci 136:178–186

Tibary A, Anouassi A (1997a) Reproductive physiology in female camelidae. In: Tibary A, Anouassi A (eds) Theriogenology in camelidae; anatomy, physiology, pathology and artificial breeding. Abu Dhabi Printing Press, Abu Dhabi, United Arab Emirates, pp 169–241
Tibary A, Anouassi A (1997b) Breeding soundness examination of the female Camelidae. In: Tibary A, Anouassi A (eds) Theriogenology in camelidae; anatomy, physiology, pathology and artificial breeding. Abu Dhabi Printing Press, Abu Dhabi, United Arab Emirates, pp 243–316
Tibary A, Anouassi A (1997c) Reproductive physiology of the male. In: Tibary A, Anouassi A (eds) Theriogenology in camelidae; anatomy, physiology, pathology and artificial breeding. Abu Dhabi Printing Press, Abu Dhabi, United Arab Emirates, pp 49–78
Tibary A, Anouassi A (1997d) Management of reproduction in camelidae. In: Tibary A, Anouassi A (eds) Theriogenology in camelidae; anatomy, physiology, pathology and artificial breeding. Abu Dhabi Printing Press, Abu Dhabi, United Arab Emirates, pp 459–479
Tibary A, Rodriguez J, Sandoval S (2008) Reproductive emergencies in camelids. Theriogenology 70:515–534
Tingari MD, El Manna MMM, Rahim ATA, Ahmed AK, Hamad MH (1986) Studies on camel semen. I: Electro-ejaculation and some aspects of semen characteristics. Anim Reprod Sci 12: 213–222
Wardeh MF (2004) The nutrient requirements of the dromedary camel. J Camel Sci 1:37–45
Wilson RT (1984) The camel. Longman, London, pp 83–102
Yagil R, Etzion Z (1984) Enhanced reproduction in camels (*Camelus dromedarius*). Comp Biochem Physiol 79A:201–204
Ziapour S, Niasari-Naslaji A, Mirtavousi M, Keshavarz M, Kalantari A, Adel H (2014) Semen collection using phantom in dromedary camel. Anim Reprod Sci 151(1–2):15–21. https://doi.org/10.1016/j.anireprosci.2014.09.016. Epub 2014 Sep 28. PMID: 25449548

Stress Transport in the Dromedary Camel 10

Mohammed El Khasmi

Abstract

Domestic animals are frequently exposed to transport stress which could alter their welfare, since it affects behavioural homoeostasis, health, meat quality and antioxidant status. During all stages of the journey, including preparation handling, loading, transport, unloading, new environmental housing and waiting, camel welfare is significantly impacted by the stressors presented in each of those stages. In dromedaries transport by road, journey length, loading density and waiting time have been identified as potential risk factors for poor welfare, using physiological, hormonal, haematological and biochemical parameters. Muscle glycogen depletion associated with high ultimate pH and hyperglycaemia has been observed. High cortisol blood levels observed after transport was positively correlated with behavioural and physiological stress responses, and negatively correlated with meat quality parameters. Respect for the well-being of the dromedary throughout road transport requires good transport conditions minimizing journey length, water and food deprivation, loading density and waiting period. This chapter indicates that there is an urgent need for legislation on the welfare and protection of dromedaries during transport according to international welfare standards.

Keywords

Stress · Oxidant stress · Transport · Physiological parameters · Hormones · Blood · Meat · Camel

M. El Khasmi (✉)
Laboratory of Physiopathology and Molecular Genetics, Ben M'Sik Faculty of Sciences, Hassan II University of Casablanca, Casablanca, Morocco
e-mail: mohammed.elkhasmi@univh2c.ma; www.fsb.univh2c.ma

B. Padalino, B. Faye (eds.), *Dromedary Camel Behavior and Welfare*, Animal Welfare 24, https://doi.org/10.1007/978-3-031-48600-5_10

10.1 Introduction

Camels are of great social and economic importance and are mainly employed for the production of milk, meat and fibre, sports (racing), beauty contests, recreation, tourism, transport and as beasts of burden and trade in arid countries (Faye 2013; Khalafalla and Hussein 2021). Meat production at the world level reached 653,000 tons in 2019 (FAOstat 2021) for an estimated camel population of 37.5 million heads, a probably underestimated number (Faye 2020; Khalafalla and Hussein 2021). With a mean slaughtering rate close to 8%, around 3 million camels are slaughtered annually. The handling of camels on farms and in markets and the conditions of their loading on trucks, transport, unloading and housing during rest periods are potentially stressful and expose camels to so-called pre-slaughter stress (Nazifi et al. 2009a; Saeb et al. 2010; El Khasmi et al. 2010, 2013, 2015; Lemrhamed et al. 2018, 2019a, b; Tabite et al. 2019a).

Farm animals face several potential welfare issues before, during and after transport, including withdrawal of food and water, injury, illness and stress associated with handling, mixing, change of environment, loading into vehicles, duration and conditions of transport, unloading and waiting time (Kannan et al. 2003; Kadim et al. 2006; Terlouw et al. 2008; Akin et al. 2018; Roadknight et al. 2021). In addition to handling operations prior to transport, animals are exposed during transport to other potential stressors, such as climatic conditions, acceleration, vibrations, noise, space restrictions, road topography, vehicle design and air pollutants (Knowles 1999; Hartung 2003; Mitchell 2009; Brown et al. 2011; Padalino and Raidal 2020). Given the impact of stress, the general aim of this chapter is to describe how transport conditions may affect camel welfare and suggest corrective and preventive measures to protect the welfare of camels during transport.

10.2 History of Transport and Mobility of the Dromedary

The first dromedary introduced into Australia arrived in Port Adelaide on the Apollon on October 12, 1840. Subsequently, dromedaries were imported into Australia between 1893 and 1896 and then in 1907 for use as a means of transport (Blancou and Parsonson 2008). In the United States, with the support of President Jefferson Davis in the mid-nineteenth century, for the American army based in the arid areas of the country, 75 dromedaries were imported from North Africa and put to work packing military supplies. Camel caravans were a common activity between San Antonio and Los Angeles just before the Civil War beginning in 1861. At the same time, commercial shipments totalling up to 400 camels were imported into Texas, Alabama and California (Young 1982).

The first camel exports from Somalia and Sudan to the Arabian Peninsula are documented since 1884 (Castiello et al. 2012; Younan et al. 2016). In addition, Sudanese racing dromedaries have been traded to Saudi Arabia (KSA) and the United Arab Emirates (UAE) via Egypt since the 1950s (Nawata 2005).

In Morocco, camel mobility over large distances (>50 km) is practised by 8% of herders towards the southern regions (Dakhla, Guelmim). While 79% of herders opt for low-amplitude movements (<50 km) which are generally limited to rangelands (Moutik 2018; Julien et al. 2021). This mobility can be pendulum (transhumance) or random (nomadism) based on the short distance (between Ethiopia and Djibouti) (Faye 1994) or the long distance (between Chad and the Central African Republic) (Aubague et al. 2011). These animals are used in peri-urban or even urban areas, either to ensure the supply of milk and meat to urban centres (Kamili et al. 2020) as is the case in other regions of Africa (Faye et al. 2003), or solely for leisure purposes in tourist areas (Kamili et al. 2020; Julien et al. 2021). A large number of camels are marketed directly on the routes, because the very poorly developed road network infrastructure in pastoral areas does not facilitate access by means of transport to the heards on these routes. These conditions favour the clandestine introduction of camels from other countries (Mauritania, Algeria, Mali and Senegal).

10.3 Camel Transport Topography

Camels are transported frequently from Mauritania, Mali, Niger, Chad, Sudan, Kenya, Ethiopia, Somalia and Djibouti, to the countries of the Near and Middle East and North Africa (Egypt, Libya and Morocco), for many reasons such as participation in races and beauty shows, and production of milk and meat (Faye 2013). The most important flows are those linking the Horn of Africa to the Egyptian and Saudi markets. To satisfy the national demand for the consumption of camel meat, several ports on the Red Sea (Port Sudan, Djibouti) or the Indian Ocean (Hargeisa, Mogadishu, Berbera) have become hubs for this trade because the majority of camels are transported by boat to Aden, Jeddah, Muscat or Dubai (Faye 2013). Somalia is the world's largest exporter of dromedaries, more than 70,000 of which are transported annually by boat to Saudi Arabia, where they are slaughtered for HAJJ, the ritual pilgrimage to the holy city of Mecca (Meyerfeld 2017).

In Egypt, the large production of camel meat depends on the import of camels mainly from Sudan and smaller numbers are imported from Somalia and Ethiopia (Napp et al. 2018). Camels arrive in Egypt via the Red Sea ports of Safaga and Suez (Sayour et al. 2015). The road transport of some dromedaries by trucks from El-Obeid in Sudan to Aswan in Egypt could last 8 days, then the dromedaries must walk on foot for another 2 days. Then, the camels remain in quarantine for a day, to be transported in trucks to the market and other commercial centres, where they spend another night before their slaughter. During these long journeys, camels are deprived of rest, water and food, chained and loaded at high density in a sternal sitting position (https://www.animals-angels.de/en/projects/animal-transports/camel-transports.html; Faye 2019). Camels are also transported from Mauritania to Algeria via Mali and to Senegal, and from Niger and Chad to Nigeria (http://camelides.cirad.fr/fr/curieux/marche_viande.html). In the Sahelian countries without a port, transport is done by truck (7–10 days) or on foot (50–60 days), via

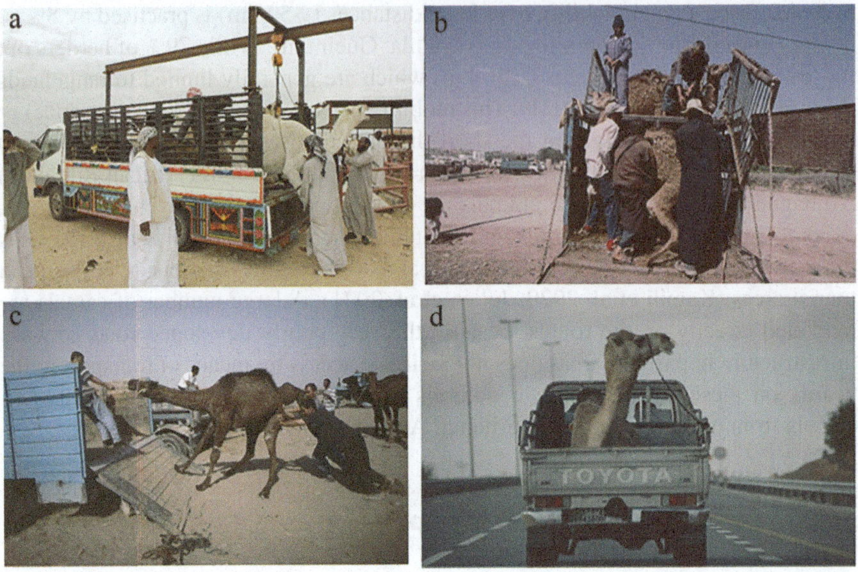

Fig. 10.1 Stress conditions linked to transport of camels in some countries: (**a**) stressful loading of camel with a crane in Saudi Arabia, (**b**) high-density transportation of camels in Morocco, (**c**) stressful manual loading of camel in Turkmenistan, (**d**) apparent quiet camel transportation in pickup at Dubai (photo: Faye Bernard)

border-crossing points such as between Abéché in Chad and Koufra in Libya. (http://camelides.cirad.fr/fr/curieux/marche_viande.html). It should be added that Somali camels destined for KSA are transported mainly from the ports of Berbera and Bosaso in northern Somalia to the Red Sea ports of KSA in Jizan and Jeddah (Younan et al. 2016). In addition to African countries, camels are also transported, but on a smaller scale, from Australia to countries in the Middle East (Kadim et al. 2013).

After arriving in the receiving countries, the dromedaries continue to be exposed to stress, because generally they are moved from one market to another on foot mainly or using small trucks (trailer or semi-trailer) not suitable for long journeys (Fig. 10.1). On the other hand, the means of transport are not suitable for these animals, the loading and unloading procedures are generally difficult and accompanied by violence, stress, fear and pain, with a lack of skills and training of the operators on humane handling (Faye 2019). These conditions could be a source of stress capable of having serious adverse effects on their well-being, their immune system (Manteca 2008) and the quality of their products (Broom 2008; Emeash et al. 2016).

10.4 Physiological Parameters

Body temperature and heart and respiratory rates depend on the metabolism of animals and are largely influenced by stress (Broom 1998). Positive correlations between these physiological parameters have been reported in dromedaries (Abdalla et al. 2011; Samara et al. 2013) and have been attributed to thermoregulatory mechanisms.

In camels, road transportation for 2 h by truck during the hot-dry season may be considered as a potent stressor which had been able to induce a significant increase in rectal temperature and heart and respiratory rates by comparison to values observed before transportation (El Khasmi et al. 2013). By comparing the effect of two loading densities in trucks on these physiological parameters, Lemrhamed et al. (2018) have found that rectal temperature and heart and respiratory rates were significantly higher when the loading density was high (1 camel/1.44–1.80 m^2) than those observed when the loading density was low (1 camel/2–3.6 m^2).

Concerning the evolution of physiological parameters after road transport, Hamad et al. (2021) found that the rectal temperature and heart and respiratory rates, which were higher at the end of the transport, approached basal values after 3 h of rest indicating a reduction in the dromedary's reactivity to stress. On the other hand, in the same species, the results obtained in the study by Lemrhamed et al. (2019a) showed that after a 2-h road transport, these same physiological parameters were returned to their basal values after a 12–16-h rest period. These differences could be due to the different conditions of transport, loading, unloading and stabling after travelling and especially the time of measurement, therefore to circadian fluctuations in the physiological functions of the dromedary (El-Allali et al. 2013; Bouâouda et al. 2014; Al-Haidary et al. 2016).

10.5 Hormonal Parameters

Stress stimulates the sympathetic adrenal medullary system (SAMS) and then the hypothalamic-pituitary-adrenal cortex (HPA) axis to release catecholamines and corticosteroids from the adrenal glands, which is why measuring the concentrations of these hormones in the blood, especially cortisol, is commonly used as a reliable biomarker of stress (Ferguson and Warner 2008; Zimerman et al. 2013). However, during stress induced by transport for example, the catecholamines released first by the sympathetic nerve endings and then by the adrenal medulla are transformed into adrenaline, while cortisol remains active in the body longer than these catecholamines (Gregory 1998). Thus, for the evaluation of stress, the quantification of cortisol levels (the most common glucocorticoid molecule) or of its metabolites is the most direct and reliable indicator of the physiological state of an animal, and an index of its response following activation of High Performance Liquid Chromatography (HPC) axis and an environmental threat of its homoeostasis (Gayrard et al. 1996; Reeder and Kramer 2005; Manteca 2009).

The difficulties in obtaining blood samples and the recognition of the stressful effect of blood sampling are the main drivers of the use of minimally invasive sampling media as biomarkers of adrenal cortex responses to animal stress. In dromedaries, instead of assessing cortisol in blood (Ziv et al. 1997; Kataria et al. 2000; Zia-ur-Rahman et al. 2007; Saeb et al. 2010), urine (El Khasmi et al. 2010), faeces (Sid-Ahmed et al. 2013) and saliva (Majchrzak et al. 2015) have been sampled to determine the effects of different stressful situations such as road transport, dehydration and environmental heat. In addition, circulating cortisol levels had been measured during lactation and milking in camels (Atigui et al. 2015; Brahmi et al. 2021) and other species (Marnet and Negrão 2000; Lupoli et al. 2001; Negrão and Marnet 2006) to assess the response of animals to environmental challenges. Atigui et al. (2015) defined dromedary camels as stressed when higher blood cortisol levels between 40 and 60 ng/mL are observed accompanied by obvious signs of fear (restlessness, moaning, kicking, defecation) while normal range of values is between 3 and 30 ng/mL (Faye and Bengoumi 2018). During more physiological adaptations, cortisol increased in camels from baseline levels of about 21.9 ± 1.0 ng/mL to over 121.6 ± 5.4 on the day of parturition or increased from 37.1 ± 1.4 ng/mL 1 day before weaning to 48.0 ± 1.5 and 69.5 ± 1.9 ng/mL at weaning and third day after weaning (Mohamed 2006).

However, cortisol levels may fluctuate also during seasons, mainly due to thermal stress. In a previous study, we determined the values of cortisol in serum, hair and faeces at two distinct periods, in 20 male dromedaries aged 3–8 years, from semi-extensive farms in the region of Essaouira and intended for slaughter at municipal slaughterhouses in Casablanca, Morocco. On average, cortisol levels in serum (ng/mL), hair and faeces (ng/g) were significantly higher in winter than in summer (respectively, 66.01 ± 13.19 vs 25.71 ± 6.71; 0.93 ± 0.26 vs. 0.61 ± 0.08 and 2.74 ± 0.14 vs. 1.42 ± 0.35; $p < 0.05$) (Bargaâ et al. 2016a; Farh et al. 2018). However, in dromedaries, a higher cortisolemia (ng/mL) in summer than in winter was observed by Baraka (2012) (38.6 ± 5.3 vs. 28.5 ± 4.8), Elias and Weil (1989) (45.0 ± 11.9 vs. 8.0 ± 1.3) and Zia-ur-Rahman et al. (2007). These authors explained this hyperactivity of the adrenal cortex observed in summer by the stress induced by the external heat load and the bodily dehydration following intense sweat losses, as a major thermolytic pathway. The cortisol levels in females are higher than males (Baraka 2012) and in young more than adult camels (El Khasmi et al. 2009; Baraka 2012).

In the dromedary camel, in addition to the levels of cortisol measured in blood, those analysed in the hair and faeces could be useful for a reliable retrospective assessment of long-term stress (Davenport et al. 2006; Sid-Ahmed et al. 2013). Collection of these samples minimizes stress on the animal and are easy to transport and store. In fact, it has been shown in dromedaries that intravenous injection of adrenocorticotrophic hormone (ACTH) (0.5 mg/animal) increased the level of cortisol in the blood and that of glucocorticoid metabolites in the faeces (Sid-Ahmed et al. 2013). Thus, 24 h after this injection, the level of cortisol increased from 0.6–10.8 to 10.9–42.2 ng/mL in the blood, and from 286.7 to 2559.7 ng/g in the faeces (Sid-Ahmed et al. 2013). Furthermore, concentrations of cortisol in blood were positively correlated with those in saliva and/or hair or faeces concentrations in dromedaries (Bargaâ et al. 2016a), cattle (Tallo-Parra et al. 2015) and ewes (Yates et al. 2010).

Cortisol and thyroid hormones are regarded as indicators of stress in several species, such as camel (Nazifi et al. 2009a, b; El Khasmi et al. 2011; Saeb et al. 2010; Baraka 2012; El Khasmi et al. 2013, 2015), alpacas (Anderson et al. 1999) and cattle (Grandin 1997). In dromedaries, road transport was able to increase blood levels of cortisol (Saeb et al. 2010; El Khasmi et al. 2010, 2013) and thyroid hormones (El Khasmi et al. 2010; Lemrhamed et al. 2019a, b) and a decrease in circulating testosterone concentrations (Mohamed et al. 2021), without any change in blood levels of parathyroid hormone and 25-hydroxyvitamin D (El Khasmi et al. 2010). If in the dromedary the stress induced by transport (El Khasmi et al. 2010, 2013, 2015) or physical exercise (distance = 4 km, speed = 6.1 m/s, Ta = 25 °C) (Riad 1995) had induced a significant rise in plasma cortisol levels, according to Dahlborn et al. (1992) these rates showed no significant variation under the effect of stress induced by starvation for 4 days in the same species.

In camels, high transport distance and stocking density are potential factors of stress, inducing a significant secretion of thyroid and glucocorticoids hormones (El Khasmi et al. 2010, 2013, 2015; Lemrhamed et al. 2018, 2019a). El Khasmi et al. (2015) reported very high elevations of cortisol in dromedaries (between 88.32 ± 19.4 and 152.4 ± 25.18 ng/mL) depending on the length of the journey (between 72 and 170 km). The animal could restore normal value after a rest period, but a lairage time not respecting the good conditions of the welfare of camel could increase concentrations of these hormones again (Lemrhamed et al. 2019b). The return to the basal physiological and behavioural state, under conditions of food deprivation but with access to water, has been observed after a stall period of 12–24 h in dromedaries (Kadim et al. 2013), 24–48 h in cattle (Tadich et al. 2005; Mounier et al. 2006) and more than 17 h in pigs (Jama et al. 2016).

Finally, in a recent investigation, we found that frequencies of some behavioural reactions of dromedaries, to stress recorded during their unloading at the end of transport and during their conduction to the waiting area (slips, falls, blather, urination and defecation) were positively and significantly correlated with serum cortisol and malondialdehyde (MDA) levels analysed in the same animals (Lemrhamed 2020; El Khasmi et al., 2021). These behavioural reactions to stress could indeed be interpreted in terms of pain, agitation and fear in the presence of conditions that do not respect the well-being of animals during transport and unloading (Grandin 1997; Bourguet et al. 2010).

10.6 Haematological Indicators of Transport Stress

Transport constitutes a source of stress, capable of reducing immunity and disturbing the homoeostasis of farm animals (Kannan et al. 2003; Minka and Ayo 2007; Ishizaki and Kariya 2010; Adenkola and Ayo 2010; De la Fuente et al. 2012). The leukocyte count by blood smear has been widely used as a reliable indicator to assess stress in many domestic ruminants (Idrus et al. 2010; Wickham et al. 2015; Stockman et al. 2012) including the dromedary (Bargaâ et al. 2016b; Lemrhamed et al. 2019b). In this species, haematocrit, erythrocytes number, and haemolysis increased under transport stress (El Khasmi et al. 2013) and gradually with transport distance, thus, over longer distances (350–360 km) these parameters were significantly higher compared to short distances (72–80 km) (El Khasmi et al. 2015). The increase in haematocrit may be explained by a water loss by thermoregulation and urination, and/or a splenic contraction (Carlson 1990). In fact, after activation of the SAMS under stress (McCarty et al. 1988), the secretion of catecholamines into the circulation could induce a splenic contraction through α-adrenergic receptors and then a release of erythrocytes into the circulation (Montane et al., 2002; Tauler et al. 2003). In addition, the hyperthermia observed in camels during transport under heat (El Khasmi et al. 2013) may induce a water loss caused by sweat evaporation, then contribute to dehydration and increase of haematocrit.

Camels have more resistant red blood cells (RBC) than other mammalian species (Livne and Kuiper 1973; Mirgani 1992; Al-Qarawi 1999; Arikan 2003); however a decrease of osmotic resistance of these cells by road transportation stress under heat had been reported (El Khasmi et al. 2013). In camel, the osmotic fragility test of RBC could be used as a diagnostic tool in road transportation stress (El Khasmi et al. 2013) suggesting oxidant alterations of plasmic membrane of these cells (El Khasmi et al. 2015), following a free radical production and damage of the membrane proteins and lipids (Tauler et al. 2003).

Numerous studies have reported in farm animals that transport stress dramatically increased the neutrophil/lymphocyte ratio (NLR) (Nwe et al. 1996; Kannan et al. 2000; Rajion et al. 2001; Minka and Ayo 2007; Sowińska et al. 2020). This effect could be due to the release of adrenaline and corticosteroids, which are capable of inducing an increase in the number of neutrophils and a decrease in the number of lymphocytes (Stanger et al. 2005; Emeash et al. 2016), in especially helper T lymphocytes involved in the response to foreign substances, resulting in a decrease in all cell-mediated immunity (Earley et al. 2011, 2017). In addition, the values of the NLR were positively correlated with blood levels of cortisol, glucose and lactate under the effect of transport stress in camels (El Khasmi et al. 2015; Lemrhamed et al. 2018, 2019a, b) and cattle (Grigor et al. 2004; Mounier et al. 2006; Chulayo et al. 2016). The high NLR could also be associated with free radical generation and immunosuppression under stressful conditions (Klokker et al., 1993; Flerov and V'iushina, 2011), making the animal vulnerable to infections and inflammatory diseases.

In dromedary camels, road transport induced a significant decrease in the number of leukocytes (Tadich et al. 2005; El Khasmi et al. 2013; Emeash et al. 2016; Hamad et al. 2021). In this same species, the NLR and haemolysis showed considerably high values at the end of a 2-h transport with a high loading density of animals in the truck (1.44–1.80 m^2/camel) (Lemrhamed et al. 2018). Under the same conditions, this increase persisted from 12 to 16 h after the end of transport (Lemrhamed et al. 2019a). According to Hamad et al. (2021), in dromedaries, the increase in haematocrit and RBC count and the decrease in lymphocytes and granulocytes count observed after a waiting time of 3 h after transport were lower than those recorded after 10 h rest. However, in the same species, Liotta et al. (2007) observed that leukocytes count after a long period of rest after transport was lower than that observed after a short period of rest.

In dromedaries, a 2-h road transport associated or not with environmental heat induced significant hyperglycaemia (El Khasmi et al. 2010, 2013). A significant gradual increase in blood glucose with increasing transport distance was reported in the same species (El Khasmi et al. 2015), goats (Minka et al. 2009), cattle (Malena et al. 2006), pigs (De Silva and Kalubowila 2012) and horses (Stull and Rodiek 2000). In dromedaries, hyperglycaemia was also manifested under the effect of increased loading density in trucks (Lemrhamed et al. 2018) and during stabling after transportation (Lemrhamed et al. 2019a). Hyperglycaemia is considered to be a reliable indicator of transport stress in farm animals (Nwe et al. 1996; Kannan et al. 2007; Sowińska et al. 2020). In cattle, hyperglycaemia was maintained for 24 h after 3-h transport (Tadich et al. 2005) or 31 h (Mounier et al. 2006). It could be due to the elevated blood levels of cortisol following the activation of the sympathetic and HPA axes during the exposure of the animals to stress, causing a release of cortisol and catecholamines which directly stimulate the muscular and hepatic mobilization of the glycogen, resulting in increased circulating glucose levels (Grandin 1997; Kannan et al. 2000; Kuo et al. 2015) and muscle glycogen depletion (Terlouw et al. 2008). In addition, cortisol plays a very important role in gluconeogenesis because it stimulates the liver to convert fats and proteins into indirect metabolites. These metabolites are ultimately converted to glucose as an energy source (Saeb et al. 2010). On the contrary, glucose concentrations in transported camels were lower than those of non-transported controls (Mohamed et al. 2021), and which could be explained by very prolonged food deprivation (Kataria and Kataria 2004; Kataria et al. 2007) and/or much more stressful conditions that have depleted energy reserves (Badakhshan and Mirmahmoudi 2016). Transport stress was able to induce an increase in blood lactate in dromedaries (El Khasmi et al. 2015), suggesting the possibility of certain lesions of the muscles or other tissues of the animal (Knowles et al. 2014).

Domestic ruminants transported in vehicles without feed prior to transport were highly stressed and exhibited an increase in unesterified fatty acids (NEFA) in the blood, suggesting greater muscle activity and instability (Kreuzer et al. 1998; Carnovale et al. 2021). In dromedaries, neither the stress induced by a 2-h road transport (El Khasmi et al. 2010), nor that induced by the density of loading (Lemrhamed et al. 2018) were able to modify the blood levels of cholesterol and triglycerides.

Total plasma proteins may provide a better indication of the hydration state of the animal since they are not affected by transport (Jarvis et al. 1996). So, in dromedaries, the transport as well as the density of loading did not cause any significant variation in the plasma levels of total proteins, urea and creatinine (El Khasmi et al. 2010; Lemrhamed et al. 2018). However, in the same species, an increase in circulating levels of creatinine and a decrease in those of total proteins have been noted (Baghshani et al. 2010; Saeb et al. 2010; Tharwat et al. 2013;

Emeash et al. 2016; Mohamed et al. 2021). These differences could be attributed to transport conditions and environmental differences.

The increase in haptoglobin and fibrinogen by transport stress observed in dromedaries (Mohamed et al. 2021) had been reported in cattle (Marti et al. 2017; Kang et al. 2017) and could indicate food and water deprivation activating the acute phase response (Earley et al. 2011). In contrast, camels transported over long distances showed no increase in fibrinogen after transport and unloading (Baghshani et al. 2010).

The metabolism of the liver is a vital physiological process essential for maintaining the body's homoeostasis under stressful conditions. Fluctuations in hepatic transaminase values in animals may indicate an intensification of metabolic processes involved in the conversion of carbohydrates, proteins and fats or metabolic disorders (Hrkovic-Porobija et al. 2017). Transport stress induced higher blood levels of hepatic transaminases in dromedaries (Mohamed et al. 2021) indicating liver damage. Likewise, long-distance road transport by truck in horses had increased these rates (Padalino et al. 2017). On the contrary, the levels of transaminases did not show any significant variation in camels having been subjected to stress induced by 2-h transport (El Khasmi et al. 2010) or by a high density of loading during transport (Lemrhamed et al. 2018).

Creatine kinase is an enzyme involved in the production of Adenosine Triphosphate (ATP) in skeletal muscle and its activation increases during muscular efforts requiring a large amount of energy (Volfinger et al. 1994). This enzyme is used as an indicator of muscle fatigue during stress induced by transporting animals (Warriss et al. 1995; Zhong et al. 2011) and could be affected by driving style during transport. In 25 healthy camels, a significant increase in blood creatine kinase levels was detected 2 h after the end of 5-h road transport, these levels returning to their basal level 24 h after transport (Tharwat et al. 2013).

Lactate dehydrogenase (LDH) catalyses the interconversion of pyruvate and lactate, the end-product of glycolysis in muscle, when oxygen is absent or insufficient. In farm animals, blood concentration of LDH could increase under the effect of transport (Warriss et al. 1984; De et al. 2021), suggesting greater muscle and metabolic activity (Bórnez et al. 2009; Carnovale et al. 2021).

Finally, in dromedaries, the stress induced by road transport decreased the blood levels of iron, copper (Mohamed et al. 2021) and magnesium (El Khasmi et al. 2013), reflecting water and food restriction (Kataria et al., 2007). So in the same species, calcium and phosphorus levels were not impacted either by transport (El Khasmi et al. 2010) or by stocking density (Lemrhamed et al. 2018).

10.7 Postmortem Muscle Metabolism

Stress hormones released under transport stress may influence muscle metabolism parameters, such as glycogen levels, ultimate pH (24-h postmortem pH), proteolytic processes, water holding capacity, tenderness and flavour (Ferguson and Warner 2008). According to Benaissa et al. (2014), the evolution of the postmortem pH of camel meat is characterized by a rapid drop during the first 8 h after slaughter, followed by a slowing down and then stabilization at stage 24 hour. In camels, high circulating levels of cortisol after transportation had been associated with a decrease in glycogen levels and a high pH value in meat (El Khasmi et al. 2010).

According to Barka et al. (2016), the ultimate pH (pHu) values of muscles (*Triceps brachii*, *Musculus obliquus* and *Diaphragma*) had been higher in camels transported for 160 km compared with camels transported for 72 km only (respectively, 6.4 ± 0.2 vs. 5.7 ± 0.2; 6.5 ± 0.2 vs. 5.6 ± 0.1 and 6.3 ± 0.1 vs. 5.6 ± 0.2). In farm animals, pHu is the main key to meat quality which is closely related to their transport before slaughter (Warriss 1990; Ferguson and Warner 2008; Mounier et al. 2006). Long transport distances are stressful factors that could affect the pHu. Kadim et al. (2006) observed that generally young camels tend to produce meat with a higher pH than that of older ones due to the low glycogen content in the latter.

In addition, the glycogen levels in camel muscles (*Triceps brachii*, *Musculus obliquus* and *Diaphragma*) (mg/100 g) decreased significantly at high transport distance compared with short ones (respectively, 170 ± 20 vs. 226 ± 25; 191 ± 21 vs. 241 ± 27 and 180 ± 23 vs. 237 ± 25) (Barka et al. 2016). In the dromedary camel, the circulating levels of cortisol were positively correlated with transport distance (El Khasmi et al. 2015) and with the values of postmortem pH, drip loss, cooking loss, dimensional shrinkage and total haem pigment in meat (Tabite et al. 2019a). It had been reported that high road transport distance had induced higher pHu, expressed juice, cooking loss and shear force in goats (Kadim et al. 2006) and sheep meat (Ruiz-De-La-Torre et al. 2001). Prolonged food deprivation during transport and stabulation could significantly reduce muscle glycogen levels (Terlouw et al. 2015), and the duration of the fast was positively correlated with the value of postmortem pH of camel meat (Benaissa et al. 2021). In addition to the impact of the length of food deprivation, the waiting time after transport influences the quality of meat.

In camel meat, the 25-hydroxyvitamin D levels were negatively correlated with drip loss and cooking loss (Tabite et al. 2019b) suggesting a possible role of this vitamin in the meat quality in transported animals.

10.8 Oxidant Stress Indicators

Oxidative stress (OS) groups together interrelated phenomena that increase the generation of free radicals, in particular reactive oxygen species (ROS), reactive nitrogen species (RNS) and damage associated with cellular constituents. The increased production of ROS can induce the formation of lipid peroxidation products and oxidized proteins, and oxidative damage to DNA and RNA (Storey 1996; Aschbacher et al. 2013). The quantification of malondialdehyde (MDA) is widely used as an indicator of the peroxidation of polyunsaturated fatty acids, which results in structural disruption and impaired function of cell membranes (Simsek et al. 2006).

In mammalian species, cells could protect themselves against free radical damage by using a variety of enzymatic and non-enzymatic antioxidant systems. Catalase, glutathione peroxidase (GSH-Px), superoxide dismutase (SOD), glutathione reductase and thioredoxin are the main enzymatic antioxidants that can react directly or indirectly with ROS and whose activities have been correlated with OS (Nielsen et al. 2011; Martinez-Haroa et al. 2011). Non-enzymatic antioxidants like vitamins C and E, glutathione, uric acid, albumin, bilirubin, *N*-acetylcysteine, ubiquinol-10, carotene, methionine, reduced glutathione and melatonin act as free radical scavengers (Noori 2012).

Oxidant stress had been evaluated in serum and plasma of clinically healthy adult camels (Salar-Amoli et al. 2009) and stressed adult camels by road transport (Nazifi et al. 2009a; El Khasmi et al. 2013, 2015; Mohamed et al. 2021). In the dromedary camel, transport stress seems to activate antioxidant defence systems, leading to an increase in oxidative metabolites and antioxidant capacity and a decrease in endogenous antioxidants. The high levels of antioxidants help reduce the cellular damage generated by OS induced by adverse conditions (Kataria et al. 2012; Kumar et al. 2019). Under the effect of road transport stress, dromedaries showed an imbalance of the oxidant-antioxidant status, marked by a slight increase in the activity of SOD and a slight decrease in that of GSH-Px (Kataria et al. 2010; Mohamed et al. 2021). In addition, a significant increase in blood levels of MDA for a distance travel greater than 350–360 km has been noted (El Khasmi et al. 2015). According to Nazifi et al. (2009a), camels transported during the hot summer season showed a slight increase in GSH-Px activity and a significant increase in plasma concentrations of MDA up to 24 h after the end of transport (Nazifi et al. 2009a) and susceptibility to infections such as pneumonia (Werner and Kaaden 2002). If the camel can adapt to transport stress for a long time, the imbalance between antioxidant enzymes and ROS may be delayed a little until the concentration of non-enzymatic antioxidants like ascorbic acid becomes reduced and unable to fight against tissue damage induced by OS (Mohamed et al. 2021). In addition, ROS are implicated in numerous disorders (Halliwell and Gutteridge 1999; Agarwal et al. 2003), and are associated with an activation of the enzymatic antioxidant system and with low concentrations of ascorbic acid in the body tissue and/or blood level in dromedaries (Mousa et al. 2006). In addition, blood levels of some antioxidants such as vitamins A, C and E, and glutathione had been significantly lower in a cold environment in dromedaries (Kataria et al. 2010), suggesting a low antioxidant status of these animals during the winter season (Lektib et al. 2016b).

El Khasmi et al. (2015) evaluated the OS related to road transport distance in three groups of camels that were transported at short (72–80 km), medium (160–170 km) and long (350–360 km), respectively. Plasma cortisol and MDA levels and catalase activity increased progressively and significantly with distance of transport, and these indicators of OS were positively correlated with each other (El Khasmi et al. 2015). The increase in oxidative damage revealed by a significant lipid peroxidation of the membranes of camel RBC could be explained at least in part by the higher levels of serum cortisol under the effect of transport (Saeb et al. 2010; El Khasmi et al. 2010, 2013, 2015). In fact, administration of glucocorticoid hormones has been reported in rats to induce lipid peroxidation by reducing the levels of enzymatic and non-enzymatic antioxidants in erythrocytes (Orzechowski et al. 2000) and hypothalamus (Flerov and V'iushina 2011). In addition, a significant positive correlation between serum MDA and cortisol levels was demonstrated in calves that had been transported by truck (Wernicki et al. 2006).

On the other hand, in the muscle (*musculus abdominis obliquus externus*) of camels which had high circulating cortisol levels (80.29–107.21 ng/mL) after transport, the levels of MDA were higher and the activities of Catalase (CAT) were lower than those of camels which had low cortisol (13.07–67.9 ng/mL) (Tabite et al. 2019a). So, the analysis of circulating cortisol levels after transport just before slaughter (El Khasmi et al. 2013, 2015; Lemrhamed et al. 2018, 2019a, b) may predict the quality and OS status of meat in the camel (Barka et al. 2016; Tabite et al. 2019a). In addition, in the same species, Barka et al. (2016) had reported that the levels of MDA increased significantly while the Catalase activity (CATa) decreased significantly in meat (*Triceps brachii, Musculus obliquus and Diaphragma*) when the transport distance before slaughter increased (72 km vs. 160 km). This OS could be amplified by oxidation of oxymyoglobin and lipids, microbial contaminations leading to the production of free short-chain fatty acids and unstable lipid hydroperoxide (Silva et al. 1999; Aidani et al. 2014), and slaughter conditions.

In view of the impact of transport OS on camel homoeostasis and the metabolism of its muscle, finding food supplements with effective antioxidant properties will be of great use in this species. In camels, incubation of RBC with vitamin E or vitamin C may attenuate oxidative alterations of these cells by hydrogen peroxide (Chakir et al. 2013) and reduce MDA production in the incubation medium (Lektib et al. 2016a). According to Adenkola et al. (2011), in goats, supplementation of the diet with antioxidants before transport was able to significantly reduce the impact of stress and increase the total antioxidant capacity of the animal during transport. Thus, the supplementation by exogenous antioxidants like vitamin E and vitamin C (Alvarado et al. 2006; Niki 2010; Moskowitz et al. 2018) seems to be very helpful during road transportation of dromedary camels by improving cellular immune activity, cytokine production, inflammatory responses and antioxidant status. However, the use of antioxidants as therapy remains controversial (Lykkesfeldt and Svendsen 2007).

Some practices such as transport, handling or poor feeding of animals increase their stress, which can make the muscle more sensitive to lipid and protein peroxidation (Gobert et al. 2013). Camel meat, compared to red meat from other farm animals, is relatively rich in polyunsaturated fatty acids (Sahraoui et al. 2014; Kadim et al. 2020), in myoglobin and in other hematinic compounds, which could act as prooxidants and thus promote the oxidation of the meat (Maqsood and Benjakul 2011), especially when this meat came from camels having a more pronounced stress transport before slaughter (Tabite et al. 2019a).

10.9 World Organization for Animal Health Standards

The World Organization for Animal Health (WOAH) standards on animal welfare, include a chapter on the transport of animals by land and thus play an important role in international trade, as they are the only global standards based on the science agreed upon by the trading nations of the world. According to these standards, plans must be drawn up in relation to the preparation, resting, watering, feeding, observation and housing of animals, transport, travel time, design and maintenance of vehicles and weather conditions. The OIE has 167 member countries, of which 120 are developing countries without any animal welfare standards. All countries in the Middle East and other African countries such as Morocco, Tunisia and Egypt are members of the OIE. Unfortunately, while Oman and the United Arab Emirates have animal welfare rules, including provisions on the transport of animals, many countries in the Middle East and Africa have not yet adopted national rules and do not propose any provisions for the protection of animals during transport based on a legislative framework and OIE standards. Moreover, even if the countries of the Gulf Cooperation Council (GCC) have a legal framework for the movement of dromedaries, and laws on animal quarantine, these rules remain limited to the health of animals without taking into account their protection and their well-being.

10.10 Conclusion and Recommendations

As for other domestic animals, camels cannot escape the stressful conditions which begin at the farm, the place of breeding and the market, and continues with loading, transport, unloading, reception and conduction to the area where the animal should wait several hours. This species is more susceptible to stress and OS induced by road transport, depending on journey conditions, such as stocking density and microclimatic parameters during transport, and consequently journey duration and lairage waiting time (in case of travel towards slaughterhouse). These different sources of stress and OS can alter the homoeostasis of the dromedary and its welfare and health status.

During the transport stages of camels in the Middle East and Africa, the animal welfare standards of the OIE for animals during transport are not respected. Indeed, too often, the vehicles are multi-purpose and are not adapted or designed for the transport of dromedaries, are not equipped with loading and unloading devices and offer no protection against the sun and sandstorms. Animals are roughly handled by untrained and unskilled operators, and transported without bedding and at very high ambient temperatures. However, up to date, there are no welfare standards for camels during transport, so we can conclude that the transport of dromedaries is not sufficiently regulated by law and is not subject to any official control on the welfare of these animals during transport.

As poor handling practices during camel transport can endanger animal health and therefore public health and food safety, these practices must be carried out by experienced and knowledgeable personnel. Thus, our recommendations are as follows:

It is necessary to allow people handling or transporting camels to have the practical experience and/or sufficient training on humane handling based on animal learning theory (see Chap. 8), allowing them to know and comply with national and international legislation on animal welfare. Moreover, it would be better if camels are trained to be handled, loaded and unloaded.

Minimize the duration of transport and feed withdrawal, transport at an appropriate stocking density with comfortable bedding, transport only healthy and fit camels and optimize their management prior to transport (Broom 2008).

Camels must be transported in suitably equipped vehicles with a roof, a minimum height of 2.3 m, a resistant floor that minimizes the risk of slipping and falling, sufficient bedding, adequate ventilation inside the vehicle and vehicle should be equipped with non-slippery and not too steep loading ramps. During loading, some space should be provided between the animal and the wall of the vehicle to prevent injury to the animal during loading/unloading procedures. In addition, experienced camels should be loaded first, to be demonstrators for the others, taking advantage of the fact the camels are follower animals. Finally, loading should never be performed using cranes and camels should never be transported in a forced and restrained sternal position.

Camels must be inspected before departure, during transport, at each stop or rest period and at the unloading stage.

When handling camels, avoid any procedure capable of causing pain or fear to the camels, consequently whipping, tail twisting, nasal twitching, loud noise and pressure on the eyes, lips, ears or external genitalia should be avoided.

During transport on long journeys, camels should be provided with sufficient and adequate food and water, at least twice a day. The dromedaries must be distributed in the vehicle in such a way that they can all lie down on their sternum.

Dromedaries can be tired after transport, so they should be unloaded from the vehicle at recommended facilities as soon as possible after arrival at the destination, waiting time before unloading should be minimized. In housing areas, these animals must have a favourable environment with good ventilation and sufficient space to allow them to rest, drink, feed, protect themselves from extreme weather conditions and have enough place to stand, lie down and turn around.

Finally, the transport practices of dromedary camels urgently need legislation to protect their welfare during all transport steps according to international standards.

References

Abdalla EB, Zeidan AEB, Abd El-Salaam AM et al (2011) Thermoregulation and blood biochemical changes in male dromedary camels during hot-humid and hot dry environments under Egyptian conditions. J Camel Pract Res 18:297–304

Adenkola AY, Ayo JO (2010) Physiological and behavioural responses of livestock to road transportation stress. A review. Afr J Biotechnol 9:4845–4856

Adenkola AY, Ayo JO, Sackey AK et al (2011) Modulatory role of ascorbic acid on behavioural responses of pigs transported by road during the harmattan season. Niger J Physiol Sci 26:61–65

Agarwal A, Saleh RA, Bedaiway MA (2003) Role of reactive oxygen species in the pathophysiology of human reproduction. Fertil Steril 79:829–843

Aidani E, Aghamohammadi B, Akbarian M et al (2014) Effect of chilling, freezing and thawing on meat quality: a review. Int J Biosci 5:159–169

Akin PD, Yilmaz A, Ekiz B (2018) Effects of stocking density on stress responses and meat quality characteristics of lambs transported for 45 minutes or 3 hours. Small Ruminant Res 169:134–139

Al-Haidary AA, Abdoun KA, Samara EM et al (2016) Daily rhythms of physiological parameters in the dromedary camel under natural and laboratory conditions. Res Vet Sci 107:273–277

Al-Qarawi AA (1999) Phospholipids of the camel erythrocyte membrane. J Camel Pract Res 6:39–40

Alvarado C, Álvares P, Puerto M et al (2006) Dietary supplementation with antioxidants improves functions and decreases oxidative stress of leukocytes from prematurely aging mice. Nutrition 22:767–777

Anderson DE, Grubb T, Silveira F (1999) The effect of short duration transportation on serum cortisol response in alpacas (*Llama pacos*). Vet J 157:189–191

Arikan S (2003) A comparison of the effect of methyl-β-cyclodextrin on the osmotic fragility of ovine, bovine and human erythrocytes. Turk J Vet Anim Sci 27:383–387

Aschbacher K, O'Donovan A, Wolkowitz OM et al (2013) Epel E. Good stress, bad stress and oxidative stress: insights from anticipatory cortisol reactivity. Psychoneuroendocrinology 38: 1698–1708

Atigui M, Marnet PG, Barmat A et al (2015) Effects of vacuum level and pulsation rate on milk ejection and milk flow traits in Tunisian dairy camels (*Camelus dromedarius*). Trop Anim Health Prod 47:201–206

Aubague S, Mannany AA, Grimaud P (2011) Difficultés de transhumance des chameliers dans le Tchad central liées aux aléas climatiques. Science et changements planétaires/Sécheresse 22(1): 25–32

Badakhshan Y, Mirmahmoudi R (2016) Blood metabolites of one-humped camel (*Camelus dromedarius*) versus sheep during summer heat stress. Iran J Vet Med 10:65–71

Baghshani H, Nazifi S, Saeb M et al (2010) Influence of road transportation on plasma concentrations of acute-phase proteins, including fibrinogen, haptoglobin, serum amyloid A, and ceruloplasmin, in dromedary camels *Camelus dromedarius*. Comp Clin Patholol 19:193–198

Baraka TA (2012) Clinical evaluation of vitamin A, ß-carotene, vitamin E and cortisol levels in health and selected diseases in camels (*Camelus dromedarius*) in Egypt. J Am Sci 1s:106–111

Bargaâ R, Lektib I, Chakir Y et al (2016a) Corrélation des taux de cortisol dans le sérum, les poils et les fèces chez le dromadaire (*Camelus dromedarius*). Rev Elev Med Vet Pays Trop 69:87–91

Bargaâ R, Lektib I, Barka K et al (2016b) Seasonal variations of haematological profile and its correlations with thyroid and adrenal gland hormones in male Moroccan camel (*Camelus dromedarius*). IJAES 1:8–14

Barka K, Farh M, Riad F et al (2016) Impact of transport distance on some stress biomarkers levels in camel meat. MOJ Food Process Technol 2:00043

Benaissa A, Ould El Hadj-Khelil A, Adamou A et al (2014) Quality of camel meat in the slaughterhouses of Ouargla in Algeria. I. Some physico-chemical characteristics of meat during maturation. Rev Elev Med Vet Pays Trop 67:223–228

Benaissa A, Babelhadj B, Touhami I et al (2021) Study of some parameters influencing the *post mortem* PH of carcass's muscles of dromedaries slaughtered at the Touggourt slaughterhouse Algeria. Int J Environ Agric Biotechnol 6. https://doi.org/10.22161/ijeab

Blancou J, Parsonson I (2008) Historical perspectives on long distance transport of animals. Vet Ital 44(1):19–30

Bórnez R, Linares MB, Vergara H (2009) Haematological, hormonal and biochemical blood parameters in lamb: Effect of age and blood sampling time. Livestock Sci 121:200–206

Bouâouda H, Achâaban MR, Ouassat M et al (2014) Daily regulation of body temperature rhythm in the camel (*Camelus dromedarius*) exposed to experimental desert conditions. Physiol Rep 2: e12151

Bourguet C, Deiss V, Gobert M et al (2010) Characterising the emotional reactivity of cows to understand and predict their stress reactions to the slaughter procedure. Appl Anim Behav Sci 125:9–21

Brahmi M, Atigui M, Hammadi I et al (2021) Oxytocin and cortisol release during suckling, hand-milking and machine milking in camels. J Dairy Res. https://doi.org/10.1017/S0022029921000522

Broom DM (1998) Welfare, stress and the evolution of feelings. Adv Stud Behav 27:371–403

Broom DM (2008) The welfare of livestock during road transport. In: Cussen M, Garcés V, Lambert L, Turner LJ (eds) Long distance transport and the welfare of farm animals. CABI, Wallingford, UK, pp 157–181

Brown JA, Samarakone TS, Crowe T et al (2011) Temperature and humidity conditions in trucks transporting pigs in two seasons in Eastern and Western Canada. Trans ASABE 54:2311–2318

Carlson GP (1990) Clinical chemistry tests. In: Smith BP (ed) Large animal internal medicine. Mos CV by Co, St Louis, MO, p 393

Carnovale F, Xiao J, Shi B et al (2021) The effects of vehicle type, transport duration and pre-transport feeding on the welfare of sheep transported in low temperatures. Animals (Basel) 11:1659

Castiello M, Innocente S, Wamalwa K et al (2012) Sustainable livelihood: potential role and quality assurance of camel export trade in Somalia. Sci Rep 1:227

Chakir Y, El Khasmi M, Farh M et al (2013) Effects of vitamin E and vitamin C on hydrogen peroxide-induced hemolysis in Moroccan dromedary camels (*Camelus dromedarius*). Greener J Med Sci 3:111–120

Chulayo AY, Bradley G, Muchenje V (2016) Effects of transport distance, lairage time and stunning efficiency on cortisol, glucose, HSPA1A and how they relate with meat quality in cattle. Meat Sci 117:89–96

Dahlborn K, Benlamlih S, Zine-Filali R et al (1992) Food deprivation and refeeding in the camel (*Camelus dromedarius*). Am J Physiol 262:R1000–R1005

Davenport MD, Tiefenbacher S, Lutz CK et al (2006) Analysis of endogenous cortisol concentrations in the hair of rhesus macaques. Gen Comp Endocrinol 147:255–261

De AK, Sawhney S, Ponraj P et al (2021) Physiological and immune responses to long road transportation in Andaman local pigs. Trop Anim Health Prod 53:247

De la Fuente J, González de Chávarri E, Sánchez M et al (2012) The effects of journey duration and space allowance on the behavioural and biochemical measurements of stress responses in suckling lambs during transport to an abattoir. Appl Anim Behav Sci 142:30–41

De Silva PHG, Kalubowila A (2012) Relationship of transport distance, sex on live weight loss of pigs during transit to slaughter house. Vet World 5:150–154

Earley B, McDonnell B, Murray M et al (2011) The effect of sea transport from Ireland to the Lebanon on inflammatory, adrenocortical, metabolic and behavioural responses of bulls. Res Vet Sci 91:454–464

Earley B, Buckham Sporer K, Gupta S (2017) Invited review: relationship between cattle transport, immunity and respiratory disease. Animal 11:486–492

El-Allali K, Achaâban MR, Bothorel B et al (2013) Entrainment of the circadian clock by daily ambient temperature cycles in the camel (*Camelus dromedarius*). Am J Physiol Regul Integr Comp Physiol 304:R1044–R1052

Elias E, Weil S (1989) Serum cortisol levels in camels (*Camelus dromedarius*) during the reproductive cycle. Comp Biochem Physiol A 94:787–790

El Khasmi M et al (2009) Effects of age on transportation and preslaughter stress responsiveness in Moroccan dromedary camels. In: Chilliard Y et al (eds) Proceedings of the XIth international symposium on ruminant physiology, Clermont-Ferrand, France

El Khasmi M, Riad F, Safwate A et al (2010) Effects of preslaughter stress on meat quality and phosphocalcic metabolism in camels (*Camelus dromedarius*). J Camelid Sci 3:33–38

El Khasmi M, Riad F, Safwate A et al (2011) Seasonal variation of circulating levels of 25-hydroxyvitamin D and testosterone during the rutting and non-rutting periods in Moroccan dromedary camels (*Camelus dromedarius*). Emir J Food Agric 23:368–374

El Khasmi M, Chakir Y, Riad F et al (2013) Effects of transportation stress during the hot-dry season on some haematological and physiological parameters in Moroccan dromedary camels (*Camelus dromedarius*). J Life Sci 7:13–25

El Khasmi M, Chakir Y, Bargaâ R et al (2015) Impact of transport distance on stress biomarkers levels in dromedary camel (*Camelus dromedarius*). Emir J Food Agric 27:507–512

El Khasmi M, Farh M, Iddar A (2021) Correlation between indicators of stress and oxidative stress and behavioral responses before slaughter in dromedary camel. Final meeting of the Caravan Project, 9–10 June 2021

Emeash HH, Mostafa AS, Karmy M et al (2016) Assessment of transportation stress in Dromedary camel (*Camelus dromedarius*) by using behavioural and physiological measures. J Appl Vet Sci 1:28–36

FAOstat (2021) The state of the world's land and water resources for food and agriculture—Systems on the brink. Summary report 2021. Rome. https://doi.org/10.4060/cb7654fr

Farh M, Lektib I, Bargaâ R et al (2018) Hormonal profile in rutting and non-rutting periods in dromedary camel (*Camelus dromedarius*). Rev Mar Sci Agron Vét 6:390–398

Faye B (1994) Systèmes pastoraux, agro-pastoraux et agricoles d'Ethiopie. In: Chantal B-P, Jean B (eds) Dynamique des systèmes agraires. A la croisée des parcours. Pasteurs, éleveurs, cultivateurs. ORSTOM, Paris, pp 269–287

Faye B (2013) Camel meat in the world. In: Kadim IT et al (eds) Camel meat and meat products. CABI, Wallingford, UK

Faye B (2019) TADs in the dromedary. In: Kardjadj M et al (eds) Transboundary animal diseases in Sahelian Africa and connected regions. Springer, Cham. https://doi.org/10.1007/978-3-030-25385-1_6

Faye B (2020) How many large camelids in the world? A synthetic analysis of the world camel demographic changes. Pastoralism Res Pol Pract 10:25. https://doi.org/10.1186/s13570-020-00176-z

Faye B, Bengoumi M (2018) Camel clinical biochemistry and hematology. Springer, New York, p 346

Faye B, Bengoumi M, Barkat A (2003) The development of peri-urban dairy camel systems in Africa. In: International workshop on camel milk in Africa. FAO-CIRADKARKARA, Niamey (Niger), 5–8 November, pp 115–125

Ferguson DM, Warner RD (2008) Have we underestimated the impact of preslaughter stress on meat quality in ruminants? Meat Sci 80:12–19

Flerov MA, V'iushina AV (2011) Free radical oxidation of lipids in the rat hypothalamus under stress after cortisol pre-treatment. Ross Fiziol Zh Im I M Sechenova 97:898–902

Gayrard V, Alvinerie M, Toutain P (1996) Interspecies variations of corticosteroid-binding globulin parameters. Domest Anim Endocrinol 13:35–45

Gobert M, Damon M, Durand D (2013) Stress oxydant et qualité nutritionnelles des produits animaux. Cah de Nutr et de Diet 48:225–232

Grandin T (1997) Assessment of stress during handling and transport. J Anim Sci 75:249–257

Gregory NG (1998) Physiology of stress, distress, stunning and slaughter. In: Animal welfare and meat science. CABI, Wallingford, UK

Grigor PN, Cockram MS, Steele WB et al (2004) A comparison of the welfare and meat quality of veal calves slaughtered on the farm with those subjected to transportation and lairage. Livest Prod Sci 91:219–228

Halliwell B, Gutteridge JMC (1999) Free radicals in biology and medicine. Oxford University Press, Oxford
Hamad B, Hadef L, Aggad H (2021) Effect of lairage time on the physiological and hematological parameters of dromedary camel (*Camelus dromedarius*). Biol Rhythm Res 52:1438–1447
Hartung J (2003) Effects of transport on health of farm animals. Vet Res Commun 27:525–527
Hrkovic-Porobija A, Hodzic A, Hadzimusic N (2017) Functional liver stress in dairy sheep. Indian J Small Rumin 23:194
Idrus Z, Bahyuddin N, Wai CY et al (2010) Physiological responses in goats subjected to road transportation under the hot, humid tropical conditions. Int J Agric Biol 12:840–844
Ishizaki H, Kariya Y (2010) Road transportation stress promptly increases bovine peripheral blood absolute NK cell counts and cortisol levels. J Vet Med Sci 72:747–753
Jama N, Maphosa V, Hoffman LC et al (2016) Effect of sex and time to slaughter (transportation and lairage duration) on the levels of cortisol, creatine kinase and subsequent relationship with pork quality. Meat Sci 116:43–49
Jarvis AM, Harrigton DWJ, Cockram MS (1996) Effect of source and lairage on some behavioural and biochemical measurements of feed restriction and dehydration in cattle at a slaughterhouse. Appl Anim Behav Sci 50:83–94
Julien L, Moutik FE, Haloui C et al (2021) Paramètres démographiques et économie de l'élevage camelin : une étude au Maroc. Cah Agric 30(1)
Kadim IT, Mahgoub O, Al-Kindi A et al (2006) Effects of transportation at high ambient temperatures on physiological responses, carcass and meat quality characteristics of three breeds of Omani goats. Meat Sci 73:626–634
Kadim IT, Farouk MM, Mahgoub O et al (2013) Slaughtering and processing of camels. In: Kadim IT (ed) Camel meat and meat products. CABI, Wallingford, UK, pp 54–72
Kadim A, Sunardi N, Husain T (2020) The modeling firm's value based on financial ratios, intellectual capital and dividend policy. Accounting 6:859–870
Kamili A, Faye B, Tligui NS et al (2020) Typology of camel farming systems in the south of Morocco. Rev Elev Med Vet Pays Trop 73(2):71–80. https://doi.org/10.19182/remvt.31862
Kang HJ, Lee IK, Piao MY et al (2017) Effects of road transportation on metabolic and immunological responses in Holstein heifers. Anim Sci J 88:140–148
Kannan G, Terrill TH, Kouakou B et al (2000) Transportation of goats: effects on physiological stress responses and live weight loss. J Anim Sci 78:1450–1457
Kannan G, Kouakou B, Terrill TH et al (2003) Endocrine, blood metabolite and meat quality changes in goats as influenced by short-term, preslaughter stress. J Anim Sci 81:1499–1507
Kannan G, Saker KE, Terrill TH et al (2007) Effect of seaweed extract supplementation in goats exposed to simulated preslaughter stress. Small Rumin Res 73:221–227
Kataria N, Kataria AK (2004) Use of blood analytes in assessment of stress due to drought in camel. J Camel Pract Res 11:129–133
Kataria N, Kataria AK, Agarwal VK et al (2000) Effect of water restriction on serum aldosterone and cortisol in dromedary camel during winter and summer. J Camel Pract Res 7:1–7
Kataria N, Kataria AK, Agarwal VK et al (2007) Solute loads and transfer function of kidney in dromedary camel during dehydration and rehydration in winter and summer. Vet Arh 77:237–246
Kataria AK, Kataria N, Maan R (2010) Correlation of serum IgE with stress in Indian dromedaries affected with skin wounds. J Stress Physiol Biochem 6:17–24
Kataria N, Kataria AK, Joshi A et al (2012) Serum antioxidant status to assess oxidative stress in brucella infected buffaloes. J Stress Physiol Biochem 8:5–9
Khalafalla AI, Hussein MF (2021) Infectious diseases of dromedary camels. Springer, Cham. https://doi.org/10.1007/978-3-030-79389-0
Klokker M, Kharazmi A, Galbo H (1993) Influence of in vivo hypobaric hypoxia on function of lymphocytes, neutrocytes, natural killer cells, and cytokines. J Appl Physiol 74:1100–1106
Knowles TG (1999) A review of the road transport of cattle. Vet Rec 144:197–201

Knowles TG, Warriss PD, Vogel K (2014) Stress physiology of animals during transport. In: Grandin T (ed) Livestock handling and transport, 4th edn. CABI, Oxfordshire, p 406

Kreuzer M, Langhans W, Sutter F et al (1998) Metabolic response of early-lactating cows exposed to transport and high altitude grazing conditions. Anim Sci 67:237–248

Kumar P, Giri A, Bharti VK et al (2019) Evaluation of various biochemical stress markers and morphological traits in different goat breeds at high altitude environment. Biol Rhythm Res 52: 261–272

Kuo T, McQueen A, Chen TC et al (2015) Regulation of glucose homeostasis by glucocorticoids. Adv Exp Med Biol 872:99–126

Lektib I, Bargaâ R, Chakir Y et al (2016a) Study of incubation conditions for erythrocytes osmotic fragility testing in dromedary camel (*Camelus dromedarius*). Int J Res Environ Sci 2:1–11

Lektib I, Bargaâ R, Farh M et al (2016b) Blood oxidant stress in relation to seasonal activity of thyroid and adrenal glands in *Camelus dromedarius*. Int J Res Environ Stud 3:101–108

Lemrhamed A (2020) Evaluation de la réactivité au stress de pré-abattage et d'abattage chez le dromadaire (*Camelus dromedarius*). Thèse de Doctorat National, Université Hassan II de Casablanca, Faculté des Sciences Ben M'Sick, Maroc, p 175

Lemrhamed A, Farh M, Riad F et al (2018) Evaluation of stress responses induced by the loading density in dromedary camel "*Camelus dromedarius*". Emir J Food Agric 30:803–808

Lemrhamed A, Tabite R, Farh M et al (2019a) Evaluation of preslaughter stress responses during waiting time at lairage in the dromedary camel (*Camelus dromedarius*). J Camel Pract Res 26: 149–156

Lemrhamed A, Tabite R, Abbadi N et al (2019b) Evaluation of slaughter stress responses in the dromedary camel. Int J Livest Res 9:78–90

Liotta L, Costam LN, Chiofalo B et al (2007) Effect of lairage duration on some blood constituents and beef quality in bulls after long journey. Ital J Anim Sci 6:375–384

Livne A, Kuiper PJ (1973) Unique properties of the camel erythrocyte membrane. Biochim Biophys Acta 318:41–49

Lupoli B, Johansson B, Uvnas-Moberg K et al (2001) Effect of suckling on the release of oxytocin, prolactin, cortisol, gastrin, cholecystokinin, somatostatin and insulin in dairy cows and their calves. J Dairy Res 68:175–187

Lykkesfeldt J, Svendsen O (2007) Oxidants and antioxidants in disease: oxidative stress in farm animals. Vet J 173:502–511

Majchrzak YN, Mastromonaco GF, Korver W et al (2015) Use of salivary cortisol to evaluate the influence of rides in dromedary Camels. Gen Comp Endocrinol 211:123–130

Malena M, Voslarova E, Tomanova P et al (2006) Influence of travel distance and the season upon transport-induced mortality in fattened cattle. Acta Vet 75:619–624

Manteca X (2008) Physiology and disease. In: Appleby MC et al (eds) Long distance transport and welfare of farm animals. CABI, Wallingford, UK

Manteca X (2009) Conceptos generales de bienestar animal. In: Manteca X (ed) Etología Veterinaria. Multimédica Ediciones Veterinarias, Barcelona, Spain, pp 225–243

Maqsood S, Benjakul S (2011) Retardation of haemoglobin-mediated lipid oxidation of Asian Sea bass muscle by tannic acid during iced storage. Food Chem 124:1056–1062

Marnet PG, Negrão JA (2000) The effect of a mixed-management system on the release of oxytocin, prolactin and cortisol in ewes during suckling and machine milking. Reprod Nutr Dev 40:271–281

Marti S, Wilde RE, Moya D et al (2017) Effect of rest stop duration during long-distance transport on welfare indicators in recently weaned beef calves. J Anim Sci 95:636–644

Martinez-Haroa M, Greenb AJ, Mateoa R (2011) Effects of lead exposure on oxidative stress biomarkers and plasma biochemistry in water birds in the field. Environ Res 111:530–538

McCarty R, Horwatt K, Konarska M (1988) Chronic stress and sympathetic-adrenal medullary responsiveness. Soc Sci Med 26:333–341

Meyerfeld B (2017) L'hécatombe des dromadaires d'Afrique de l'Est. Le Monde, https://www.lemonde.fr/planete/article/2017/04/18/l-hecatombe-des-dromadaires-d-afrique-de-l-est_5112875_3244.html

Minka NS, Ayo JO (2007) Physiological responses of transported goats treated with ascorbic acid during the hot-dry season. Anim Sci J 78:164–172

Minka NS, Ayo JO, Sackey AK et al (2009) Assessment and scoring of stresses imposed on goats during handling, loading, road transportation and unloading, and the effect of pretreatment with ascorbic acid. Livest Sci 125:275–282

Mirgani T (1992) Lipid composition of camel erythrocytes. J Arid Environ 22:401–405

Mitchell MA (2009) Chick transport and welfare. Avian Biol Res 2:99–105

Mohamed HE (2006) Factors affecting CORT status in camels (*Camelus dromedarius*). J Anim Vet Adv 5:307–309

Mohamed RH, Abo El-Maaty AM, Abd El Hameed AR et al (2021) Impact of travel by walk and road on testicular hormones, oxidants, traces minerals, and acute phase response biomarkers of dromedary camels. Heliyon 7:e06879

Montane J, Marco I, Lopez-Olvera J et al (2002) Transport stress in roe deer (*Capreolus capreolus*): Effect of the short acting antipsychotic. Anim Welf 11:295–303

Moskowitz A, Andersen LW, Huang DT et al (2018) Ascorbic acid, corticosteroids, and thiamine in sepsis: a review of the biologic rationale and the present state of clinical evaluation. Crit Care (London, England) 22:283. https://doi.org/10.1186/s13054-018-2217-4

Mounier L, Dubroeucq H, Andanson S et al (2006) Variations in meat pH of beef bulls in relation to conditions of transfer to slaughter and previous history of the animals. J Anim Sci 84:1567–1576

Mousa HM, Omer OH, Ali BH et al (2006) Antioxidant levels in tissues of young and adult camels (*Camelus dromedarius*). J Physiol Biochem 62:213–218

Moutik FE (2018) Estimation des paramètres démographiques d'un cheptel de dromadaires et leurs effets sur ses performances (lait et croît). Thèse en Agronomie, IAV Hassan II de Rabat, Maroc, p 109

Napp S, Chevalier V, Busquets N et al (2018) Understanding the legal trade of cattle and camels and the derived risk of Rift Valley Fever introduction into and transmission within Egypt. PLoS Negl Trop Dis 12(1):e0006143

Nawata H (2005) Historical Socio-economic Relationships between the Rashāyda and the Beja in the Eastern Sudan: The production of racing camels and trade networks across the Red Sea. In: Ikeya K, Fratkin E (eds) Pastoralists and their neighbors in Asia and Africa, Senri ethnological studies 69. Senri Ethnological, Osaka, pp 187–213

Nazifi S, Mansourian M, Nikahval B et al (2009a) The relationship between serum level of thyroid hormones, trace elements and antioxidant enzymes in dromedary camel (*Camelus dromedarius*). Trop Anim Health Prod 41:129–134

Nazifi S, Mahdi S, Hasan B et al (2009b) Influence of road transportation during hot summer conditions on oxidative status biomarkers in Iranian dromedary camels (*Camelus dromedarius*). Afr J Biochem Res 3:282–287

Negrão JA, Marnet PG (2006) Milk yield, residual milk, oxytocin and cortisol release during machine milking in Gir, Gir × Holstein and Holstein cows. Reprod Nutr Dev 46:77–85

Nielsen F, Mikkelsen BB, Nielsen JB et al (2011) Plasma malondialdehyde as biomarker for oxidative stress: reference interval and effects of life-style factors. Clin Chem 43:1209–1214

Niki E (2010) Antioxidants in relation to lipid peroxidation. Chem Phys Lipids 44:227–253

Noori S (2012) An overview of oxidative stress and antioxidant defensive system. Open Access Sci Rep 1:413–421

Nwe TM, Hori E, Manda M et al (1996) Significance of catecholamines and cortisol levels in blood during transportation stress in goats. Small Rumin Res 20:129–135

Orzechowski O, Ostaszewski P, Brodnicka A et al (2000) Excess of glucocorticoids impairs whole-body antioxidant status in young rats. Relation to the effect of dexamethasone in soleus muscle and spleen. Horm Metab Res 32:174–180

Padalino B, Raidal SR (2020) Effects of transport conditions on behavioural and physiological responses of horses. Animals 10:160

Padalino B, Raidal SL, Carter N et al (2017) Immunological, clinical, haematological and oxidative responses to long distance transportation in horses. Res Vet Sci 115:78–87

Rajion MA, Saat IM, Zulkifli I et al (2001) The effects of road transportation on some physiological stress measures in goats. Asian Australas J Anim Sci 14:1250–1252

Reeder DM, Kramer KM (2005) Stress in free-ranging mammals: integrating physiology, ecology, and natural history. J Mammal 86:225–235

Riad F (1995) Regulation endocrinienne du metabolisme hydroéléctrolytique et phosphocalcique chez le dromadaire. Thèse de Doctorat d'Etat, Université Hassan II de Casablanca, Faculté des Sciences Ben Msik, Maroc, p 174

Roadknight N, Mansell P, Jongman E et al (2021) Invited review: The welfare of young calves transported by road. J Dairy Sci 104:6343–6357

Ruiz-de-la-Torre JL, Velarde A, Manteca X et al (2001) Effects of vehicle movements during transport on the stress responses and meat quality of sheep. Vet Rec 148:227–229

Saeb M, Baghshani H, Nazifi S et al (2010) Physiological response of dromedary camels to road transportation in relation to circulating levels of cortisol, thyroid hormones and some serum biochemical parameters. Trop Anim Health Prod 42:55–63

Sahraoui N, Olivier Dotreppe M, Errahmani B et al (2014) Caractérisation des acides gras de la viande cameline en Algérie. Cah Nutr Diet 49:231–234

Salar-Amoli J, Hejazy M, Ali Esfahani T (2009) Comparison between some oxidative Stress Biomarkers values in serum and plasma of clinically healthy adult camels (*Camelus dromedarius*) in Iran. Vet Res Commun 33:849

Samara EM, Ayadi M, Al-Haidary AA et al (2013) Thermophysiological study in lactating and dry camels (*Camelus dromedarius*) under summer conditions. Emir J Food Agric 25:308–313

Sayour AE, Elbauomy EM, El-Kholi MK et al (2015) Brucellosis prevalence and serologic profile of male one-humped camels reared in Somaliland and Eastern Ethiopia for meat production. Global Veterinaria 14:67–76

Sid-Ahmed OE, Sanhouri A, Elwaseela BE et al (2013) Assessment of adrenocortical activity by non-invasive measurement of faecal cortisol metabolites in dromedary camels (*Camelus dromedarius*). Trop Anim Health Prod 45:1453–1458

Silva JA, Patarata L, Martins C (1999) Influence of ultimate pH on bovine meat tenderness during ageing. Meat Sci 52:453–459

Simsek S, Yuce A, Utuk AE (2006) Determination of serum malondialdehyde levels in sheep naturally infected with Dicrocoelium dendriticum. Firat Univ Saglik Bil Dergisi 20:217–220

Sowińska J, Milewski S, Witkowska D et al (2020) Effect of dietary supplementation with β-hydroxy-β-methylbutyrate on stress parameters in goat kids. Revista Brasileira de Zootecnia 49:e20200035. https://doi.org/10.37496/rbz4920200035

Stanger KJ, Ketheesan N, Parker AJ et al (2005) The effect of transportation on the immune status of *Bos indicus* steers. J Anim Sci 83:2632–2636

Stockman CA, McGilchrist P, Collins T et al (2012) Qualitative Behavioural Assessment of Angus steers during pre-slaughter handling and relationship with temperament and physiological responses. Appl Anim Behav Sci 142:125–133

Storey KB (1996) Oxidative stress: animal adaptations in nature. Braz J Med Biol Res 29:1715–1733

Stull CL, Rodiek AV (2000) Physiological responses of horses to 24 hours of transportation using a commercial van during summer conditions. J Anim Sci 78:1458–1466

Tabite R, Lemrhamed A, El Abbadi N et al (2019a) Relationship between circulating levels of cortisol at slaughter and meat physico-chemical and oxidant stress parameters in dromedary camels. Emer J Food Agric 31(11):874–883

Tabite R, Lemrhamed A, El Abbadi N et al (2019b) Relationship between 25-hydroxyvitamin D content and quality characteristics and lipid oxidation in raw and cooked camel meat during cold storage. JALS 6(2). https://doi.org/10.30845/jals.v6n2p2

Tadich N, Gallo C, Bustamante H et al (2005) Effects of transport and lairage time on some blood constituents of Friesian-cross steers in Chile. Livest Prod Sci 93:223–233
Tallo-Parra O, Manteca X, Sabes-Alsina M et al (2015) Hair cortisol detection in dairy cattle by using EIA: protocol validation and correlation with faecal cortisol metabolites. Animal 9(6): 1059–1064
Tauler P, Aguilo A, Gimeno I et al (2003) Pons, Influences of vitamin C diet Supplementation on endogenous antioxidant defences during exhaustive exercise. Eur J Physiol 446:658–664
Terlouw EMC, Arnould C, Auperin B et al (2008) Pre-slaughter conditions, animal stress and welfare: current status and possible future research. Animal 2:1501–1517
Terlouw EMC, Cassar-Malek I, Picard B et al (2015) Stress en élevage et à l'abattage: impacts sur les qualités des viandes. In: Picard B, Lebret B (eds) Numéro spécial, Le muscle et la viande, vol 28. INRA Prod Anim, pp 169–182
Tharwat M, Al-Sobayil F, Buczinski S (2013) Cardiac biomarker changes in camels (*Camelus dromedarius*) secondary to road transportation. J Vet Cardiol 15(1):15–22
Volfinger L, Lassourd V, Michaux JM et al (1994) Kinetic evaluation of muscle damage during exercise by calculation of amount of creatine kinase released. Am J Physiol 266(2 Pt 2):R434–R441
Warriss PD (1990) The handling of cattle pre-slaughter and its effects on carcass and meat quality. Appl Anim Behav Sci 28:171–186
Warriss PD, Kestin SC, Brown NS et al (1984) The time required for recovery from mixing stress in young bulls and the prevention of dark cutting beef. Meat Sci 10:53–68
Warriss PD, Brown SN, Knowles TG et al (1995) Effects on cattle of transport by road for up to 15 hours. Vet Rec 136(13):319–323
Werner W, Kaaden OR (2002) Nonpathogenic viral infections. In: Infectious diseases in camelids. Blackwell, Berlin, pp 209–211
Wernicki A, Urban-Chmiel R, Kankofer M et al (2006) Evaluation of plasma cortisol and TBARS levels in calves after short-term transportation. Rev Méd Vét 157:30–34
Wickham SL, Collins T, Barnes AL et al (2015) Validating the use of qualitative behavioral assessment as a measure of the welfare of sheep during transport. J Appl Anim Welf Sci 18(3):269–286
Yates DT, Ross TT, Hallford DM et al (2010) Technical note: Comparison of salivary and serum cortisol concentrations after adrenocorticotropic hormone challenge in ewes. J Anim Sci 88(2): 599–603
Younan M, Bornstein S, Gluecks IV (2016) MERS and the dromedary camel trade between Africa and the Middle East. Trop Anim Health Prod 48(6):1277–1282
Young JA (1982) Camels on the Western range. Rangelands 4(6):248–251
Zhong RZ, Liu HW, Zhou DW et al (2011) The effects of road transportation on physiological responses and meat quality in sheep differing in age. J Anim Sci 89:3742–3751
Zia-Ur-Rahman NA, Bukhari SA, Akhtar N et al (2007) Serum hormonal, electrolytes and trace elements profiles in the rutting and non-rutting one humped male camel. Anim Reprod Sci 101: 172–178
Zimerman M, Domingo E, Grigono G et al (2013) The effect of pre-slaughter stressors on physiological indicators and meat quality traits on Merino lambs. Small Rumin Res 111:6–9
Ziv A, Sod-Moriah UA, Creveld C et al (1997) Is water restriction a stress for camels (*Camelus dromedarius*)? J Camel Pract Res 4(2):217–221

The Welfare of Dromedary Camels at Slaughter

11

Isam T. Kadim, Issa S. Al-Amri, Abdulaziz AlKindi, and Dawood Sulaiman Al-Ajmi

Abstract

The welfare of camels during slaughter is the communal responsibility of abattoirs, meat industry members, transporters, and operators. Current camel slaughtering practices frequently fail to follow ethical animal slaughtering practices, breach welfare protection laws, and will commonly induce stress-related physiological changes in animals. When camels move from the farm to the slaughterhouse, they will try to use a wide range of physiological mechanisms and behaviors to survive the new situation that often cause distress. There is also little in the way of legislation and supporting regulations and guidelines to ensure camel welfare, which is basically concerned with preventing or minimizing thirst, hunger, distress, disease, pain, injuries, and the inability to express normal behavior. The most common camel slaughtering methods in the world are currently unnecessarily cruel and painful because of ways camels are unloaded, handled, restrained, and walked to the abattoirs. Important areas that need to be addressed to improve camel welfare include significant changes to the infrastructure of abattoirs and working practices, facility management, regulations, and enforcement. It is important to adopt a holistic approach to solving current camel-welfare related problems and to adopt appropriate technologies and practices that will be sustainable into the future.

I. T. Kadim (✉) · I. S. Al-Amri · A. AlKindi
Department of Biological Sciences and Chemistry, College of Arts and Sciences, University of Nizwa, Birkat Al-Mouz, Nizwa, Sultanate of Oman
e-mail: isam@unizwa.edu.om

D. S. Al-Ajmi
Department of Integrative Agriculture, College of Agriculture and Veterinary Medicine, United Arab Emirates University, Al-Ain, UAE

B. Padalino, B. Faye (eds.), *Dromedary Camel Behavior and Welfare*, Animal Welfare 24, https://doi.org/10.1007/978-3-031-48600-5_11

Keywords

Dromedary camel · Welfare legislation · Slaughtering methods · Transportation · Stunning · Handling

11.1 Introduction

Many abattoirs have not been designed for camels due to animal size, limited camel meat production, and low per capita consumption of camel meat compared to meat from other livestock (Kadim et al. 2013). Recently, camel meat is consumed in certain places in Africa, Asia and other arid regions where alternative forms of protein is limited or where camel meat has had a long cultural history (Kadim et al. 2008). The slaughter of camels is an enterprise that creates a significant amount of emotion for the people involved. There are various religious and cultural customs involved in the production of camel meat for human consumption (Kadim et al. 2013). Dromedary camels have been slaughtered to fulfill local meat consumer requirements. Several pre-slaughter stress responses are induced by the handling and slaughtering methods employed (Kadim et al. 2013; El Khasmi et al. 2013, 2015), stocking density, and waiting time before slaughtering (Lemrhamed et al. 2018a, b). This is because camels use a wide range of physiological mechanisms and behaviors to cope with situations that are different from what they are accustomed to (Velarde et al. 2016).

From the animal rights viewpoint, camel welfare is an important concern and everything possible should be done to ensure that camels are protected from stressful elements during their life, from farm to fork. Most dromedary camels are reared in developing countries where animal welfare is often poor due to a lack of awareness, knowledge, facility quality, and funds (Koei 2009). There are also cultural differences across countries and regions that can worsen the conditions. Improving animal welfare has a positive effect, not only on the health of the animals, but also on the quality of its products (Gottardo et al. 2003), on the safety of abattoir workers, and the incidence of diseases (Brown-Brandl et al. 2008). Salmonellae in livestock increase in stressed animals (Corrier et al. 1990; Galland 1997). According to Koei (2009), in addition to poor meat quality, animals who experienced stress before and during slaughter produce pathogen levels that were at least ten times higher than the normal levels. This is an important issue for disease control and public health, particularly in camel-raising countries where many diseases are prevalent. Furthermore, camel welfare is poor due to a lack of education and training, poor infrastructure at older abattoirs, and poor pre-slaughter stunning and slaughter practices. Most if not all camel slaughter facilities require significant improvement, with major changes also to the unloading and lairage facilities. There is often no clean water available at libitum for camels to avoid dehydration. Slaughter staff needs education to minimize improper handling of the camels, preventing further fear and injury to both them and the animals (Koei 2009).

Camels respond to stress in their environment with a mixture of physiological, biochemical and behavioral mechanisms, which include increased release of hormones such as adrenaline, corticosteroids, glucagon, prolactin, and vasopressin (Lemrhamed et al. 2019). The hormonal changes increase metabolic rate in addition to causing behavioral changes that help camels to adapt to new conditions. With low levels of stress, recovery to baseline is possible which improves product quality and enhances animal welfare. Conversely, if the level of stress and its duration increase, distressed animals may express noticeable signs that harm the animal and its produce. Continued production of various hormones related to stress changes the activity of the secreting glands and affects the regular functioning of the body resulting in lower productivity and resistance to infection (Appleby and Hughes 1997). Moreover, the concentrations of glycogen in muscles are lowered by stress leading to poorer meat quality and a decreased shelf life of meat products (Andriessen 2004). This chapter discusses aspects and concerns about camel welfare both during the pre-slaughter period that involves unloading, mixing of unfamiliar groups, handling at lairage, as well as during stunning and bleeding. In addition, a consideration of requirements for legislation and regulation is included along with recommendations for protecting the welfare of camels at slaughter.

11.2 Pre-slaughter Treatments and Handling

The World Organization for Animal Health (WOHA) considers pre-slaughter and handling treatments to be important for animal welfare and treats it as a multi-faceted public policy issue that includes important scientific, ethical, economic, and political dimensions (Koei 2009). In 2006, Animal Welfare in Livestock Operations (IFC guidance note) confirmed that WOHA is giving priority to animal welfare during pre-slaughter handling. Two IFC guides provide informative case studies of the economic profits from improving pre-slaughter treatments, supported by scientific literature that provides evidence to endorse further action in this area (IFC 2006). However, every living camel faces a wide range of different challenges that keep it under “stress.”

Camels are often stressed as they walk to the slaughterhouse. This is exacerbated by entrances to slaughter areas that are dark, narrow, and slippery. It is not natural for camels to enter such stressful spaces; therefore, it is not unusual for the camels to be forced in and often animals panic and become injured in the process. Camel slaughter practices vary, but often are inadequate in terms of animal welfare and worker safety. This leaves the camel in serious pain and distress until the slaughter cut is made, which is often not carried out immediately (Guya and Neme 2015). The same authors stated that batch slaughtering camels is practiced and they are in close distance to others being slaughtered and in distress. This practice can be extremely stressful for camels.

Animals are usually suffering stress while waiting to be slaughtered (Gregory 1996). The responses to pre-slaughter stress in the dromedary camel, from the arrival at the slaughterhouse, were evaluated by Lemrhamed et al. (2019). The study

Table 11.1 Blood parameters in dromedary camels ($n = 8$) at unloading (after 90 min transportation), at lairage (after 20 h rest), and after bleeding. Neutrophile/Lymphocyte ratio, and concentrations of hypotonic salt solution, glucose, cortisol, triiodothyronine, thyroxine, malondialdehyde, and catalase activity in eight dromedary camels (Lemrhamed et al. 2019)

Parameter	Unloading	Lairage	Bleeding
Hematocrit%	34	34	35
Neutrophile/Lymphocyte ratio	14	8	13
Hypotonic salt solution (mosm/dL)	13	9	14
Glucose (mmol/L	8	7	8
Cortisol (nmol/L)	7	4	7
Triiodothyronine (nmol/dL)	4	2	5
Thyroxine (μg/dL)	22	8	23
Malondialdehyde (UI/dL)	4.2	6.4	3.5
Catalase activity (nM/L)	3.8	1.8	5.5

analyzed the neutrophil/lymphocyte ratio (NLR), hemolysis 50 (H50), hematocrit (Hct), glucose, cortisol and thyroid hormones and oxidant stress (MAD and CATa) after unloading from transport vehicle and resting for 20 h. The results showed that stress responses observed after transport and unloading were significantly decreased after a rest period of 20 h (Table 11.1). The loading/unloading, waiting duration, environmental situations during the transport, starvation/dehydration, and lairage before slaughter are the most common stressors for camels (El Khasmi et al. 2015; Lemrhamed et al. 2018a, b). These factors should be considered because they compromise welfare and increase susceptibility to diseases (Broom 2014). Camels can suffer from bruising and injury caused by rough pre-slaughter handling, such as beating them with sticks when they refuse to move forward or dragging them along the ground when they fall (Kadim et al. 2013). Although dromedary camels are less vulnerable (Kadim et al. 2013) to mishandling practices than other livestock, there are specific mishandling practices that can have a considerable impact on stress (Cortesi 1994).

11.2.1 Unloading

Walking is the ideal form of transport when camels are in urban locations with minimal distance between farms, markets, and slaughterhouses. Travel over longer distances will require transportation to slaughterhouses (Kadim et al. 2013). If long distances are covered, ramps or cranes and trucks are often used to load or unload camels to slaughterhouses, which can cause severe stress to camels (Fig. 11.1). It is important to recognize that animals which have become habituated to a particular regime and handling technique are more resilient and adaptable compared to animals reared or managed in an inconsistent method. In addition, pickup and unloading sites rarely have suitable ramps and ancillary equipment. The absence of guidelines covering camel loading and unloading is due to a lack of understanding and recognition of a present issue.

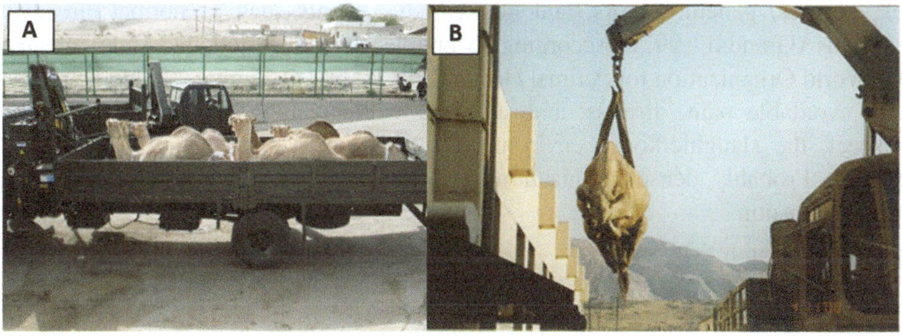

Fig. 11.1 Camels transported to slaughter. Image **a** showing camels that have had their legs tied "sitting" on a truck or trailer, image **b** showing the use of a crane to load or unload camels from the truck. Camels in either of these situations are likely to be under significant stress

The handling of camels during unloading and lairage involves inhumane treatments, such as crowded tie-ups, heat stress, a lack of water troughs during holding, pathways of access to slaughter sites that cause the animals to panic, and abuse from the handlers who use wooden prods and other brutal methods to move animals forward (Koei 2009). The mishandling of camels is considered one of the principal stressful factors for camels leading to hypercortisolemia (El Khasmi et al. 2013) and activation of free radical groups (Nazifi et al. 2009). According to El Khasmi et al. (2013, 2015), hemolysis and concentration of cortisol, thyroid hormones, and glucose in camels after unloading were significantly higher than those before disembarking. Researchers have reported that mishandling of dromedary camels during unloading played a significant role in blood stress indicators (El Khasmi et al. 2013, 2015) and, as a result, their welfare is compromised and their susceptibility to diseases is increased (Broom 2014).

11.2.2 Mixing of Unfamiliar Groups

At many slaughterhouses unloaded camels are held in pens where they are often exposed to a variety of stressors that may result in high levels of physiological and physical discomfort (Lemrhamed et al. 2019). In many countries, camels are kept in an open area without shade and are mixed with different species and classes of stock. Furthermore, mixing camels with cattle before slaughter leads to extensive bruising, especially when the cattle have large horns. The stress responses to the lairage conditions could be influenced by unfamiliar environments such as slaughterhouse design (Miranda de la Lama et al. 2010), slaughterhouse environment (Njisane and Muchenje 2017), slaughterhouse workers (Terlouw et al. 2011), stocking density (Lemrhamed et al. 2018a), and different animal species (Vimiso et al. 2012). Moreover, Lulietto et al. (2018) stated that the loud noise at the slaughterhouse might also affect the animals' response, as opposed to the quiet environment on the farm. An animal's stress response is a compound interaction that depends on its past experiences, genetic factors, and non-aversive events such as normal circadian rhythms (Grandin 1997). According to animal slaughter guidelines implemented by World Organization for Animal Health (OIE 2009a, b), animals should be spared any avoidable pain, distress, and suffering while waiting for slaughter. In this respect, the slaughterhouse environment is playing a significant role in camel welfare. Probably, dense urbanization around many camel slaughterhouses prevents their expansion.

11.3 Camel Stress Mitigation Strategies

Camels are often moved into the slaughter area by force, and animals may be aware of their impending fate, possibly by smelling blood and hearing the distress of other camels. It is recommended that camels should be handled properly before and after arrival at the slaughterhouse and should be calmly guided into the lairage for an adequate period before slaughter with access to water (Kadim et al. 2013). Resting before slaughter reduces stress and improves camel welfare. All aspects of camel handling should be carried out by well-experienced personnel aware of domestic and international legislation on animal welfare. Moreover, pre-slaughter stress can be reduced by preventing the mixing of different groups of dromedary camels, keeping them cool with adequate ventilation, and avoiding overcrowding. Holding and restraint of dromedary camels can pose an occupational health and safety risk as they are declared dangerous and hazardous animals. They can deliver a kick in all directions and their teeth can cause severe injury. Recommendations to minimize the camel stress at slaughter are presented in Table 11.2.

11.4 Slaughter of Sick and Diseased Camels

Sick and injured camels should be treated and not sent for slaughter until fully recovered, but unfortunately sick and diseased camels are often sent for slaughter to avoid losing the animal or monetary returns (Kadim et al. 2008). Sickness and disease exacerbate the suffering of camels and transporting seriously ill animals imposes additional stress that the ill animal is less able to withstand because of lowered immunity and a decreased ability to resist the challenge of pathogens (Koei 2009). Moreover, animals affected by some diseases are more likely to defecate larger amounts of infectious organisms to infect people or other animals. If the sick camels are exposed to various stresses including dehydration, starvation, transportation, overcrowding, mental stresses, slippery flooring, badly designed housing, poorly maintained equipment, excess noise, or poor lighting, these will reduce the immunity system further and will increase the spread of diseases among other livestock and workers in the slaughterhouse. Probably, the worst examples of poor camel welfare occur during the pre-slaughter period with diseased camels where the potential for disease dissemination is at its most dangerous.

Table 11.2 Recommendations regarding conditions, procedures, and facilities involved in pre-slaughter transportation and holding of dromedary camels in order to minimize stress

Aspect	Recommendation
Temperature	The temperature in the holding pens should not be extremes of heat or cold with recommendation range from 10 to 40 °C
Shed	The shedding pens must be positioned and designed to protect camels from harsh weather
Mixing	If the holding pens are used for mixing species, they must be large enough to accommodate any territorial needs of camels and avoid aggressive behaviors to minimize the risk of injury
Water	Camels must have easy excess to clean water ad libitum
Space	The holding design should provide enough space for lying down/resting area and allow the camel full movement in all directions, including turning, standing, or stretching. A good rule for this would be of approximately the length of the camels by 1½ times the width
Location	The holding pens should be located nearby the slaughterhouse and allow for the camels to easily move to the slaughter area
Transport	Camels must be in crouch positions for the duration of the journey so enough space is needed to allow all camels to sit comfortably and not lie or trip over one another
Time of transportation	Loading onto the transport vehicles must occur as close to departure as possible, preferably within 30 min, and unloading must begin immediately after arrival. For long and day time transportation, adequate stops must be made to allow the watering of camels
Health	Health checks must be performed upon arrival at the slaughterhouse to avoid the slaughter of sick animals

11.5 Pre-slaughter Stunning

Some religions demand that exsanguination of camels should be completed without pre-slaughter stunning, citing their religious interpretations, which vary between schools of Islam and also by location (Koei 2009). In developed countries, pre-slaughter stunning of livestock is normal practice and is in most cases mandatory, whether religious practices are required to be observed or not, mainly due to the need to meet high animal welfare standards. In these countries, however, facilities are of a very high standard, practices and conditions are strictly regulated, and funds are available to meet the high processing costs which are passed on to consumers, who themselves demand high-quality meat (Koei 2009). Even though pre-slaughter stunning is normally practiced in developed countries, scientific research on different species has, so far, returned equivocal results about the welfare of animals slaughtered with and without stunning (Grandin 1997). The results of mainstream research, such as that by the European Food Safety Authority (EFSA 2004), suggest that stunning before slaughter is the most humane method and that slaughter without pre-stunning can place undue pain and discomfort on the animal. However, the level of pain during slaughter without pre-stunning (Halal method) is minimal provided that the procedure is performed correctly and in the right conditions (Al-Amri et al. 2022). Various methods are available that can render camels insensible, but the level of pain or distress caused by different combinations can be difficult to quantify. Some of the slaughter practices and conditions reported certainly require modification if they are to meet even basic animal welfare standards.

Traditionally, camel stunning is not practiced at slaughter in countries consuming camel meat products. However, stunning a camel can be carried out like cattle with a captive bolt pistol on the intersection of the medial corner of the eye and upper ear attachment (Herrmann and Fischer 2004). The purpose of stunning is to reduce the animal's sensibility and improve its welfare (Gregory 1998). The basic principle is the transfer of kinetic energy from a moving object to the brain, which results in neuronal dysfunction and/or destruction, and subsequent insensibility. Captive bolt stunning can cause animals to lose their evoked potentials immediately and they do not return, which can be used in camel pre-slaughter to improve its welfare. EFSA (2004) described the drive for stunning as follows: the majority of the slaughtered livestock for human consumption are bled by cutting the major blood vessels in the neck or thorax so that rapid blood loss occurs. Without stunning, the animal becomes unconscious only after a certain degree of blood has been lost, which may take several seconds, and they will experience pain, fear, panic and inhale blood as a result of the cut. Therefore, camels should be stunned into unconsciousness before their bleeding to ensure death with less suffering. In this respect, pre-slaughter stunning of camels should be encouraged to minimize the experience of pain during bleeding while not stunned. Acceptance of stunning in some way may depend on demonstrating and convincing those responsible for traditional cultural and religious practices that it is preferable for the camel. To encourage the use of stunning before bleeding camels, educational materials should be disseminated to demonstrate how stunning works and why it is a humane practice. Electrical stunning is the most common method before slaughter due to its low cost (Daly and Simmons 1994), and it can be used in camels before slaughter. Electrical stunning involves passing a sufficient current through the brain to depolarize neurons and causes insensibility (Simmons and Daly 2004). The stunning can be reversible (head-only) or irreversible (head-to-body) by inducing cardiac arrest (Grandin 2003). Head-only electrical stunning causes the animal to be unconscious and insensible to pain, yet the animal can fully recover if the slaughter cut is not made. This type of stunning can be applied to camels (halal meat). Moreover, Gilbert (1993) stated that head-only electrical stunning is accepted as humane to the animal, safe for the workers, and virtuous. There is enough evidence to conclude that head-only electrical stunning can be used with camels to improve camel welfare because it does not kill the animal before it is bled and the procedure is painless for the animal.

11.6 Bleeding Procedures and Method of Constraint

Slaughterhouse practices can significantly affect the welfare of camels such as being dragged from holding yards and being slaughtered without stunning within sight of other camels. Legs of the un-stunned slaughtered camels are usually roped so they topple (Fig. 11.2). They may be repeatedly struck with sticks to effect movement, causing unnecessary pain and injury. Lack of restraint makes slaughter difficult and causes more discomfort to the animal when they are brought to the ground physically. Most slaughtermen are usually untrained workers who are highly skilled in knife work but do not have the technical knowledge or background to perform their tasks with due consideration to animal welfare and food safety issues (Koei 2009).

Camels are most commonly slaughtered in the crouching position with the head held in a caudal position (Fig. 11.2, image b). A rapid cut is made with a sharp blade at the base of the neck between the neck and the thorax to induce fast bleeding. This

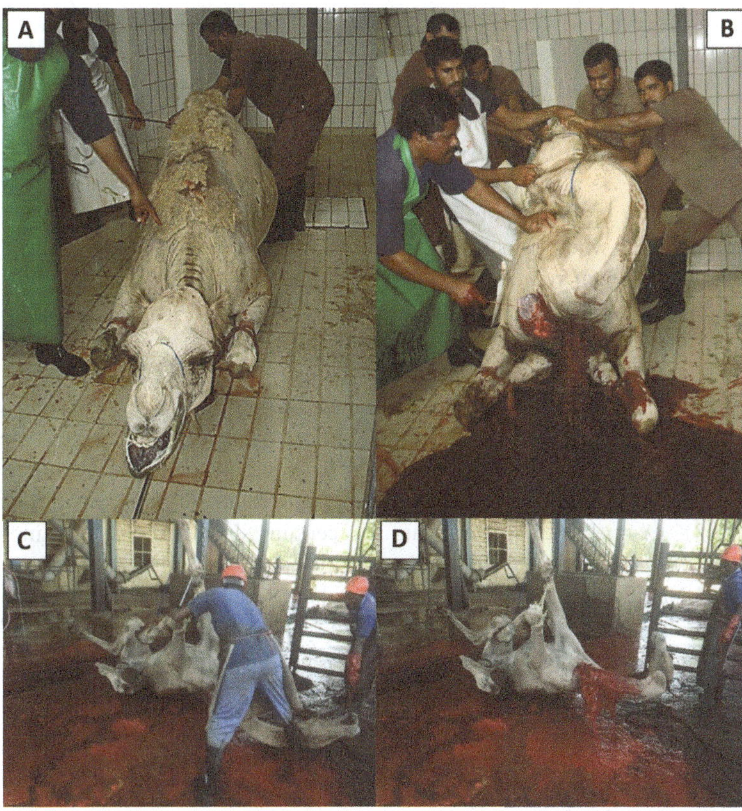

Fig. 11.2 Images **a** and **c** show a restrained dromedary camel in the crouching position or hanging from their legs, respectively, with obvious stress from the position and smell of the blood. Images **b** and **d** illustrate how the head is commonly turned toward the tail to limit physical movement and maximize extension of the neck to facilitate bleeding (© photo B. Faye)

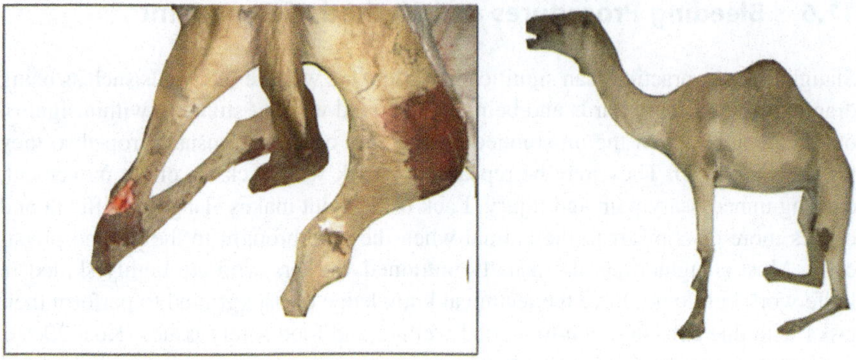

Fig. 11.3 A camel dragging its hind legs after the Achilles' tendon has been cut at the hock as a means of animal restraint

is because major blood vessels are more exposed at this point than further up the neck where the transverse processes of the cervical vertebrae conceal the carotid arteries. Animals should be allowed to completely bleed out to reduce meat contamination. Approximately 40–60% of the total blood is usually lost at slaughter, which is highly desirable to ensure early brain death and minimize bacterial growth (Kadim et al. 2013).

Camels find being forced into a crouched position a particularly aversive experience, in particular when camels are restrained, they struggle vigorously and make escape attempts when rotated, making a bad situation a lot worse (Kadim et al. 2013). When the restraint is not done properly, the camel may right itself, sustain an injury, to itself and possibly the staff, and prevent effective slaughter. Proper restraint is desirable because camels can be dangerous to people, but restraint must be carefully done in such a way that the animal is in a safe but controlled position that is acceptable for welfare. Restraint procedures that do not invert the camel will generally improve welfare and can be used for slaughter without pre-stunning. It provides for an effective transverse cut when the head is turned toward the tail (Fig. 11.2). Figure 11.3 shows that in certain slaughterhouses, camels are slaughtered in cruel ways by cutting the Achilles tendon to immobilize the animals (Guya and Neme 2015). In some cases, camels are restrained by holding the neck and putting the animal in a squat position before severing the neck (Guya and Neme 2015), meaning that the camel was conscious and apparently suffering from excruciating pain. Procedures must be developed to prevent such avoidable suffering.

Unfortunately, many attempts have been made to restrain animals with sheer force instead of using behavioral principles that have been outlined by Grandin (1995). Grandin (1995) has also explained how improvements in the design of restraining devices enhance animal welfare and reduce stress and injuries. Changing the design of a squeeze chute can reduce injuries to cattle. Under the best conditions, cattle can become bruised or injured in a conventional squeeze chute. Excessive hydraulic pressure can also cause severe injuries (Grandin 1995). Good management can prevent many bruises and injuries but there is still a great need for improved restraint devices for use in slaughterhouses. Grandin (1995) listed some key principles of low-stress restraint of cattle including solid sides or barriers around the animals to prevent them from seeing people, provision of non-slip flooring for animals, use of sufficient (optimum) pressure, and minimize noise. In general, beef slaughter plants have been using the V restrainer system for restraining cattle during stunning and shackling (Edwards 1971; Schmidt 1972; Willems and Markey 1972). The V restrainer was a major humane and safety improvement over old-style restraining devices. Revised restraining devices need to be developed for camels to improve their welfare. They would help to keep camels calmer prior slaughter. To reduce stress camels must willingly enter a restraint device. Stress levels are greatly increased if camel balks and refuses to enter or if several attempts are required to restrain it. In this respect, adequate lighting is required so that the camel can see where it is going. In indoor facilities, lamps must be used to illuminate the restrainer entrance. Animals have a basic behavioral tendency to move from a darker area toward a more brightly illuminated area (Lambooij and Von Putten 1993). When a restraining device is developed, a small window in the front gate will encourage camels to enter.

Some of the undesirable slaughter practices described above have been observed in developing countries where slaughterhouse facilities and procedures were inadequate leading to significant concerns regarding camel welfare. Camel slaughterhouses need to develop regulations and facilities to protect slaughter animals from procedures that threaten their welfare. The Humane Method of Livestock Slaughter Act (HMLSA 1902) states that either all animals are rendered insensible to pain before being slaughtered or are slaughtered in accordance with the formal requirements of the Islamic and Jewish faiths whereby the animal suffers loss of consciousness due to the simultaneous and instantaneous severance of the carotid arteries with a sharp knife. Ethical concerns for animal welfare are based on the assessment of their mental capacity to feel pain consciously (Allen 1998). Although, the slaughtering procedures of livestock have been significantly improved over time (Gregory 1998), there is no authority responsible for policies that protect the rights of dromedary camels in slaughterhouses. Although camel slaughtering practices differ from region to region and from country to country, the basic elements of camel welfare have not been practiced. In this respect, Gregory (1998) stated that animals should move quietly walking to the slaughterhouse with a minimum of visible excitement or agitation. The importance of reducing stress during slaughter is clear and has been pointed out by many animal welfare scientists. In animals, reducing excitement and agitation during handling will improve welfare and increase the likelihood of high meat quality (Grandin 1994; Voisinet et al. 1997). However, little information is available on camel welfare in relation to mishandling at the slaughterhouse. Traditionally, camels were slaughtered, dressed, and cut up on the often-dusty concrete floor due to a lack of hygienic facilities in most of the slaughterhouses.

11.7 Animal Welfare Legislation

Camel welfare legislation needs to be further developed in many countries by governments and private-sector bodies such as veterinary services. Countries slaughtering camels do not have a common formal animal welfare unit, and there are no common formal regulations protecting camel welfare. With the exception of very few articles such as the work done by Lemrhamed et al. (2019), no country has conducted trials or procedures to monitor or control camel welfare during transportation, sale, or slaughter, either through official veterinary services or by an animal welfare society. In addition, no country has any formal animal welfare training protocols for those working with handling or slaughtering camels. Specific criteria must be developed for inclusion in welfare assessments at the slaughterhouse based on their validity, reliability, repeatability, and feasibility (Velarde and Dalmau 2012). Monitoring systems and legislation mostly rely on inspection of inputs: "what" or "how much" of different resources are given to camels. Assessments of camel welfare at slaughter should include the number or proportion of animals slipping, falling, turning around, and moving backward. Poor camel welfare due to poor pre-slaughter management practices is linked with quantitative and qualitative losses in the value of carcasses and meat quality (Lemrhamed et al. 2019). Training programs for pre-slaughter management practices and a focus on improving staff competency can mitigate these losses. Therefore, camel handling training programs along with improvements in yard and race design should be implemented for fast and effective prevention of animal suffering during pre-slaughter and slaughter management. Such training can be assessed in terms of camel reactions, behavior, and evaluating progress in its welfare outcomes. Training programs in the past have resulted in a significant reduction in proportion of downgraded animal carcasses due to bruising and carcass damage (Paranhos da Costa et al. 2014).

Outlined below are a few examples of ways in which certain countries have set up standard regulations related to livestock welfare. In Europe, the measures for sanitary checks, animal welfare protection, and slaughtering procedures are harmonized throughout the European Union, and are detailed by the European Commissions' regulations CE 853/2004, 854/2004 and 1099/2009 (Koei 2009). In Canada, the handling and slaughter of livestock is a shared responsibility of the Canadian Food Inspection Agency (CFIA), industry, stakeholders, transporters, operators, and every person who handles live livestock. Canadian law requires that all federally registered slaughter establishments ensure that all species of food animals are handled and slaughtered humanely (Koei 2009). The CFIA's humane slaughter requirements take effect when the animals arrive at the slaughterhouse including the conditions for the humane slaughter of livestock. The regulations included guidelines and procedures for the proper unloading, holding, and movement of livestock in the slaughterhouse. Requirements for the segregation and handling of sick or injured animals are included. Livestock slaughter in the United Kingdom (UK) is governed by regulations issued by the Department for Environment, Food and Rural Affairs (Defra), which is the main governing body responsible for legislation and codes of practice covering animal slaughter in the UK. In the

United States, the United States Department of Agriculture (USDA) specifies the approved methods of livestock slaughter (Koei 2009). Each of these methods is outlined in detail, and the regulations require that inspectors identify operations which cause "undue" "excitement and discomfort" to animals. In New Zealand, the National Animal Welfare Advisory Committee within the Ministry of Primary Industries (MPI) is responsible for setting out the regulations and acceptable procedures associated with the welfare of animals. This is done through the production of a range of codes. The best example of protection of animal welfare at slaughter is a 46-page code entitled "Code of Welfare—Commercial Slaughter" (2018) which sets out the standards that must be achieved to meet the requirements under the Animal Welfare Act 1999.

11.8 Feral Camels Slaughtering Procedure and Welfare Considerations

The Australian government has commissioned the Department of Agriculture and Water Resources (DAWR) to develop a set of guidelines "The Australian Animal Welfare Standards and Guidelines (www.animalwelfarestandards.net.au)" to assist state and territory governments in implementing state-specific legislation about livestock welfare (Table 11.3). These guidelines in concert with the Commonwealth government's Biosecurity Act 2015 and Environmental Protection and Biodiversity Conservation Act 1999 provide a framework for the humane management of feral pest species in Australia. In the absence of reference to animals (except fisheries) in the Australian Constitution, the guidelines are not legally binding with the enforcement of animal welfare laws being governed by local state governments and the

Table 11.3 State and territory of Australian governments' legislation about livestock welfare

State/territory	Animal welfare act	Animal welfare regulation
Australian Capital Territory	Animal Welfare Act 1992	Animal Welfare Regulations 2001
New South Wales	Prevention of Cruelty to Animals Act 1979	Prevention of Cruelty to Animals Regulation 2012
Northern Territory	Animal Protection Act 2018	Animal Protection Act 2000
Queensland	Animal Care and Protection Act 2001	Animal Care and Protection Regulation 2012
South Australia	Animal Welfare Act 1985	Animal Welfare Regulation 2012
Tasmania	Animal Welfare Act 1993	Animal Welfare (General) Regulation 2013
Victoria	Prevention of Cruelty to Animals Act 1986	Prevention of Cruelty to Animals Regulation 2019
Western Australia	Animal Welfare Act 2002	Animal Welfare (General) Regulations 2003

Source: www.RSPCA.org.au (Home → Legislation → Animal Welfare Legislation)

Royal Society for the Prevention of Cruelty to Animals (RSPCA) (*Parliamentary Briefing Book—Key Issues for the 45th Parliament 2016*, page 190).

The acts and regulations cover, inter alia, cruelty, abuse, confinement, administering/laying poison, transport of livestock, trapping and snaring, aerial shooting, control of feral animals, and commercial pest control. The various legislations also specify aspects of unloading, pre-slaughter handling and the slaughter process with intentions to encourage the humane treatment of livestock and to minimize stress. The acts and regulations establish the fundamental care required for livestock and feral pest management and slaughter. State and territory animal welfare legislation may refer to standards and guidelines or codes of practice such as the "Model Code of Practice" for the welfare of animals at slaughtering establishments. General guidelines for slaughtering operations are contained in the "Australian Model Code of Practice" for handling and slaughtering of camels (Model Code of Practice for the Welfare animals 2006).

The population of feral camels (*Camelus dromedarius*) in Australia is estimated to be between 500,000 and one million according to the sources (Saalfeld and Edwards 2010; Lethbridge et al. 2016; Al-Jassim and Lisle 2016) with numbers increasing around 8% per annum. The loading, unloading, holding, and slaughter of feral camels in Australia are performed in a manner that elicits minimal stress, pain, or suffering. Best practice guidelines require that camels are given 24 h to roam and explore their new environment to acclimatize to their new surroundings for reducing stress and for helping in their handling. Feed and hydration troughs are required for camels that are contained for greater than 24 h. In the 24 h before transfer to slaughter establishments, the camels must be assessed for signs of injury, disease, inappetence, illness, late pregnancy, or distress to protect the animal's wellbeing.

During loading, short straight races and ramps with a minimal incline are to be used and any metal loading ramps should be covered with dirt/sand to reduce the sound which unsettles the caravan and cause camels to baulk or resist during loading. The operators must be patient and remain calm to reduce animal stress and anxiety to aid with their handling. Within 24 h before slaughter, animals are examined by meat safety inspectors to ensure they are healthy, and that their meat quality is suitable for human consumption. Just before slaughter, animals are led up a raceway into the abattoir where they enter the stunning box to separate individual camels from the caravan. Within seconds of entering the stunning box, an operator stuns the camel with a captive-bolt pistol ensuring the camel has been pitched or bled by severing major neck vessels. The primary purpose of stunning is to render the animal unconscious and insensible to pain before being bled out. Due to loss of blood the camel should not regain consciousness or sensibility before dying.

To comply with animal welfare legislation, the Australian meat industry has developed its industry standards for livestock processing establishments, covering regulatory and customer requirements, best practices, and codes to ensure animal welfare during transport, processing, and slaughter at abattoirs. Industry standards must be observed by the abattoir to receive Australian Animal Welfare Certification, a voluntary scheme jointly owned by the Australian Meat Industry Council (AMIC) and the Australian Meat Processor Corporation (AMPC). Admission to the scheme demonstrates the abattoir's compliance with the industry's animal welfare standards through annual audits conducted by AUS-MEAT Limited.

11.9 Recommendations

To improve camel welfare at slaughter, there are a number of important areas that need to be addressed. To attain satisfactory levels of camel welfare, significant changes will often be required in the infrastructure of the slaughterhouses, working practices, facility management, regulations, and enforcement. It is important, therefore, to adopt a holistic approach to solving current camel welfare-related problems by adopting appropriate technologies and practices that will be sustainable in the future.

Below is a list of recommendations and requirements to ensure satisfactory and acceptable camel welfare. In some situations, some or all of these requirements are already in place, but where they are not fully implemented, they need to be addressed as a matter of urgency, often through training, stakeholder engagement, and most importantly political willingness.

11.9.1 Pre-slaughter Transport and Holding

- Develop handling protocols requiring that camels are treated fairly so they are not injured, stressed, or overly excited before slaughter, in particular during the transportation to the slaughterhouse.
- Institute driver training and accreditation working toward compulsory licensing for transportation of camels to slaughter.
- Feed camels adequately during the pre-slaughter period so they are not subjected to prolonged periods of hunger and thirst.
- Provide suitable housing including thermal comfort and ease of movement. Lairage at slaughter permits camels to recover from transport stress.
- Ensure that camels are in good condition, which includes the absence of injuries, disease, and pain. Injuries can cause acute and/or chronic pain that can be a consequence of rough handling. Fighting with other camels can also cause injury; this is more common when camels are mixed with unacquainted individuals and when animals have to compete for access to resources.
- Provide facilities that permit appropriate behavior, which includes the expression of social behavior, expression of other behaviors, and good human-animal relationships. Unloading at the lair or moving to the slaughter area may cause fear that involves physiological and behavioral changes that prepare the camel for coping with the danger.
- Provide detailed requirements for the segregation and handling of sick or injured animals.

11.9.2 Handling, Stunning and Bleeding at Slaughter

- Prepare guidelines and procedures for the proper unloading, holding, and movement of animals in slaughter facilities.
- Avoid that camels are slaughtered in front of other camels, and that knives are sharpened in front of camels to be slaughtered.
- Require that the bleeding trap used must be properly cleaned before introducing a camel to limit the presence of blood.
- Improve the efficiency of existing slaughterhouses by extending operating hours and introducing refrigeration. This requires a change in perception of what is considered "fresh meat" through public awareness campaigns.
- Investigations into more welfare-friendly methods of moving camels to the site of slaughter such as through the use of moving V races based on those that have been used successfully with cattle and sheep.
- Develop pre-bleeding stunning methods that are effective with camels, fulfill halal requirements, and are acceptable to all groups involved, as has been done in several countries for animals other than camels.

11.9.3 Legislation, Regulations, and Training

- Establish national legislation that includes comprehensive and detailed regulations, guidelines, and penalties.
- Urge governments to recognize the reality and magnitude of the problems.
- Develop public awareness programs covering such issues as food hygiene and safety, animal welfare, safe and humane slaughter.
- Encourage government bodies to focus on the enhancement of services and collaboration in the areas of animal welfare, veterinary services, public health, food safety, and disease control.
- Develop veterinary practices and meat inspection services appropriate for developing countries, including training materials.
- Publish animal welfare requirements appropriate to developing countries, including training materials.
- Prepare detailed guidelines for stunning and slaughter practices appropriate for developing countries, including training materials and home slaughter.
- Develop public awareness programs covering such issues as food hygiene and safety, animal welfare, safe and humane slaughter. Some of these could be linked to existing water, sanitation, and waste initiatives and could even be introduced at junior school level.
- Improve camel welfare and slaughter practices through education and training of slaughterhouse staff.
- Reduce and eventually eradicate informal slaughtering through public awareness campaigns, improved legislation, and strengthening inspection and regulation.

- Establish a national camel welfare section within the Ministry of Agriculture (or equivalent) to develop a national strategy for establishing welfare standards, application, staff education, and public awareness.
- Establish animal welfare training systems, initially to motivate veterinarians and officials in its concepts.
- Establish written national standards and guidelines with transport police assistance.
- Establish written standards and guidelines for all slaughterhouse activities.
- Develop training programs for slaughterhouse staff with respect to good camel welfare, stunning and slaughtering practices, and hygiene and disease risks.
- Institute public awareness campaign about public health issues associated with animal slaughter for meat production.
- Develop guidelines that ensure slaughtering is rapid to avoid unjustified pain, suffering, injury, or fear for the camel.

11.10 Conclusions

Animal ethics and welfare standards associated with camel pre-slaughter handling and slaughtering must be established in legislation and regulations so that procedures are followed preventing the slaughtering of camels in a cruel manner. The management of camel slaughterhouses must enable the consistent achievement of acceptable standards of camel welfare. This will require the provision of comprehensive legislation, standards, and guidelines to follow. Currently, very poor standards of knowledge and skill were on frequent display in all parts of many facilities. Camel handling practices were generally aggressive and ill-refined, leading to camel welfare concerns. Facilities were generally of poor design, in light of modern knowledge and senior personnel with adequate specialist knowledge to correct fundamental flaws were seldom present. The widespread inadequacy of restraint at slaughter significantly reduces the efficiency of the technique, dramatically increasing the level of stress on a camel, and allowing increased brutality and associated safety problems for the workers themselves. Similarly, the possibility of one camel to watch another is totally avoidable and design faults could be easily corrected with little expense. An extensive informal camel-slaughtering sector remains unchecked in many countries due in part to a lack of enforcement of regulations (where they exist), along with traditional cultural and religious practices.

References

Al-Amri IS, Kadim IT, Alkindi AY, Haq QMI, Al-Ajmi DS, Haemd A, Quibol R, Al-Magbali RS, Khalaf SK, Al Hosni KS (2022) The effect of pre-slaughter electrical stunning on bleeding efficiency, meat quality, histology and microbial count of several goat muscles. S Afr J Anim Sci 52(5)

Al-Jassim R, Lisle A (2016) Prediction and management of feral camel population in Australia. In: Baxter G, Finch N, Murray P (eds) Advances in conservation through sustainable use of wildlife. Proc. Conf. held in Brisbane, Australia, 30th August–1st September 2016, pp 74–77
Allen C (1998) Assessing animal cognition: ethological and philosophical perspectives. J Anim Sci 76:42–47
Andriessen EH (2004) Meat safety, quality and veterinary public health in Australia. Powerhouse, Peterhead, Australia. ISBN 0646-30939-0
Appleby MC, Hughes BO (1997) Management and welfare of farm animals. CABI, Wallingford, UK
Broom DM (2014) Welfare of transported animals: Factors influencing welfare and welfare assessment. In: Grandin T (ed) Livestock handling and transport, 4th edn. CABI, Wallingford, pp 23–38
Brown-Brandl T, Berry E, Wells J, Arthur T, Nienaber J (2008) The impact of stress level on fecal bacteria and pathogen shedding in feedlot cattle. American Society of Agricultural and Biological Engineers, St. Joseph, MI
Corrier DE, Purdy CW, Deloach JR (1990) Effect of marketing stress on fecal excretion of Salmonella spp. In feeder calves. Am J Vet Res 51:866
Cortesi M (1994) Slaughterhouse and human treatment. Sci Tech Rev Off Int Epizooties 13:171
Daly CC, Simmons NJ (1994) The use of high frequency currents for the pre-slaughter stunning of livestock. In: CAPAA Engineers' conference, Gunnedah, NSW, Australia
Edwards E (1971) Apparatus for preparing animals for slaughter. U.S. Patent No. 3,599,276. Washington, DC
EFSA (2004) European Food Safety Authority Welfare aspects of the main systems of stunning and killing the main commercial species of animals. EFSA J 45:1–29
El Khasmi M, Chakir Y, Riad F, Safwate A, Tahri E, Farh M, El Abbadi N, Abouhafs R, Faye B (2013) Effects of transportation stress during the hot-dry season on some haematological and physiological parameters in Moroccan dromedary camels (Camelus dromedarius). J Life Sci (USA) 7(1):13–25
El Khasmi M, Chakir Y, Bargaâ R, Barka K, Lektib I, El Abbadi N, Belhouari A, Faye B (2015) Impact of transport distance on stress biomarkers levels in dromedary camel (Camelus dromedarius). Emirates J Food Agric 27(6):507–512
Galland JC (1997) Risks and prevention of contamination of beef carcasses during the slaughter process in the United States of America. Sci Tech Rev Off Int Epizooties 54:391
Gilbert KV (1993) Electrical stunning and slaughter in New Zealand. MIRINZ Technical Report 908, AgResearch Ltd, Hamilton, New Zealand
Gottardo F, Ricci R, Fregolent G, Ravarotto L, Cozzi G (2003) Welfare and meat quality of beef cattle housed on two types of floors with the same space allowance. Ital J Anim Sci 2(4): 243–254. Avenue Media, Italy
Grandin T (1994) Euthanasia and slaughter of livestock. J Am Vet Med Assoc 204:1354–1360
Grandin T (1995) Restraint of livestock. In: The animal behavior and the design of livestock and poultry systems international conference proceedings, Indianapolis, Indiana, USA, April 19–21, 1995
Grandin T (1997) Assessment of stress during handling and transport. J Anim Sci 75:249–257
Grandin T (2003) Stunning options. Meat Poult October issue, 118
Gregory NG (1996) Welfare and hygiene during preslaughter handling. In: Meat for the consumer 42 international congress of MEAT science and technology, vol 43, pp 35–46
Gregory NG (1998) Animal welfare and meat science. CABI, Oxon, UK, pp 223–253
Guya ME, Neme G (2015) Camel slaughtering practices and meat production in Eastern Ethiopia. Sci Technol Arts Res J 4(3):123–128
Herrmann K, Fischer A (2004) Method for hygienic slaughter of camel. In: Farah Z, Fischer A (eds) Milk and meat from the camel; Handbook on products and processing. vdt Hochschulverlag AG an ETH Zurich, Zurich
HMSA (1902) Humane Method for Livestock Slaughter Act. Retrieved 2021-25-08

IFC Guidance Note (2006) Animal welfare in livestock operations. International Finance Corporation/World Bank Group, Washington, DC

Kadim IT, Mahgoub O, Purchas RW (2008) A review of the growth and of the carcass and meat quality characteristics of the one-humped camel (Camelus dromedarius). Meat Sci 80:555–569

Kadim IT, Mustafa MF, Mahgoub O, Bekhit A (eds) (2013) Slaughtering and processing of camels. In: Camel meat and meat products. CABI, Oxon, UK, pp 54–72

Koei N (2009) Global study of livestock markets, slaughterhouses and related waste management systems. Final report in Association with ProAnd Associates Australia, February 2009

Lambooij E, Von Putten G (1993) Transport of pigs. In: Grandin T (ed) Livestock handling and transport. CABI, Oxon, UK, pp 213–239

Lemrhamed A, Farh M, Riad F, El Abbadi N, Tahri E, Belhouari A, Faye B, El Khasmi M (2018a) Evaluation of stress responses induced by the loading density in dromedary camel (Camelus dromedarius). Emir J Food Agric 30(9):803–808

Lemrhamed A, Farh M, Riad F, El Abbadi N, Tahri E, Faye B, El Khasmi M (2018b) Impact of lairage time at slaughterhouse on physiological biomarkers of stress in dromedary camel (Camelus dromedarius). In: Proceedings the 5th conference of the International Society of Camelid Research and Development ISOCARD. Laâyoune, Morocco, November 12–15, pp 323–324

Lemrhamed A, Tabite R, Abbadi N, El Khasmi M, Faye B (2019) Evaluation of slaughter stress responses in the dromedary camel. Int J Livest Res 9(8):78–90

Lethbridge M, Keith W, Saalfeld B, G.P., and Edwards. (2016) Measured reductions in the density of camels under the Australian Feral Camel Management Project. Rangeland J 38(2):173–179

Lulietto FM, Sechi P, Mansi Gaudenzi C, Grispoldi L, Ceccarelli M, Bar-bera, S., and Terzo Cenci-Goga B (2018) Noise assessment in slaughterhouses by means of a smartphone app. Italian Journal of Food Safety, 7:7053

Miranda de la Lama GC, Villarroel M, Liste G, Escós J, María GA (2010) Critical points in the pre-slaughter logistic chain of lambs in Spain that may comprise the animal's welfare. Small Rumin Res 90:174–178

Model Code of Practice for the Welfare Animals (2006) The camel (Camelus dromedarius), 2nd ed, Primary Industries standing Committee. PISC Re-port, No. 86. CSIRO, Australia

Nazifi S, Saeb M, Baghshani, H., and Saeb, S. (2009) Influence of road transportation during hot summer conditions on oxidative status biomarkers in Iranian dromedary camels (Camelus dromedarius). African Journal of Biochemistry Research, 3(7): 282–287

Njisane YZ, Muchenje V (2017) Farm to slaughterhouse conditions, animal factors and their subsequent effects on cattle behavioural responses and beef quality. A review. Asian Australas J Anim Sci 30:755–764

OIE (2009a) Transport of animals by land: Terrestrial Animal Health Code, 18th edn. World Organization for Animal Health, Paris, France

OIE (2009b) Slaughter of animals: Terrestrial Animal health Code, 18th edn. World Organization for Animal Health, Paris, France

Paranhos da Costa MJR, Huertas SM, Strappini AC, Gallo C (2014) Handling and transport of cattle and pigs in South America. In: Grandin T (ed) Livestock handling and transport. CABI, Wallingford, UK, pp 174–192

Saalfeld WK, Edwards GP (2010) Distribution and abundance of the feral camel (Camelus dromedarius) in Australia. Rangeland J 32:1–9

Schmidt CO (1972) Cattle handling apparatus. U.S. Patent No. 3,657,767. Washington, DC

Simmons NJ, Daly CC (2004) Electrical stunning of sheep, principles and auditing methods. Meat New Zealand Milestone Report E24, Meat New Zealand, Wellington, New Zealand

Terlouw C, Bourguet C, Cohen Tannugi C, Deiss V (2011) Stress reactions of cattle in a French industrial slaughterhouse. Viandes & Produits Carnés – 28 September

Velarde A, Dalmau A (2012) Animal welfare assessment at slaughter in Europe: moving from inputs to outputs. Meat Sci 92:244–251

Velarde A, Temple D, Mainau E, Mantecca X, Raj M (2016) Chapter one: Introduction. In: Velarde A, Raj M (eds) Animal welfare at slaughter. Benchmark House, Sheffield, UK

Vimiso P, Muchenje V, Marume U, Chiruka R (2012) Preliminary study on consumers' and meat traders' perceptions of beef quality and how the beef quality is affected by animal welfare practices. Sci Res Essays 7:2037–2048

Voisinet BD, Grandin T, O'Connor SF, Tatum JD, Deesing MJ (1997) Bos indicus cross feedlot cattle with excitable temperaments have tough meat and a higher incidence of borderline dark cutters. Meat Sci 46:367–377

Willems DJ, Markey EF (1972) Animal shackling device. U.S. Patent No. 3,693,216. Washington, DC

The manufacturer's authorised representative in the EU is Springer Nature Customer Service Centre GmbH, Europaplatz 3, 69115 Heidelberg, Germany. If you have any concerns regarding our products, please contact ProductSafety@springernature.com

Printed and bound by CPI Group (UK) Ltd, Croydon, CR0 4YY
28/09/2026
02224081-0011